THE

Amateur Naturalist

THE
Amateur Naturalist
Gerald Durrell

With LEE DURRELL

Hamish Hamilton
London

This book is for

Theo
(Dr Theodore Stephanides),

my mentor and friend,
without whose guidance I would have
achieved nothing,

and for

Pa
(Wilson James Northcross Sr),

Lee's grandfather, who
encouraged her early interest in wildlife,
especially by building palatial
homes for her animals.

The Amateur Naturalist was conceived, edited and designed by
Dorling Kindersley Limited, 9 Henrietta Street, London WC2E 8PS.

Editor Stephen Parker
Art Editor Neville Graham
Editorial Director Christopher Davis
Art Director Stuart Jackman

ISBN 0 241 10841 1

First published in Great Britain in 1982 by
Hamish Hamilton Limited
Garden House, 57–59 Long Acre, London WC2 9JZ.

Printed and bound in Italy by Arnoldo Mondadori, Verona

CONTENTS

Foreword 6
On Becoming a Naturalist 9

FOREWORD

In writing this book, Lee and I have tried to produce the sort of work which we ourselves, as young aspiring naturalists, would have liked to possess—a book that gives a little guidance in the somewhat bewildering and complex job of becoming a naturalist. Of course, no one book is in any way the be-all and end-all of naturalist books. Look upon our work more as the hors d'oeuvre at the beginning of a splendid meal, as an appetizer which we hope will encourage you to read the numerous brilliant books on natural history. We have tried to open a few windows on the extraordinary and beautiful living world, in the hope that this will encourage you to explore more thoroughly the magical environment around you and so add to the pleasures of your life.

It has been one of our aims to show that the wonders of nature are not confined to exotic places like the tropical forests of the world. They are just as accessible in your own back garden, if you search for them. Through the naturalist's eyes a sparrow can be as interesting as a bird of Paradise, the behaviour of a mouse as intriguing as that of a tiger, and a humble lizard as fascinating as a crocodile. In addition, we have tried to stress that to be a naturalist you do not need a lot of expensive equipment. This is very nice if you can afford it, but you can still study successfully and discover amazing things with the simplest of tools. A proper scalpel is not essential for dissection—a razor blade will do. A high-powered microscope is a wonderful tool, but a hand lens can reveal extraordinary things. It is nice to have special collecting containers for your specimens, but a matchbox will suffice.

Never be ashamed to use the words "I don't know". This is one of the most useful phrases in the English language, since if you admit your ignorance you will find the world full of people who are only too delighted to teach you. True naturalists love to share their knowledge. The first time I was in Western Australia, I spent three days being taken round by a local ornithologist. She showed me all the beautiful birds that lived in the area, about which I was woefully ignorant. When I left, I thanked her most sincerely for her help and patience. "Not at all," she replied, "I should thank you. I have had three lovely days, and, apart from the pleasure I have had from watching your pleasure, I had never realized before how much I know about our birds."

We greatly enjoy the world, and we have always done our best to respect and protect it; we hope that you will do the same. Our planet is beautifully intricate, brimming over with enigmas to be solved and riddles to be unravelled. As a naturalist you will never suffer from that awful modern disease called boredom—so go out and greet the natural world with curiosity and delight, and enjoy it.

ON BECOMING A
NATURALIST

All of us are born with an interest in the world about us. Watch a human baby or any other young animal crawling about. It is investigating and learning things with all its senses of sight, hearing, taste, touch and smell. From the moment we are born we are explorers in a complex and fascinating world. With some people this may fade with time or with the pressures of life, but others are lucky enough to keep this interest stimulated throughout their lives.

I can never remember the time when I was not intrigued and excited by everything about the world and, in particular, the other animals that inhabited the world with me. One of my very first recollections was when I was two years old. I went for a walk along a mountain road in India accompanied by my ayah. There had been heavy rain some time during the day, and the earth smelt rich and moist. At a bend in the road my ayah met two friends, a man and a woman, and I remember that the woman was wearing a brilliant magenta-coloured sari that shone like an orchid against the green undergrowth alongside the road. I soon lost interest in what they were talking about and made my way to a ditch nearby where I discovered, to my delight, two huge khaki-coloured slugs brought out by the rain. They were slowly wending their way along the ditch, leaving glittering trails of slime behind them. I remember squatting down and watching them, enraptured, seeing how they slid over the earth without any legs to propel them. I remained captivated by these two creatures until my ayah came over to see what I was doing and pulled me away from them, saying that I should not watch such dirty things. To me, they were not only fascinating but, in their own way, as pretty as my ayah's friend in her beautiful sari.

Throughout my life—and I have seen a great number of creatures all over the world—it has always astonished me that people can look at an animal or plant and say "Isn't that loathsome?" or "Isn't it horrible?" I think that a true naturalist must view everything objectively. No creature is horrible. They are all part of nature. You may not want to curl up in bed with a rattlesnake or a stinging nettle, and you may find it irritating (as I once did) when giant land snails invade your tent and eat all your food, but these organisms have just as much right in the world as we have.

When I was five or six, the next step in my career as a naturalist was collecting woodlice or ladybirds and carrying them around in matchboxes in my pocket. Then when my family and I went to live on the Greek island of Corfu my interest in natural history really blossomed. I found the island had a whole array of wonderful wildlife, and so I not only made collections of things, like butterflies and beetles, but I also kept them alive and studied them. At any one time I would have in my room or in the garden a varied assortment of creatures ranging from eagle owls to scorpions and from tortoises to sea horses. I was extremely lucky in

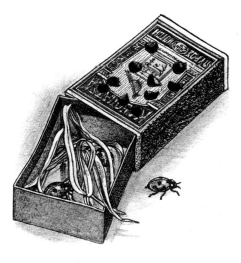

The young naturalist
In the photograph opposite, taken when I was ten years old, I am proudly holding one of my most treasured possessions—a barn owl.

9

Creatures great and small
Every animal and plant is a marvel of nature and should be viewed with interest and admiration. The enormous elephant seal (top) and the deadly black mamba (centre) seem exotic and exciting creatures, but for the naturalist a familiar house sparrow (bottom) holds just as much wonder and fascination.

having as a mentor and friend Dr Theodore Stephanides, who had, and still has, an encyclopaedic knowledge of the natural world. It was he who gave me my first "professional" piece of equipment when I was eight years old—a pocket microscope. A whole new world was opened up for me and I discovered that ordinary ponds and ditches were full of minute creatures, so that every puddle became a teeming jungle of tiny life.

Later I became a professional animal collector for zoos, and I travelled in places as far removed from each other as Patagonia and the Cameroons, from Guyana to Malaysia. Over the years I have been lucky enough to meet a magnificent selection of wildlife. I have lain on a beach in South America surrounded by huge carunculated elephant seals, some of them snoring loudly, others—if you got too close—rearing up two metres high in a shower of shingle, threatening you with open mouths and deep bellows. In the tropical forests of South America I have captured electric eels a metre and a half long, capable of stunning prey almost their own length with a charge of electricity. In the cloud forests of Costa Rica I have watched the beautiful flight of the quetzal, with his metre-long golden-green tail glittering like a banner as he flew across the clearings between the trees, and on the other side of the world I have spent an exciting—if somewhat dangerous—half-hour at night trying to capture a black mamba, one of the deadliest of African snakes, which had escaped from one of my cages. I have had great fun being introduced to and playing with a duck-billed platypus that looked like Donald Duck in a fur coat, but at the same time I get enormous pleasure by simply looking out of my kitchen window and watching the sparrows bustling about in the hedge beneath.

A naturalist is lucky in two respects. First, he enjoys every bit of the world about him and has a much more enriched life than someone who is not interested in nature. Second, he can indulge his hobby in any place at any time, for a naturalist will be fascinated to watch nature struggling to exist in the midst of a great city as well as observe its riotous splendour in a tropical forest. He can be equally interested and moved by the great herds on the African plains or by the earwigs in his back garden.

The first naturalists

Mankind has always had an interest in the creatures that share the world with him. Because some animals were a source of food and others were dangerous predators, primitive men became the first naturalists. They were forced to observe nature as a means of survival, and we can tell how beautifully and carefully they did so from the early cave paintings found in different parts of the world.

Then came the time, maybe 20,000 years ago, when man, instead of being merely a hunter, started to domesticate animals. The dog helped in his hunting activities and geese and wild ungulates were kept and bred as a food source—it was easier than having to go out and hunt them. Once humans had domesticated animals they turned their attention to plants. Instead of being gipsy-like nomads, drifting from place to place following the game animals and the seasonal change of plants, they began to create farms and thus enter a more settled way of life. Conglomerations of people sprang up into villages and towns. Now animals and plants were kept not merely as food sources but for interest's sake or for their beauty. An early forerunner of the modern zoological garden was founded in about 1100 BC in China and called, sensibly enough, the "Garden of Intelligence". The Pharoahs of ancient Egypt and most Kings and Queens of that time, including King Solomon, had huge animal collections.

One of the first serious books written about nature was *Historia animalium*, the great encyclopaedia of animal life compiled by Aristotle in 335 BC. It described at least 300 species of vertebrates accurately enough for modern naturalists to identify them. It was largely from this work that Pliny the Elder, in Italy, drew information for his momentous 37-volume work of 75 AD, *Historia naturalis*. Although there was plenty of good sound zoology in this, it did show how gullible Pliny was for he included such unlikely beasts as winged horses, unicorns and mermaids.

For many hundreds of years following Pliny the subject of natural history, in common with many other areas of knowledge, progressed little. For the most part such zoos as existed were in the hands of the dilettante nobility and were no more than second-rate menageries, lacking in any scientific purpose. As late as the 16th and 17th centuries, books on natural history consisted chiefly of illustrated bestiaries and herbals. These were, in the main, concoctions of myth, folklore and fact. Herbals were pretty unreliable, dealing mainly with plants as medicine and often as magic. In the bestiaries some real animals were given extraordinary attributes: the stag, for example, was supposed to suck snakes from their holes and eat them, and weasels could bring their dead young to life.

Widening horizons in natural history

Developments during the Renaissance and the work of Galileo during the early 1600s gave naturalists two great tools, the microscope and the telescope, so that during the 17th century there was a gradual change in attitudes from the credulous to the scientific. The bestiaries and herbals slowly gave way to proper books on natural history. As this happened, naturalists began to realize that they needed a system for classifying living things since, as more and more of the world was explored and more and more animals and plants were discovered, it was difficult to keep track of them all. A young Swedish student, Carl von Linné, finally solved the problem. In the disorganized offices of the naturalist he created in the mid-1700s a filing system and this system (though like most things it has been improved) still forms the basis of how we cope with the enormous array of organisms in the world.

Von Linné's idea was this: he first looked at all living organisms and those that were similar he assigned to one of several *classes*; the organisms in one class were assigned to several *orders* on the basis of more detailed similarities; an order was broken down yet again into *genera* and finally, a genus into *species*. This system of *classification* is still in use today (see page 287). But probably the most important thing von Linné did was to give every organism a unique scientific name in Latin which tells you at once its genus and its species in descriptive terms. Take, for example, the greater horseshoe-nosed bat, whose scientific name is *Rhinolophus ferrumequinum*, from *rhinos* (the Greek word for nose), *lophos* (Greek for crest), *ferrum* (Latin for iron) and *equinum* (Latin for horse). This animal is one of the leaf-nosed bats who have fleshy protruding decorations on the nose, in this case shaped like a horseshoe. A bird or fish or tree or mushroom may have a different common name in, say, Britain or France or Timbuktu, but its scientific name, telling the naturalist exactly what it

The manticora
"A beast is born in the Indies called a manticora. It has a threefold row of teeth meeting alternately; the face of a man, with gleaming, blood-red eyes, a lion's body, a tail like the sting of a scorpion, and a shrill voice which is so sibilant that it resembles the notes of flutes. It hankers after human flesh most ravenously." The purely mythical manticora, seen here from Topsell's *Historie of Foure-Footed Beastes* (1607), was still appearing in serious books in the late 17th century.

The powers of plants
Like the animal bestiaries, the herbals were curious books, being mixtures of astute observation and outlandish myth. In *The Compleat Herbal* of 1719, Tournefort warned of the dangers of herb sniffing: "A Certain Gentleman of Siena being wonderfully taken and delighted with the Smell of Basil, was wont very frequently to take the Powder of the dry Herb, and snuff it up his nose; but in a short time he turn'd mad and died; and his Head being opened by Surgeons, there was found a Nest of Scorpions in his Brain."

Carolus Linnaeus (1707–78)

Gilbert White (1720–93)

Charles Darwin (1809–82)

is, will be the same wherever the organism is found. Von Linné even gave himself a scientific name, calling himself Carolus Linnaeus, and his great work *Systema naturae* of 1758 is a major landmark of natural history.

At the same time in Hampshire lived an extraordinary man, Gilbert White, a clergyman who has been called the first English naturalist. Rather than involving himself in the scientific basis of natural history, as Linnaeus was doing, he simply observed nature with a sharp eye and wrote about it lovingly. White never strayed far from the village of Selborne where he was born, and all his observations were made in the surrounding countryside. Yet his delightful book *The Natural History and Antiquities of Selborne* (1788) is a wonderful model of clear and fascinating natural history writing, and with its details of the relationships between plants and animals it was far ahead of its time.

In the 19th century came a new wave of exploration, and naturalists travelled far and wide making vast collections of specimens. The richness and variety of the world, particularly the tropics, excited and astonished them. Von Humboldt, Bates and Alfred Russell Wallace worked in the rich tropical forests of South America, and later Wallace went to the East Indies. In Africa Selous, Emin, Johnson and others were penetrating deeper and deeper into what they called the "dark continent" and making equally astounding natural history discoveries. In North America Wilson and Audubon were busy exploring and cataloguing, and painting birds.

In Victorian England it was the heyday of the amateur naturalist. Ladies and gentlemen searched for shells, plants, fossils, rocks, marine organisms, birds' eggs, butterflies, moths and beetles, and in a world which had not, up to that point, become as depleted as it is today, it was the thing for naturalists to make large collections of whatever they found.

Evolution—the key to understanding nature

It was at this point and in this atmosphere that a young English naturalist named Charles Darwin was asked if he would like to sail on a voyage of discovery around the world on HMS *Beagle*. It was to be a momentous voyage, not only for the man himself but for the whole of science. Up to that moment most people believed that all the different forms of life had come into being simultaneously during the Creation as if God had tipped them out of a box ready-made, like a lot of tin soldiers, and that an elephant, say, or a pine tree had always looked exactly the same. But there were many curious facts coming to light. There were fossils of bizarre creatures no longer in existence and rocks that were millions of years old. There was man's ability to breed strange forms of domestic animals and plants. These facts were difficult to reconcile with the theory of the Creation, which stated that the world and all living things (and all fossils) had been created by God in six days in the year 4004 BC. After many years of patient study, Darwin provided a brilliant solution.

He was a perfect naturalist—observant, patient and ever curious, and nothing escaped his notice or his questioning mind. In his diary of the voyage you can see how everything was fascinating to him, from the fossils he found in Patagonia to the spiders in the Brazilian jungle to the fact that his flannel waistcoat glowed with phosphorescence during a lightning storm in the Andes. When Darwin reached the Galapagos Islands, his great theory began to take shape. There he found many different species of finch. They were all basically similar to one species found on the South American mainland, 900 kilometres to the east, but they differed slightly from each other in size and plumage, and especially in the size and shape of the beak. The species with the largest beak fed on the largest seeds; another species ate fleshy cacti, tearing into the soft succulent bits with a hooked beak; while a third species with a long thin beak probed for insects. Darwin saw that each kind of beak was suited to the feeding habit of each kind of finch. But why should the Galapagos finch species be so similar to, yet slightly different from, both each other and the mainland species?

The notion of the evolution of life—that species change over time so that their descendants become new species—is as old as the Greek philosophers, and the great naturalists of the 18th and 19th centuries, like Cuvier, Lamarck and Buffon, were aware that the theory of the Creation would not long stand up to the incoming evidence for evolution. But it was Darwin and, on the other side of the world, Alfred Russell Wallace who realized just *how* evolution takes place. As keen naturalists they had noted that within a group of similar organisms, no two individuals are exactly alike. Furthermore, they were both reminded by the work of Thomas Malthus, an economist of the early 19th century, that most organisms produce many more offspring than can possibly live long enough to become parents themselves. Some must die. Those that differ from their fellows, even slightly but in such a way that makes them *better* suited to their environment, have the best chance of survival. In a sense, nature is "choosing" or "selecting" which individuals will be the parents of the next generation. If the offspring inherit the "life-saving" characteristics or *adaptations* from their parents, they too will stand a better chance of surviving the hazards of the environment and so will pass on the adaptations to their offspring. This process continues, generation after generation. The great number of species in the world have evolved by the gradual accumulation of different adaptations brought about by this process of *natural selection* in different and changing environments.

What had happened in the Galapagos Islands? It is now believed that a few finches were blown by a storm from the South American mainland to one of the islands about a million years ago. Their descendants became adapted to different types of food and there was enough variety for 13 species to evolve.

As you can imagine, Darwin's theory, set out in great detail in his book *On the Origin of Species* (1859), caused an uproar. The Church thought it was blasphemous, but to the majority of naturalists the theory of evolution by natural selection was wonderfully illuminating for it explained so many puzzling things, like fossils that were clearly ancestral to modern forms and fossils of bizarre creatures long extinct because they could not adapt quickly enough to a rapidly changing environment. The theory also explained the existence of creatures of similar build and habits, but which are found in quite separate regions with no apparent connections in the past. Such creatures are the result of *convergent evolution*. They show similar adaptations because of their similar roles in nature, yet have evolved from unrelated ancestors. The work of Darwin and Wallace has probably had more impact than any other in shaping our understanding of the natural world.

There was another great theory being developed at this time, unknown to Darwin and Wallace and most of the world. The Austrian monk

Natural and artificial selection
When Charles Darwin visited the Galapagos Islands in 1835, his perceptive naturalist's mind saw a parallel between selective breeding done by man and selection done by nature. For example, pigeon fanciers have created all sorts of weird and decorative variations by selective breeding from the original rock dove. In a sense, nature is doing the same thing through the "struggle for existence". The various beak shapes belonging to the Galapagos finches have been adapted by natural selection to eating different kinds of food, so avoiding competition between the finch species. Darwin suspected that the original model for the Galapagos finches was a South American species blown to the islands by a storm.

Natural selection in Galapagos finches

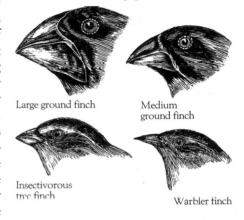

Large ground finch

Medium ground finch

Insectivorous tree finch

Warbler finch

Artificial selection in pigeons

Rock dove

Show racer

English pouter

English fantail

Gregor Mendel (1822–84)

The workroom of a master naturalist
The house of the great French naturalist Henri Fabre, which is near Orange in southern France, has been made into a museum in his honour. In Fabre's workroom (shown in the photograph opposite) is preserved much of his specimen collection and equipment. You can see that, besides having cases of specimens, Fabre set a new trend among Victorian naturalists by keeping live creatures and studying the reasons why they behaved as they did.

Mendel was in his monastery garden doing experiments with garden peas. He was particularly interested to find out if the seed produced by a certain plant resembled the seeds from which the plant itself grew. He made hundreds of different crosses of different varieties of pea plants and kept massive detailed notes of the types of seed that came from these crosses. His theory had to do with the way organisms inherit characteristics from their ancestors—the study of *genetics*. The great significance of his work was not fully appreciated until 40 years later, when scientists realized how it helped support Darwin's theory of natural selection, but Mendel's story shows that you don't have to go to the ends of the earth to discover something interesting and important.

The modern approach to natural history

The science of life is being put together like a giant jigsaw puzzle—a piece added here by the ancient Chinese, a bit there by Aristotle, and sections contributed by people like Linnaeus, Darwin and Mendel. We now have some important areas complete, but there is still a long way to go.

From a naturalist's point of view the progress of knowledge can be divided into four stages: what, where, how and why. We are still working on the "what" and "where", of course, though nowadays we have a pretty good idea of which living organisms inhabit the earth with us and where they are found. With Darwin's revolutionary theory we come to the "why" stage in our studies. Why are there so many different kinds of finch in the Galapagos Islands? Why are orchid flowers shaped so curiously? Darwin, too, asked "How?"—how does a plant climb a trellis, and how do tiny corals build whole tropical islands?

One of the foremost naturalists in Victorian times was the French entomologist Fabre. As far as he was concerned, although it was important to make collections of insects it was also important to know how and why they behaved as they did. He was one of the first entomologists to go out into the field and study how insects lead their lives in nature. Not only that, he could write about his discoveries simply and beautifully so that even people who did not understand anything about entomology could appreciate them. I remember when I was ten years old, my brother Lawrence introduced me to Fabre's writings. They opened up a magical world. Here was a man who could tell me why dung beetles so busily made little balls from horse or cow manure and buried them and how the glow-worm overpowers and consumes snails; who could tell me what a praying mantis did on her wedding night and who could describe exciting experiments to prove his theories—he once borrowed a cannon and fired it off in front of his house (breaking all the windows) to see if cicadas could hear. Though his main work was with insects, Fabre was interested in the whole of nature from mushrooms to fossils and his writings meant that you were suddenly transported out into the open air instead of, as with so many Victorian naturalists, into a museum. This tradition has been carried on by modern naturalists such as Konrad Lorenz, through whose studies of animal behaviour has emerged the science of *ethology*.

Ecology—the tapestry of life

The modern naturalist is particularly interested in how the life style of an organism lets it fit in with the other organisms living around it. He

Who eats what?

A food web is a diagram showing feeding relationships, with arrows indicating the diet of the various members. In a typical food web, shown below, plants are at the base while large carnivores such as the fox are at the top. This web is of course greatly simplified and, being a "consumer" food web, is only half the story. The other half—the "decomposer" web—involves organisms such as carrion beetles, fly larvae, fungi and bacteria, all of which live on dead animal and plant remains and help to return raw materials to the soil.

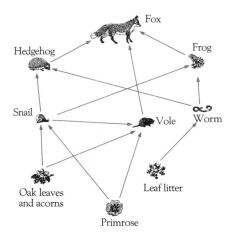

Fox

Hedgehog

Frog

Snail

Vole

Worm

Oak leaves
and acorns

Leaf litter

Primrose

Cooperation between species

The little blue tropical fish called the barber gets itself a meal by nibbling off and eating the tiny creatures living on the skin and gills of larger fish, who in return are relieved of their irritating parasites. The large fish, which could easily eat a barber, sits quietly while the little barbers swim round agitatedly and perform their cleaning duties.

observes everything that goes on in the *community*, the collection of all living things in a natural area—be it forest or coral reef—and he sees that each species has a certain role to play in nature. This role is called the species *niche* and is defined by the profession and homelife and the friends and enemies of the individuals that make up the species.

One of the chief ways in which living things in a community depend on each other is in obtaining food. There are the plants, then the animals that eat the plants, and the animals that eat animals and then, finally, the fungi and bacteria of the soil that feed on the dead bodies of both plants and animals. This network of feeding relationships, of eating and being eaten, is called a *food web*. The main decomposers—fungi and bacteria—process dead bodies like a factory into the chemicals they are made of, and so enrich the soil with compounds of carbon, oxygen, hydrogen, nitrogen and phosphorus. These chemicals enter the cycle again when plants use the nutrients in the soil plus energy from sunlight for their growth. The chemical building blocks of life leave and enter the bodies of organisms at many other points as well. A panting fox will produce moist breath, for example, that may take two million years to come back to an oak tree: the moisture enters the atmosphere and becomes part of a snowflake; this falls and adds to a glacier that slowly melts into a river and runs into the sea; this eventually evaporates into clouds that rain on the soil in which the oak tree lives. Nutrients travel in the *nutrient cycles* from the non-living to the living world and back again.

Organisms depend on each other for things other than food—things such as homes, places to hide in and the right conditions for their growth. A tick makes its home in the fur of a squirrel or rabbit; the squirrel raises its young in a tree, and the rabbit rears its young in burrows among the roots. The tree gives shade to allow the growth of such things as ferns or bluebells and it helps vines to climb up into the sunlight.

Living things are also very dependent on the physical aspects of their environment. A whale cannot live in the middle of a desert, nor a palm tree at the bottom of the sea. It is very important from the organism's point of view whether its environment is warm or cold, wet or dry, and even the slant of the sunlight or the precise slope of the ground may make a great deal of difference.

The availability of the things necessary for an organism's life—food supply, a proper home and the right physical conditions—naturally puts a limit on where living things are found and how many can be supported there. A community together with its physical environment is called an *ecosystem*. The study of what controls the numbers and distribution of living things is called *ecology*, coming from the Greek word *oikas* meaning, appropriately enough, a house. There are several sorts of ecosystem on this planet, and each of these depends first and foremost on temperature and moisture. In areas of the world where conditions for life are harsh, ecosystems are quite simple—there are fewer species, and in consequence fewer links in the food web. In cold dry regions there is tundra and its place is taken in hot dry regions by the great deserts. In milder areas ecosystems become more complex. There are coniferous and deciduous forests in more moist areas and savannahs and plains in the drier areas. Around the very moist and warm "tummy" of the planet are tropical rain forests, where the ecosystem is at its richest—an enormous number of species in many specialized niches. The waters of the world—inland waters, shores and oceans—follow the same patterns, holding fewer species in the extreme northerly and southerly climes.

People frequently get the idea that any creature or plant living in nature has a happy-go-lucky sort of existence. In actual fact, they have to work hard to survive and to reproduce. The successful organism will be the one who wins the competition for eating more food, getting better hiding or growing or nesting places and above all raises more offspring than its fellows. Many species have evolved complex adaptations for survival, ranging from thorns and stinging hairs in some plants to *mimicry* among the animals. There are intricate adaptations to ensure reproduction—weird seed shapes that aid in seed dispersal, the special relationship between a flower and its own insect pollinator, and spectacular methods of courtship, like the fantastic feather displays of the birds of Paradise and the mating dance of the alligators.

Courtship is just one aspect of the fascinating world of animal communication. By signalling their intention to one another with flamboyant colours, by postures, sounds and scents and even by touch, animals waste less time and energy fighting between themselves over resources and mates, and thus can concentrate on the important jobs of finding food, avoiding their enemies, mating and raising their young. The importance of communication for animal life was brought home to me when I spent a few weeks filming and recording in a huge conglomeration of two million penguins in Patagonia. The nest site for the colony was about three kilometres from the sea, and to get to it the penguins had to waddle over a series of enormous sand dunes. When they came back from their fishing, their crops full of food for their babies, they had to find their way to their own home through thousands and thousands of nest burrows, each one containing babies that rushed

Courting gannets
Many birds have complicated and interesting patterns of courtship behaviour, in which various acts and gestures are performed in a set order. The movements are often borrowed from other parts of the bird's behavioural repertoire—feeding or fighting, for example—but they are performed as though in a ceremony. Julian Huxley, who studied the courtship behaviour of the great crested grebe, coined the term "ritualization" for this sort of out-of-place borrowed behaviour. In the photograph the male gannet is indulging in what is termed "neck-biting", which he would normally do to another male invading his territory. If the female responds not by pecking him back but by the submissive "facing-away" gesture then the pair can move on to the next stage in their courtship ritual.

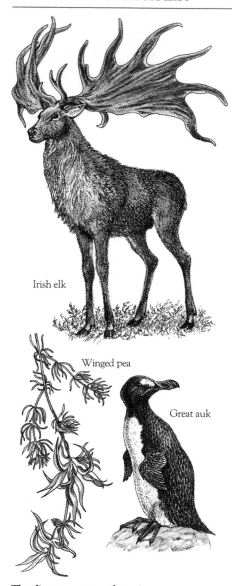

Irish elk

Winged pea

Great auk

The disappearance of species

There has always been a constant turnover of species in nature, with poorly adapted ones dying out and new ones taking their place. Since man appeared on the scene, this process has been vastly speeded up. The huge Irish elk, which once roamed much of northern Europe, has been extinct for many thousands of years and it is possible (though we can never be sure) that man had a hand in its downfall. The flightless great auk of the North Atlantic coasts was exterminated for its flesh, skin and eggs, the last birds being killed in 1844. The winged pea plant of the Canary Islands was luckier; extinct in the wild by the late 1800s, it survives in cultivation—in fact, collection for horticulture was probably the main reason for its disappearance.

out at them and begged for food. This made clear to me very sharply the idea that we did not observe carefully enough. To me, all the nest burrows and all the babies looked identical, and yet the parents never made a mistake and always went straight back to their own nest burrow and their own baby.

Strength and fragility in nature's fabric

Within a stable community, niches are usually arranged so that no species is in direct competition with another. The niche defines a unique role that a species plays in the community and there is little overlapping. Certain species might appear to have similar roles, but in many cases they are distinguished by time. For example, hawks catch rodents in the day and owls catch them at night. Frogs and toads may breed in the same pond but you will generally find that frogs breed early in the spring and toads later. In a community which has lots of different species, relationships among living things are interwoven in a complex fashion. There are so many ways to get food, avoid enemies and find a home. A lack in one requirement may be made up by using something else. An unusually dry spring may affect the insect population and so you will find that birds do not raise as many babies as usual, since there is not the food to feed them. Predators that eat baby birds, like hawks and foxes, will be forced to take other prey—maybe mice or lizards—until the following spring, when the number of birds is back to normal. Nature can easily cope with such minor natural disturbances.

Sometimes the members of a community may be forced into serious competition with each other. An invasion by a new species or just simply the changing seasons may create food and "housing" shortages. Some species simply move away until conditions become more favourable—birds migrate, for example. Some creatures hibernate in cold weather when food is scarce; others in hot climates do the equivalent (called aestivation) during food shortages in dry periods. In the long term, some species cannot cope and will die out, and other species quickly fill the empty niches. We know this has happened throughout the history of the earth from fossil records.

As you can see from all of this, the whole machinery of nature is dynamic and fluid; seasons revolve, niches are emptied and filled, there is competition and cooperation among species and the constant recycling of nutrients. The great ecosystems are like complex tapestries—a million complicated threads, interwoven, make up the whole picture. Nature can cope with small rents in the fabric; it can even, after a time, cope with major disasters like floods, fires and earthquakes. What nature cannot cope with is the steady undermining of its fabric by the activities of man.

We are eroding this world we live in. We over-exploit the natural forests, replacing them with quick-growing foreign trees in which the natural fauna does not flourish. We over-fish and pollute the seas. All over the world agricultural methods are so bad that within the next few years much productive land will be rendered useless. We are killing off some animal species and depleting others, like the big mammals of Africa. We introduce animals to countries where they have no business and where they affect the local fauna. We are scattering pesticides and herbicides around in a profligate fashion so that they not only kill the things we want to kill but wipe out a whole host of innocent organisms as well. With these activities combined with over-population we are creating grim prospects for ourselves. I will be saying more about this later in the book and describing what you, as a naturalist, can do to help.

Starting out as a naturalist

The first and most important thing a naturalist in the field has to learn is a code of behaviour. Remember you are privileged to live in this world and so you must treat it politely and with respect. When you visit friends you do not pick all their flowers, trample on their flower beds and fill their pond with empty tins and bottles, so treat the world as a very special garden. On page 320 you will find a code for naturalists; please study it well and even add items to it that apply in your particular area.

One of the great things about being a naturalist is that you are born with all your basic equipment—your eyes, your ears and the senses of smell, taste and touch. All these, of course, can be added to by man-made tools, but a naturalist should be capable of enjoying his craft naked on a desert island. Never forget that, while taking a deep interest in the world outside, you are yourself walking around inside a miracle. The human body is an extraordinary piece of adaptation and you should learn to use it in the same way as you learn to use the other adjuncts of the naturalist's art—the hand lens, binoculars, camera and so on.

Of course, sophisticated items of equipment will be needed if you want to watch birds at their nests or watch mammals without being observed; binoculars or a telescope, along with a hide, are essential. Hides can be very simple or very elaborate. The famous naturalist and photographer Cherry Kearton made a hide that was like a model cow and he crouched inside it with his camera. However, one day after he had installed himself and his brother had left him, he had the misfortune to fall over and so there he was, trapped inside his wooden cow for several hours before his brother came back and set him upright again. So, should you wish to make a model cow, you have been warned.

When I was a young student naturalist in Corfu I did not have access to naturalist shops (such as the wonderful Watkins & Doncaster near London) and even if this had not been the case, my meagre pocket money would not have allowed me to buy specially-made butterfly nets and special tins for collecting insects. My nets were home-made out of cheap butter muslin, sewed by myself or by my mother or sister (whichever was in a generous mood), and my killing jar was an old sweet bottle with a cottonwool pad soaked in ether purchased from the chemist. I had to be ever ingenious in obtaining my equipment—not a single bottle or tin in our house was thrown out, they all made containers of one sort or another for my collection. Matchboxes were invaluable and any solid container of wood or cardboard was a godsend. When my containers were full of specimens I unhesitatingly used all my pockets, my handkerchief, my shirt and on one occasion (to the horror of my mother) I came back to our villa stark naked because I had used up every bit of my clothing for wrapping up specimens.

In Greece, after my elder brother had opened a matchbox and found a scorpion inside and my other brother had found the bath full of watersnakes, they gave me a special room to keep my animals in. This room became a cross between a bedroom, a museum and a mini-zoo. It was a large room on the ground floor, with French windows leading out

A camera in sheep's clothing
Natural history photographers go to great lengths to obtain a good picture. In the early 1900s the Kearton brothers, Cherry and Richard, devised all sorts of disguises for themselves and their cameras. This sheep contained a camera which the brothers operated by a long remote control from the relative comfort of a nearby hide. Despite an attempt to herd the sheep by a local shepherd and his dog, they managed to photograph the sandpiper they were after.

The naturalist's rucksack

Delve into the amateur naturalist's rucksack and you might well find this mixture of purpose-made and improvized equipment. The precise contents will depend on the trip at hand—different nets for pond dipping, trowel for sandy shore digging, or paper and crayons for bark rubbing.

Secateurs

Butterfly net

Multi-bladed penknife

Killing bottle and fluid

Hand lens

Forceps

Examining tray

Pooter

Examining A white plastic or enamel dish provides a plain background for scrutinizing specimens under a lens.

Notebook and pen

Pocket field guides

Binoculars

Collecting The butterfly net dismantles for easy transport, and the pooter safely picks up tiny animals when you suck on the rubber tube.

String, elastic bands, sticky tape

Plastic and muslin bags

Butterfly envelopes

Rucksack

Local map

Camera

Jars and bottles

Larva tin

Rigid plastic container

General Use the indispensable notebook, pocket field guides and camera to identify and "collect" without disturbing nature.

Storage Plastic bags maintain humidity and keep plants fairly fresh; punched holes or muslin tops for containers admit fresh air for the creatures inside.

on to the garden. Along one wall I had aquaria for my fresh- and salt-water creatures such as sea horses, dragonfly larvae, hermit crabs, water beetles and frog and toad spawn. Next to them I had a series of glass-fronted wooden boxes which housed my snakes, toads and frogs.

There was a long shelf on which I kept a lot of big wide-mouthed sweet bottles in which lived various microscopic freshwater fauna, and a lot of old-fashioned mesh cheese covers under which I kept my spiders, scorpions and praying mantises. Metal containers being too expensive, I found a good cheap line in large pottery bowls which were ideal (with some butter muslin over the top) as homes for innumerable creatures. Along the wall opposite were all my wooden glass-topped cabinets which contained my butterfly, moth, beetle and dragonfly collections. These cabinets were made to my own specifications by the village carpenter, with whom I bargained for two or three days before we agreed what I considered was a fair price. I lined them with wine corks (never lacking in my family!) cut in half lengthways, to which I pinned the specimens.

At one end of the room I had my bed and, to my annoyance, my mother made me have a wardrobe for my clothes. I felt that this was a gross waste of valuable space, but I compensated for it by keeping all my fishing and butterfly nets under the bed, together with small kegs of ether and spirits for preserving and boxes full of hemp and cottonwool for taxidermy. In the wardrobe (by pushing all my clothes up tight) I managed to gain room for hanging my specially-designed collecting boxes in canvas bags. Each box had separate divisions containing test-tubes of various sizes, small tins, wide-mouthed jars, killing bottles, forceps and similar important equipment for my trips into the countryside. In the middle of the room was a large deal table on which I did my microscope work, setting of insects, dissection and taxidermy. Here I always kept containers in which were my latest acquisitions so that I could watch them while I worked. Of course there was a shelf for my small but precious collection of books, particularly the works of Darwin and Fabre.

Getting ready for a field trip
We are very lucky now to have such marvellous field guides that cover practically every aspect of natural history, enabling you to identify clearly and easily almost anything that you encounter. I never had such excellent books when I was young, and on many occasions I had to send drawings and photographs from Corfu to as far away as Vienna and Rome and Paris to get some of my specimens identified. Most of these field guides are small and will fit easily into your collecting bag or even your pocket.

If you are going far afield, take maps of the area and a compass. Maps are useful even if you know the area well because on them you can mark the places where you see different plants or capture different animals. There are generally very good topographic and vegetation maps for hikers and your local natural history society or library should be able to tell you where to obtain these. Always be sure to tell someone where you are going and approximately what time you expect to be back. You might have the misfortune to fall out of a tree and break your leg and if nobody knows you are there, the consequences could be disastrous.

Every naturalist should work out a sensible set of clothing guided by three factors—the weather, where you're going and the job at hand. You wouldn't wear gumboots to go snorkeling in the sea. Be thankful that we

Covers to keep insects in, not out
Many everyday items—especially kitchen equipment—can be adapted for collecting or keeping creatures. I keep butterflies, mantises and other insects (separately, of course) under fine-mesh covers that are designed to keep flies away from cheese. Changing food is easy; put the fresh food in and the occupant will soon climb on to it and start to feed, and then you can quietly remove the old food.

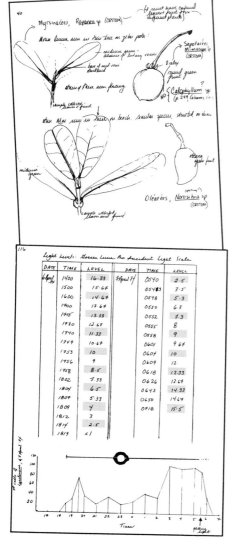

The naturalist's notebook
The uses of the notebook are many and varied. These pages from one of Lee's notebooks show detailed sketches of some plants and a table of observations with an activity graph of lemur calls. Your field notebook should contain jottings and sketches of almost everything you encounter, and these can be sifted and sorted and transferred to a proper diary when you return home from your field trip.

are not like the Victorians, who had to be "properly" dressed with collars and ties and big lace-up boots whether they were entertaining in their front parlours or catching butterflies in a meadow.

The indispensable notebook
Always remember that your best basic equipment, besides your senses, is a notebook meticulously kept. Write down your observations on the spot—don't leave them for later, because you will find your memory is not always infallible. A quick note taken at the time can act as a memory aid and be expanded upon in your proper diary when you get home.

Always try to make drawings in your diary, even if you think you can't draw. A sketch, however crude, is a great aid in reminding you exactly where the colours were on a bird, or the exact shape of a leaf or flower. In addition to the details as to what an organism looks like, you must not forget to mention such things as where and when you found it and what it was doing—eating, running, flowering or wilting—and what eventually happened to it.

Don't forget that amassing sufficient information to get a complete picture of an organism cannot be done all at once. Your work may well spread over several months or even years. Many of your researches will pose questions, the answers to which you may not discover until you have undertaken further observations. When you are observing, remember that it is not just the animal or plant you must watch, but its whole environment. So you must note the pheasant a fox eats, the grass a rabbit grazes on, the kind of tree that provides the best nesting site for a bird and so on. Nothing is too insignificant to be noted down, for there is nothing insignificant in nature. A well kept and lavishly illustrated diary will be of great help to you in the future. I had some 20 or so diaries covering the period when I lived in Corfu but, unfortunately, they were all destroyed during the war. I regret this very much, for in these diaries were such entries as the very first time I saw a goldfinch, my first dissection of a snail and my long study on the swallows that nested under our eaves. All this would have made fascinating reading for me today.

Collect with care and caution
Nature by and large is very resilient, but it is not inexhaustible and for years it has been drained by human beings. Some time ago I was in a museum in Los Angeles and there, in several large drawers, I saw numerous skins of the California condor. It was dreadful to realize that here, in front of me, carefully preserved, lay more California condors than actually existed in the wild. In the past, naturalists had the impression that nature was so prolific that the treasure trove of life was inexhaustible, so they tended to over-collect on the principle that there was "plenty more where that came from". Today we have learnt that this is untrue and nature can be easily depleted. The modern naturalist must collect with care and caution.

Making a collection of natural history specimens is fascinating and illuminating as well as being great fun, but consider what you want to get out of it and what you want to do with your specimens. Do you want to be reminded of their beauty or to identify and describe rare species? The modern naturalist, aided as he is with techniques like photography and sound recording, can "collect" without disturbing nature at all, or he can do it with a hand lens and a sketch pad. Think about live-trapping your creatures, examining them, perhaps keeping them for a time and then releasing them. Always remember, however, that the welfare of your subjects and their environment is much more important than your photograph or drawing or sound recording.

If you decide that you do want to collect, find out if permits are necessary for each species. Your local natural history society will be able to give you this information. Stay within the conditions specified on the permits and please follow the guidelines that I have set out on page 320.

Portrait of a naturalist

In my library I have the works of many naturalists, past and present. When I read these books it is not just to obtain information from them but also to try to understand what the writers have in common. What is it that links Aristotle to Mendel, or an obscure author of a bestiary to a modern ecologist? What really makes a naturalist? Well, I think that a naturalist first of all has to have a very enquiring mind. He seeks to observe every little variation in nature and to try and discover its origin and function. It was Sherlock Holmes who said, "You see, but you do not observe." That is true of most people in the world today. A naturalist must keep an open mind and be interested in many things, although he may specialize in one particular subject.

Recently I visited Evelyn Hutchinson, one of our foremost ecologists, in his laboratory at Yale University. During the course of an hour's conversation he discussed with me the breeding of giant tortoises; the animals and plants that were to be found in illuminated manuscripts from the Middle Ages; an obscure little crustacean that lives only on the gums of certain whale species; why a king crab, one of the most primitive animals alive today, should have blue blood; and whether the 18th century poet and painter William Blake could possibly have seen South American monkeys, since he had depicted some monkeys in one of his pictures as hanging by their tails, a thing only South American monkeys can do. So you see, nothing escapes Professor Hutchinson's attention or disrupts his deep interest and reverence for the world around him.

A naturalist should also be an assiduous note-taker, recording every detail of his job with accuracy and neatness. He knows that a simple note made now may be of great value in solving a problem in the future. The majority of great naturalists have been enormously modest, seeing themselves as very privileged to live in this magnificent and complex world and to be given the opportunity of unravelling some of its secrets. Towards the end of his life Darwin, having completed his mammoth works—not only on evolution but on such diverse subjects as corals, earthworms, plant fertilization and many other topics—wrote to a friend and said he wished he could have been of greater help to the world.

The work of Darwin, Mendel, Fabre, Lorenz and Hutchinson is of course being updated and expanded and applied to the problems of saving our planet. But it is well to remember that the great naturalists gained their love and appreciation for the natural world when they were youngsters dabbling about in ponds or ditches, seeing through a hand lens the minute world contained in a drop of water or walking over the moorlands or through deep woods, fascinated by the life around them—in fact doing the things that I will describe in this book.

THE NATURALIST ON
HOME GROUND

When I was young, one of the villas we lived in on Corfu was an old house built in Venetian style. It looked rather like one of those Victorian doll's houses—square, with big windows and shutters. Inside there were gigantic attics, two of which were dark and two of which had skylights so they were quite bright and sunny. The big high-ceilinged rooms had wooden floors supported on enormous beams. In summer, when it was very hot, you could hear the beams cracking like miniature muskets as they got warmer and warmer, and in winter they groaned and creaked as the heavy damp winds swayed the house. The whole building had vast cellars beneath it. Naturally all this provided an ideal habitat, not just for my family and myself, but for a multitude of other creatures as well.

Whenever I approached this villa, whether in the daytime or at night when its windows were ablaze with the glow of our oil lights, I used to wish that it really *was* an old-fashioned doll's house, so that I could open the front in two panels and thus see everything inside from top to cellar with all the animal inhabitants.

From attics to cellars
In the dark attics were all sorts of wood-boring beetles, including the deathwatch beetle which you could hear ticking in the beams—a sound it produces by beating its thorax against the wood to call a mate. This is where the bats lived, too, and on the floor there was a colony of swifts, their babies shuffling about, squeaking and calling, while the parent birds sped around and around the house in a shrieking merry-go-round, catching food for their young. In the attics with skylights there lived the *Scutigera*, a strange centipede with an amazing web of legs around its body. It was fascinating to watch it stalking the odd moth or fly that found its way up there, for it moved with incredible speed when it attacked, seeming to glide over the rafters like a stone thrown on to ice. There were also scorpions up in these attics and, in summer, ticks, fleas and spiders. In winter, earwigs prowled the beams.

We sometimes had invasions of ticks in the bedrooms and living rooms downstairs—to my mother's horror. These could be very annoying, but I found them helpful since they provided me with interesting microscopic material. In the bedrooms there were also clothes moths and silverfish and, of course, mosquitoes invaded the rooms at night to feed on us. The spiders, in consequence, found it good to set up their webbed homes there for they could eat the mosquitoes and the moths that were attracted to our lights. Our living room and dining room, which had French windows opening on to the garden, were invaded by butterflies and moths in the summer, and in the autumn some species, like tortoiseshells and peacocks, would come in to hibernate. More spiders followed them in and the geckos used to come in

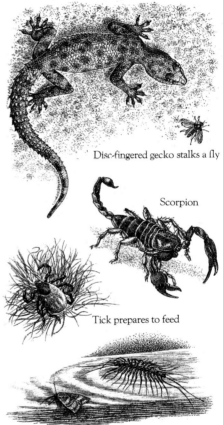

Disc-fingered gecko stalks a fly

Scorpion

Tick prepares to feed

House centipede pursues a small moth

Swift incubates eggs

The homely house martin
This member of the swallow family (opposite) was originally a cliff- and cave-nester, but protected sites on the walls of our dwellings are as good, if not better, as places to nest.

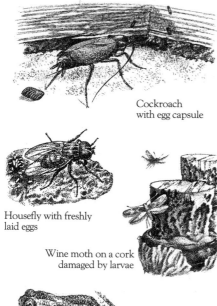

Cockroach
with egg capsule

Housefly with freshly
laid eggs

Wine moth on a cork
damaged by larvae

Common toads

at night and scuttle about the ceilings, catching the various insects that came and fluttered round the lights.

In the big stone kitchens, where my mother cooked on charcoal fires and it was warm and dry, there was a little creature resembling the silverfish called by the charming name of the firebrat. In this area of the house, we also had two species of cockroach, against which my mother waged a constant unsuccessful battle. I was fascinated by their beauty, for to me they looked as though they were carved out of tortoiseshell and their egg cases were so elegant, like the most beautiful little ladies' evening handbags. I used to collect these capsules and hatch them out in my room (unbeknownst to my mother); half I would let go and the other half I would feed to my mantis, gecko, tree frogs and so on—I felt this was fair. In the larder, we frequently had invasions of cheese mites and occasionally we would get flour mites and rice weevils. Here too, at certain times of the year, battalions of ants would invade us to carry off whatever spoils they could find, and quite often, when somebody left the larder door open, a bluebottle fly would be quick to zoom in and lay its eggs on any meat it could find. Bluebottles can smell meat from several kilometres away. Their little white upright eggs would be laid in small clusters like miniature tombstones in a microscopic cemetery.

Down in our big cool cellars, which had earth floors and grille windows looking out at ground level, the impression was more of a dungeon in some castle. As there was access through these barred windows to the open air, we got a lot of creatures down there. Toads liked the coolness of the cellars, voles invaded them and built their nests, earthworms burrowed in the soil and woodlice lived in the cracks of the walls, for it was damp enough to appeal to these little crustaceans. Here, too, were the omnivorous harvestmen (called "daddy-longlegs" in America, a name that in Britain is used for the cranefly), trembling on their long lanky legs, and shade-loving slugs that used to annoy my brothers by chewing all the labels off the wine bottles. There was another creature that threatened my brothers' precious wine and that was the wine moth, which thrived down in our damp cellars, its larvae feeding on mould growing in the walls and burrowing into the corks of the wine bottles. So you can see that a house like this holds so much of interest that it could give you almost a lifetime of study.

Creatures in residence

Human habitations provide specialized environments. There are as many hiding places as out of doors but the food supply is usually abundant and concentrated in one place, unlike it is outside, and in winter there is warmth and shelter. Many "typical" house creatures such as the familiar house mouse have been associated with humans for such a long time that they have become dependent on us, our domestic animals and our stores of food and goods.

There are many creatures of course that live on things we don't think of as food, such as the clothes moths and the various beetles that lay their eggs on wool, feathers or fur and whose larvae eat the protein keratin. Keratin itself is not sufficient, however, and so the larvae depend on "foods" like organic stains on sweaters or fur coats—sweat and food stains and so on—for the vitamins that they need. The silverfish, a very primitive creature, coming from 300 million-year-old stock, can digest paper, a very unusual diet for an animal. Other creatures eat the mould on paper, like some slugs, woodlice and the booklice.

Then there are animals that actually live in the structure of the house itself, like the wood-borers, especially the larvae of the longhorn and deathwatch beetles. In tropical and sub-tropical countries, termites are probably the worst pests for the destruction of wood, but the make-up of their societies, which are dominated by queens and have a caste system of soldiers and workers, are fascinating to study. Another extraordinary beetle you may find boring into damp wood, mainly hardwoods from the tropics, is the ambrosia beetle. The female carries on her body, on the hairs on her head, or in her digestive tract the spores of a certain species of fungus. As she gnaws out her passageway, the fungus spores brush off or are passed out in her excreta. Soon the passageway is coated with a thick growth of fungus which forms the main diet of the beetle larvae. The female tends these fungal "gardens" very carefully, fertilizing them with her excrement and keeping the temperature and humidity at the right level by blocking or unblocking the tunnel with wood dust. If the fungus lawn is not closely grazed by the larvae it quite quickly deteriorates unless it is carefully tended by the mother. If at any time she is disturbed, she will quickly swallow as much of her garden as possible so that she can "plant" it again in a different place.

Visitors

Of the many vertebrates that come into the house, the ubiquitous rats and mice come to both feed and breed, and especially to take refuge there in the winter. Other creatures like bats, owls and sparrows come in simply because they find a house a suitable area for breeding or shelter. In North America, your loft may be invaded by grey squirrels or racoons. In Europe, your loft may be home to beech martens and various species of dormice. My attic in France is home to a family of garden dormice with long fluffy tails and dark spectacled markings round their eyes like a giant panda's. These charming rodents have one irritating habit—they keep us awake by the noise they make running about on the beams. It sounds as though they are moving their furniture all night long.

In the tropics of course a wholly different, very rich fauna can be seen all the year round, so the contents of your house may be quite extraordinary. A famous naturalist in the East Indies once described how centipedes 30 centimetres long would hang like streamers, holding on to the beams with their hind pincers around the lights at night and catching moths that were attracted there. In West Africa you might find a green mamba in the house or in Madagascar it might be a mouse lemur.

In South America I remember visiting an Amerindian village where I was entertained in the chief's large palm-thatched house. During the conversation I was astonished, on looking up, to see that on several beams that ran across the hut lay draped beautiful multicoloured boa constrictors. I found this sight amazing since in most tropical countries in the world all snakes, whether poisonous or harmless, are greatly feared. I asked the chief how he came to tolerate boa constrictors in his house and he told me he had put them there, having caught them in the forest. Apparently, the Amerindians find boa constrictors infinitely superior to cats for ridding the house of annoying rodents. So not only did they look extremely decorative, like complicated and beautiful tapestries hanging

Spiders and their webs

Each species of spider found in or around the house constructs its own characteristic type of web or snare, and each web design is intended to trap a particular kind of food. Some spiders hide in a silken tube that has thin tripwires radiating from it. When a crawling insect touches a wire the spider senses the vibrations and rushes out to seize and kill the victim with its poison fangs. The orb web, spun by the garden spider, catches flying prey. The spider either waits at the hub of the web or hides nearby, behind a leaf or in a crevice. The sheet web, made by the house spider, is a flat interwoven mass of threads that ensnares any prey, flying or crawling.

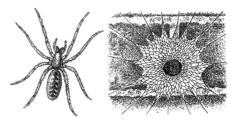

Tripwire spider with radial web

House spider with sheet web

Garden spider with orb web

Inside the home

All these specimens were garnered from dwellings in England and France, and they give a wonderful cross-section of the diversity of life that can share your house with you. Scorpions help to clean up excess insects; a wasps' nest is a masterpiece of architecture; a beautiful peacock butterfly decorates a lampshade in the vain hope of finding a dark quiet winter retreat. Of course there are a few unfriendly things that tend to undermine the fabric of your dwelling, but apart from these your house can be a living museum, full of amusement and interest.

Dry rot fungus

Woodworm

Timber destroyers Dreaded dry rot is a fungus originating in the Himalayas. Top piece shows attack, lower is actual fruiting body. Woodworm holes in plywood mark exits of adult furniture beetles.

Garden snail

Cranefly

Scorpion feeds on insects and spiders, and lives in damp corners.

Wasps' nest Found deserted in an attic, this wonderful geometrical structure was made from chewed wood glued with saliva.

Housefly

Humming-bird hawk moth

Worker wasp

Leather or hide beetle

Queen wasp

Muslin moth

Peacock

Butterflies and moths may stray indoors in search of warm winter quarters.

Old tomato stored and forgotten in a humid part of the kitchen has been almost eaten away by pin mould fungus carried there by airborne spores.

Lower jaw of young rat

An unfortunate mouse was trapped behind a deep-freezer in an outhouse.

Nests The rat's untidy mass of bedding was in an attic against a warm chimney flue. The robin's was lodged behind flowerpots on a conservatory's high shelf.

Rat's nesting material

Deserted nest of robin

over the beams, but they also put paid to any visiting mice or rats who might have ventured uninvited into the house.

Friends and enemies

There are some household inhabitants which are a help to us because they prey on the creatures that *do* destroy our goods or irritate us. These friends include spiders, centipedes and geckos that eat anything from bluebottles to mosquitoes. There are also many innocuous creatures in the house—innocuous in the sense that they do not destroy anything that we think particularly valuable. The charming musical house cricket, for example, should be encouraged and appreciated as they used to appreciate them in China. The Chinese used to keep crickets and have singing matches with them which were a very popular form of sport.

If you prefer not to share your house with other animals, that is your prerogative, for your home is your territory. On the other hand if you enjoy them—as I do—you will find that they will make your house interesting. It is fun to watch the habits of a spider, say, in her web in one corner of the room, catching insects and tending her young, or the strange antics of the craneflies that come into your house at night, or the moths and butterflies emerging from their pupae on the walls outside.

Certainly, you must have some sort of control over things that carry diseases or bite or sting you or extensively damage your food or clothes. The best approach, of course, is to try to prevent an invasion. But if you've already been invaded, try "natural" controls, like encouraging the predators of unwelcome house guests—though you may get complaints if you try to keep a boa constrictor instead of a cat!

Creatures of the walls and roof

Of course, it is not just the inside of your house that forms a habitat for creatures—a lot of them use the outside as well. In Corfu, the geckos lived under the tiles of the roof and used the walls at night as hunting grounds for moths and other insects. Spiders are particularly fond of hunting around the outside of the house. The overhanging eaves provide places where wasps can build their mud or paper nests or where swallows and house martins can found their colonies. Under the eaves you may also find pupae of butterflies and moths, protected from bad weather.

If you have shutters at the windows which you don't often use, you might find that they have been commandeered as roosts by bats, and in warmer areas the smooth cliffs of the outside of your house will be hunting grounds for praying mantises, tree frogs and diurnal lizards.

In a brick house you will find that earwigs, spiders and woodlice use the cracks in mortar as hiding or egg-laying places, and there is a certain bee—the red osmia—that actually burrows into the mortar between bricks on sunny walls to make her egg chambers. Loose and crumbly mortar is searched for food by woodpeckers and sparrows. On old buildings where there is moss between the bricks or stone you get a host of tiny creatures such as mites, millipedes and spiders, and at night snails and slugs parade up and down these miniature forests.

My French garden

Now consider the outside rim of your habitat, where you can find extraordinary things to observe whether you have a few flowerpots or are

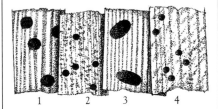

lucky enough to have a large garden. The type of garden that I really like best is one that is allowed to run wild, like my garden in the South of France. I suppose I should not really dignify it with that name since what it really consists of is about ten hectares of wilderness: hillsides covered with aromatic herbs, such as rosemary and thyme, small forests of juniper, holm-oak and umbrella pines and fig, olive, damson and walnut trees that have run wild.

It is only in the vicinity of the house that we make any attempt to "control" nature and this is very mild. It consists for the most part of training the old man's beard or mile-a-minute creeper over the bamboo-covered patio, planting a few vegetables for ourselves and rose bushes, buddleia and lavender to attract the insects and making sure that in the hot weather the tap is always dripping on to the stone-flagged patio, forming a cool oasis where the small wasps and the butterflies and beetles of the dry *garrigue* can come to drink. As the pool formed by the tap is just outside our living room door, we have to take care that we don't tread on a wasp and get stung as we come out of the house. I covered the patio with bamboo for I know that these hollow tubes form admirable hiding places for numerous insects, especially for the huge electric-blue carpenter bees, one of the biggest and stupidest but most colourful of the wild bees. All day long you can hear them scraping and scrabbling and humming to themselves as they build their nests inside the long lengths of hollow sun-baked bamboo.

Surrounding the garden are dry stone walls which make wonderful homes for many things. Slender brown wall lizards slither like quicksilver over the stones, huge eyed lizards, green as dragons, bask on top of the walls and in amongst the stones there live scorpions, natterjack toads, garden dormice and Montpellier snakes.

Everything that moves in we make welcome, whether it is a plant or an animal. To me, it is an ideal garden for a variety of reasons. Take butterflies, for example, who, although they may feed off our lavender, will need another food plant for their larvae to live on. The lovely graceful swallowtail butterflies come and drink from our lavender, pirouetting in circles, dancing their mating dance, but when they are ready to lay eggs they drift like giant snowflakes down the valley and deposit their eggs on the feathery wild fennel that grows there. The eggs hatch out and the larvae are minute hairy black dots, but as they feed and grow they shed their skins and end up as great fat green caterpillars with

Garden watering hole
Creatures in hot places will appreciate a supply of fresh water. If you have a garden tap you can attract numerous insects on hot days by leaving it dripping or turning it on at regular intervals to make a fresh clear pool. The tap on my patio draws flying visits by graceful butterflies such as swallowtails, as well as various types of wasp.

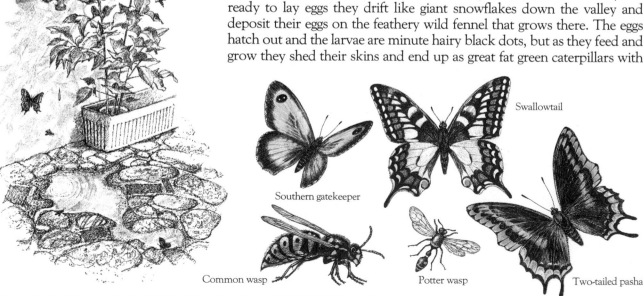

Swallowtail

Southern gatekeeper

Common wasp

Potter wasp

Two-tailed pasha

black stripes. So I can follow the whole life history of the swallowtail within a hundred metres of my house. Oddly enough, in Britain this butterfly (which is confined to the Norfolk Broads) has milk parsley as its caterpillar food plant.

Here, sitting on the patio, I can watch the praying mantis stalk butterflies or moths or spiders in the creepers above me. I can watch hunting wasps searching in the walls for spiders or see the strange decollated snails (with their elongated shells which look as though their tips have been snipped off with scissors) who live among the irises that grow around the bases of the almond trees. At twilight the longhorn beetles flying past look like witches on broomsticks. On the hillsides above and below the house I can go and lift up great slabs of rock, each embossed with yellow seals of lichen, and beneath them I might find the big creamy-white Spanish scorpions or a tarantula, while on top of the rocks there are the papery egg cases of the praying mantis, looking like elongated haystacks. Among the miniature jungle of herbs, richly aromatic as you crush them underfoot, are dozens of different grasshoppers and fragile moths and hundreds of microscopic creatures. To me, a garden like this is perfect. You can have your flowers and share them, you can have your vegetables and share them and then you have your wild hillsides which you share with whatever lives there. You feel part of nature and not divorced from it and yet, ten minutes down the road lies the large bustling city of Nîmes.

Exploring your garden

Whatever or wherever your garden is, it is worth investigating. Just sitting quietly there and waiting to see what appears can be a very exciting occupation. Early in the morning you hear the birds waking up, and then as the sun warms things a little, you see the emergence of diurnal animals. At dusk, a completely different set of creatures takes over.

Gardens old enough to have a variety of trees and plants can obviously support a much greater collection of creatures than a freshly-created one. The rapidity with which a new garden is colonized depends on how far away you are from a potential source of immigrants, and your garden visitors will be more varied the closer you live to the countryside. Don't forget that plants are at the first level of the various food chains, and so they determine what the higher links in the chain are going to be. Of course, exactly what plants grow is determined by many factors but especially by the soil—a water-loving plant won't thrive in sandy soil, for example. Also the air pollution in industrial areas has its destructive effects all along the chain of garden life.

A garden is a good place to study the life histories of plants, since you're on the spot, so to speak, and can observe their conditions of growth, when and how long they flower and set seed and what creatures help or trouble them. Even if your garden is a very carefully planted one with neat flower beds and vegetables in rows, you will still get a host of visitors and residents.

The flower garden

A well-planted flower garden can attract a great number of nectar feeders, ranging from bees and certain flies to beetles, moths, butterflies and, in tropical and Northern America, even the beautiful multicoloured

At home in the grass
The common green grasshopper is found in all grassland habitats, including that patch of rough grass at the bottom of the garden. When disturbed it leaps powerfully into the air and flies a short distance to safety. Grasshoppers can be distinguished from crickets by their antennae. The former has short thickish antennae, whereas a cricket possesses thin slender antennae which may be much longer than its body.

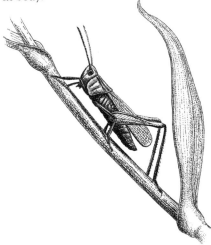

ROSE JUNGLE

Garden roses support a wealth of animal life, ranging from the sap-sucking aphids to the aphid-hunting birds. The old-fashioned types of rose are the best, for they have rich scents and plentiful nectar to attract many different animals.

Nectar feeders

The rose makes sweet nectar as bait to lure insects. As they drink the nectar the insects collect a dusting of pollen which they transfer to other rose flowers on their travels, ensuring cross-fertilization.

Red admiral

Bumble bee

Convolvulus hawk moth

Aphids and their predators

The aphid (greenfly) pierces the rose stem with its needle-like proboscis to suck the sap. Aphids breed at a phenomenal rate; one female can produce 50 offspring in a few days. Such rapid breeding is essential because aphids are virtually defenceless.

Rose aphids

Great tit

7-spot ladybird

Lacewing larva

glittering hummingbirds. Most of the nectar-feeding insects not only use the flowers in the garden as a rich food source but help, by their activities, to pollinate the plants.

Rose bushes are veritable treasure troves of insect life. Here you can find the ubiquitous sap-sucking aphids. Ants love them for their "milk" and actually herd them and build "barns" of tiny sand grains to protect them. Aphids also are a source of food and disguise for the larvae of the lacewing fly. This delicate insect with fragile transparent wings and huge golden eyes lays its eggs on long stalks among the rose leaves. When the larvae hatch out, one of their chief forms of food is the aphids. They suck them dry and then toss the drained shells onto their backs where they become entangled in their bristles. A sufficient number of aphid shells attached to the larva forms a very effective camouflage. Ladybirds, who always look to me like newly-painted clockwork toys, also feed on the aphids as do their larvae. Butterflies, beetles and hoverflies drink rose nectar and the larvae of the latter feed on the aphids. The leaf-cutter bees slice out semi-circular pieces of the rose leaf to construct cells for their young. Sometimes, you will perhaps be lucky enough to find the extraordinary treehopper, one of the bugs, which has the top part of its thorax protruded in a curved spine so that among the rose thorns it is very difficult to detect. The great haul of aphids on a rose bush often attracts the birds, and I have seen various tits, the goldcrests and even nightingales feeding on this rich bounty.

Sunflowers are a rich larder for a lot of creatures. I remember an actress friend of mine, who has a farm not far from me in a valley in Provence, once planting a huge field of sunflowers. She planted them not as a crop but because she was making a film on the famous painter van Gogh, and she needed the flowers as a backdrop to some of the scenes. It was astonishing what different species that one field of sunflowers attracted to her little valley. Tits and other birds clung to the golden flowers, feeding on the insects that the giant sunflower heads attracted. Some of these heads, bigger than a soup plate, were so heavy that they drooped over and each one contained dozens of kinds of insects among its petals. Then when the flowers died and each "soup plate" became a seed head, there was an influx of other creatures. Rats, mice, dormice and squirrels shinned up the tall stems and stole the milky seeds, and they were followed by a sounder of shaggy wild boar who, with much squealing and grunting, trampled the field of sunflowers flat and with much chomping and clattering of tusks ate up the remains of the seed heads. The stalks of the sunflowers lay and rotted gently into the soil, forming food for worms and a myriad of microscopic creatures.

Naturally, the wealth of life in flower gardens attracts predators. In warmer climates, praying mantises stalk through the flowers on the lookout for succulent butterflies, and crab spiders, who can change their colours to match the petals, crouch in the flowers themselves. Other spiders sling their webs between the flowers, and the ichneumon wasps seek out the plump caterpillars on which to lay their eggs so that, when they hatch, the caterpillars form living larders for the young ichneumons.

One of the fiercest predators is the common wasp. It stings its prey (caterpillars, flies, spiders) and then flies off with the quarry to its nest. Here the victim is used as food for the wasp grubs. It is astonishing to see what incredible weights wasps can carry—sometimes prey much bigger

Garden and back yard

This group of specimens was collected from several ordinary back gardens within commuting distance of London. It shows that anyone's garden, however small, can be a miniature nature reserve in the awful wilderness of bricks and mortar with which man surrounds himself. Get delight—and knowledge—from watching plants and creatures that share your garden with you. Simply examining a spadeful of earth or turning over a stone can be an indication of the amazing complexity of life that exists at your feet. Our specimens were collected in autumn, with the cockchafer—a zooming blundering character of spring—found naturally preserved under a spade, of all places.

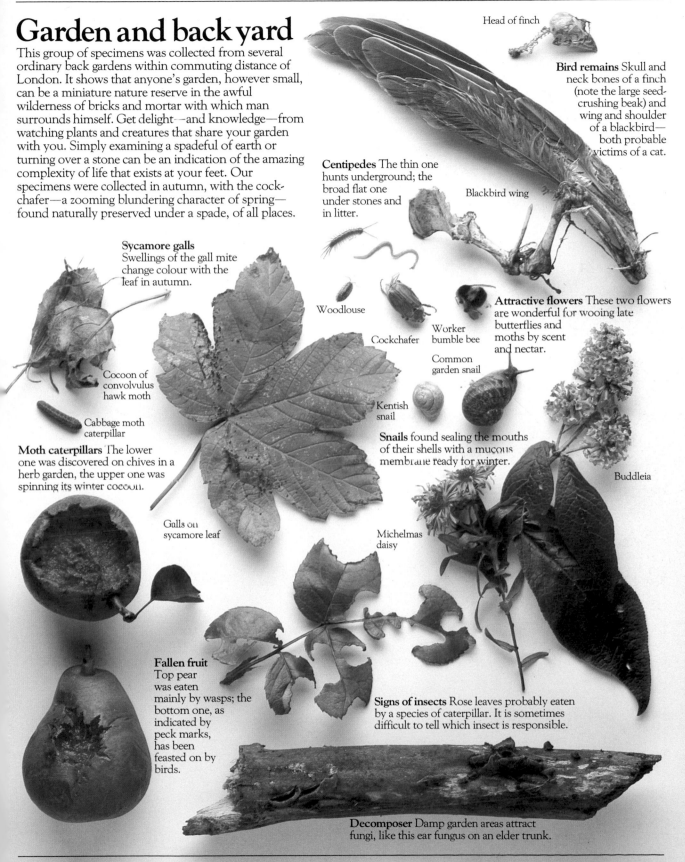

Head of finch

Bird remains Skull and neck bones of a finch (note the large seed-crushing beak) and wing and shoulder of a blackbird—both probable victims of a cat.

Blackbird wing

Centipedes The thin one hunts underground; the broad flat one under stones and in litter.

Sycamore galls Swellings of the gall mite change colour with the leaf in autumn.

Woodlouse

Cockchafer

Worker bumble bee

Attractive flowers These two flowers are wonderful for wooing late butterflies and moths by scent and nectar.

Common garden snail

Cocoon of convolvulus hawk moth

Cabbage moth caterpillar

Kentish snail

Moth caterpillars The lower one was discovered on chives in a herb garden, the upper one was spinning its winter cocoon.

Snails found sealing the mouths of their shells with a mucous membrane ready for winter.

Buddleia

Galls on sycamore leaf

Michelmas daisy

Fallen fruit Top pear was eaten mainly by wasps; the bottom one, as indicated by peck marks, has been feasted on by birds.

Signs of insects Rose leaves probably eaten by a species of caterpillar. It is sometimes difficult to tell which insect is responsible.

Decomposer Damp garden areas attract fungi, like this ear fungus on an elder trunk.

The hunting wasp
Wasps use their stings to paralyze or subdue their prey. If the victim is too large to be taken back to the wasp's nest (and a wasp can carry more than its own weight in flight) the wasp may cut it up and ferry back the pieces one at a time.

than themselves. Watching one struggle into the air with a horsefly, for instance, is like watching the lumbering take-off of a jumbo jet.

The vegetable garden

A vegetable garden can provide the naturalist with some fascinating creatures. Apart from the butterflies and moths that lay their eggs on such things as cabbages and carrots, there are the leaf and flea beetles and in Europe and North America the dreaded Colorado beetle on potato leaves. On potatoes, too, in Europe you might find the larvae of the death's-head hawk moth, one of the most spectacular of the hawk moths.

Many creatures, of course, live underground, feeding on the roots of the plants—the strange subterranean keeled slugs, for example, and some of the millipedes. There are the larvae of those strange insects, the click beetles, so called because of the noise they make when they propel themselves in leaps into the air. They do this to escape enemies and also to right themselves if they land on their backs. (Put one in the palm of your hand on its back and watch the extraordinary performance.) The

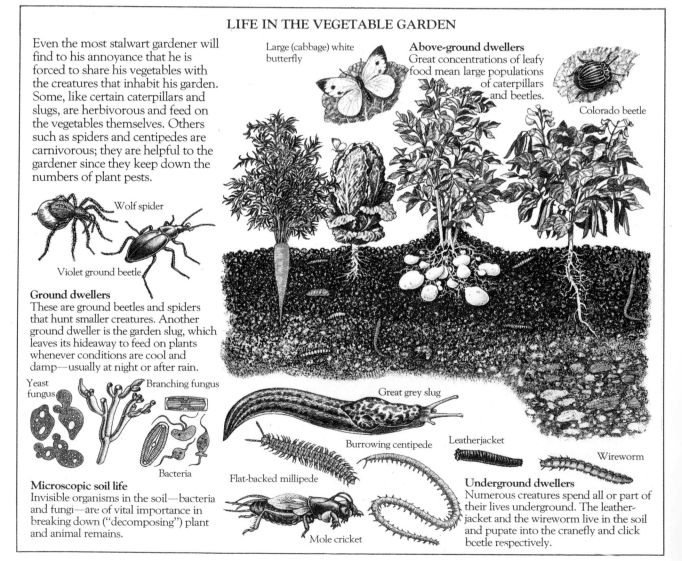

LIFE IN THE VEGETABLE GARDEN

Even the most stalwart gardener will find to his annoyance that he is forced to share his vegetables with the creatures that inhabit his garden. Some, like certain caterpillars and slugs, are herbivorous and feed on the vegetables themselves. Others such as spiders and centipedes are carnivorous; they are helpful to the gardener since they keep down the numbers of plant pests.

Wolf spider

Violet ground beetle

Ground dwellers
These are ground beetles and spiders that hunt smaller creatures. Another ground dweller is the garden slug, which leaves its hideaway to feed on plants whenever conditions are cool and damp—usually at night or after rain.

Yeast fungus

Branching fungus

Bacteria

Microscopic soil life
Invisible organisms in the soil—bacteria and fungi—are of vital importance in breaking down ("decomposing") plant and animal remains.

Large (cabbage) white butterfly

Above-ground dwellers
Great concentrations of leafy food mean large populations of caterpillars and beetles.

Colorado beetle

Great grey slug

Burrowing centipede

Leatherjacket

Wireworm

Flat-backed millipede

Mole cricket

Underground dwellers
Numerous creatures spend all or part of their lives underground. The leatherjacket and the wireworm live in the soil and pupate into the cranefly and click beetle respectively.

leap is produced by a complex mechanism which lies between the thorax and abdomen and has such a spring-like quality that the beetle can shoot rocket-like out of your hand. Then there are the eelworms, a type of roundworm whose young tunnel into the potato roots. On carrots you may find carrot fly larvae and in celery beds the larvae of the celery fly.

The lawn

Look closely at your lawn or any stretch of grass. Lawns are really rather like carpets, and if you could roll them back you would see a fascinating variety of life living in and under them. On the lawn itself, there may be wild strawberries and violets and patches of dandelions and here and there clumps of mushrooms or toadstools. Don't forget that toadstools and mushrooms have their own special inhabitants, certain species of beetle and fly which live on them, generally sheltering beneath their caps.

In the grass itself, you can find tiny leafhoppers feeding on the blades during the day and larger grasshoppers that whirr from under your feet as you approach. Feeding on the roots of the grass crop you will find the so-called leatherjackets, the larval form of the cranefly who, when they emerge from their pupae, dance over the lawns in the evening in mating flights. Also on lawns you can see the mounds covering the entrance to ants' nests like small mountains of pepper. Their size belies the extensive nests and tunnels that run beneath the soil, and you can see the ants' well-trodden roads through the grass stems along which they go foraging.

Earthworms are abundant in the soil of lawns. They tunnel near the surface, eating soil, dead grasses and other organic matter as they go. You will see their "casts", like little earthen decorations for a birthday cake, that they void on to the lawn's surface. All plants benefit from these activities, for worms aerate the soil and provide drainage with their tunnels and from their deeper burrows they bring up minerals which have been carried by the rainwater too deep for the plants' roots to reach. A few centimetres down in an old lawn you may find stones which have sunk into the fine particles of soil tilled by the worm. Fertile soil is due largely to the action of earthworms—Charles Darwin said that he doubted whether any other animal has played such an important part in world history. Worms really could be said to be the first gardeners.

In the evening, small ghost-like moths dance above the lawn attracting their mates and in the spring cockchafers hatch out and go buzzing through the trees like demented aeroplanes. Birds love lawns since the smooth surfaces provide them with an ideal terrain for spotting food. On most lawns you will see thrushes and blackbirds searching for worms and insects, accompanied by starlings and wagtails, and no lawn in America is to be seen without its attendant American robins and blue jays.

The orchard

Many creatures feed on the fruits in the orchard. You may locate the "tents" of the lackey moth on the apple and plum trees, or the tiny codling moth whose larvae eat apples and then pupate in tiny cracks in the bark of the tree. Fruit bushes attract other species, like the handsome pied magpie moth whose larvae (one of those lovely "looper" caterpillars) feed on the leaves of currant and gooseberry bushes.

Of course, when the fruit ripens and falls and splits and starts to ferment, it forms a feast for wasps and bees and various flies. Hedgehogs

Phantoms of the grass

The ghost swift, sometimes called the ghost moth, is aptly named. The white males can be seen at dusk in June or July, hovering and swinging from side to side in the air like tiny ghosts above patches of grass. They do this to attract females—which is unusual in the moth world, since it is generally the males who seek out the female by scent.

in particular are very partial to fermenting apples and grapes. In Greece and in England I have watched hedgehogs getting drunk on apples and grapes and then they stagger about, screaming abuse at each other, fighting, bumping into things and generally behaving in the vulgar way that most human drunks behave. In an apple or pear or cherry orchard, especially if it is surrounded by a hedge, you will find wood mice attracted there by the insects that feed on the fallen fruit. The fruit and the various rodents that accompany it attract badgers, foxes, weasels and stoats. The flowering of the apple, cherry and pear blossom attracts many birds for there is a great variety of insects to be had who are drawn by the sweet fragrance of the clouds of lovely blooms. Many bird species eat blackcurrants and redcurrants, gooseberries and raspberries. In North Carolina in the USA I have even seen box tortoises lumbering about among the strawberry beds, gorging themselves on the fruit (to the farmer's fury) and in Greece I have seen foxes feeding greedily off the grapes on low-growing vines.

Creating a wild garden

By letting part or all of your garden run wild and by careful planting you can do a lot to bring wildlife to you. Remember that each garden is a miniature nature reserve and you should try to manage it like that, working out desirable food, hiding places and nesting sites for all the animals that you can attract. In Great Britain over a quarter of a million hectares of gardens provide a haven for many creatures whose wild habitat is shrinking.

A good compost heap not only benefits you and your gardening activities but it is the home and larder for scores of different invertebrates—worms, snails, slugs, woodlice, earwigs, fly larvae and so on. If you want to attract the nectar feeders, try not to plant the modern type of bloom which is bred to be big, blowsy and brash, but has no scent. Try and get the more old-fashioned garden flowers, like old-fashioned roses which have rich scent and nectar. Blooms with trumpet-shaped flowers, like petunias, for example, are greatly loved by hawk moths, and if you live in America, you can plant hibiscus which will attract hummingbirds to your garden. Night-flying moths love flowers like tobacco plants and night-scented stock. In the daytime, lavender and buddleia attract many butterflies and other summer insects as does the later-flowering ice plant.

Choose your assortment of plants so that there is something flowering from spring through to late autumn, and remember to grow some wild flowers too, since if you want to keep your butterflies and moths in the garden you not only have to feed them but you have to provide them with the natural food plants for the larvae when they hatch out. So don't be too fussy about your garden—let a corner of it run wild with brambles, thistles, stinging nettles and dock and, of course, the various grasses. You will find the adult moths and butterflies will feed on the flowers and the larvae will feed on the leaves. A little section like this in your garden will also give shelter to hedgehogs or toads, ground or low-nesting birds and even slow-worms and lizards.

Be sure to encourage bushes that produce small succulent fruits. Blackberry, raspberry, blackcurrant and redcurrant will attract birds and insects from midsummer to autumn and will provide nesting sites for

The compost heap
Compost heaps are excellent homes for larvae of flies and beetles and for adult insects as well. One of the major attractions of a heap is the abundant food supply in the form of dead and decaying vegetation and other animal inhabitants. The warmth generated by rotting plant material allows the insects to go through their life cycle in a shorter time than usual, and in some cases to breed through the winter.

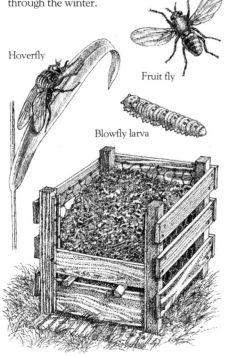

Hoverfly

Fruit fly

Blowfly larva

GARDENING FOR WILDLIFE

The butterfly garden

To attract butterflies and moths, plant a selection of highly-scented nectar-producing flowers like buddleia, primrose, lady's smock, lavender, lilac, forget-me-not, honesty and michaelmas daisy. Night-scented flowers such as petunia, evening primrose, stock and tobacco plant attract moths, and sweet rocket and honey-suckle draw butterflies and moths. You must also supply food for caterpillars as well as adults. Many common caterpillars eat stinging nettle; chop down some of the nettle bed every month to provide soft young plants. Bramble, dock, thistles and grasses are also good foods.

Buddleia

Hummingbird hawk moth

Honeysuckle

Peacock

Honesty

Apple blossom

Ivy

Primrose

Thistles and nettles

Nettle

Bramble

Song thrush

Blackbird

Robin

The bird garden

Three things will attract birds to your garden—food, water and nesting sites. Many birds like berries, seeds and nuts such as barberry, viburnum, hawthorn, elder, cotoneaster, holly, rowan, sycamore, honeysuckle, rose and sunflower. Thistles, teasels and grasses tempt small seed-eating birds, and soft fruits and berries encourage fruit-eaters, besides attracting insects which in turn lure insectivorous birds.

Water for drinking and bathing, with perches nearby, can be a shallow area in a garden pond or a simple bird bath like an up-turned dustbin lid. Daily fresh water in freezing weather will be much appreciated.

The garden pond

A pond will bring many new species of plants and animals into your garden. Stock the various depths of water with plants and animal life such as water snails and insects will soon follow, their eggs being carried on the plants by the wind and on the feet of birds and animals. Introduce small fish and frog, toad or newt spawn collected sparingly from a wild pond.

Bog asphodel

Marestail

Frogbit

Shallow area Sedges, reeds, bog asphodel, marsh ragwort, lilies.

Deep area Water starwort, water plantain, marestail, pondweeds.

Floating Duckweed, frogbit, bladderwort, floating fern.

FOOD FOR BIRDS AND OTHERS

Birds like kitchen scraps, suitable left-overs (they would not appreciate a curry but would appreciate boiled rice), cheese, cake and bread, as well as usual bird food—peanuts, maize, millet, mealworms and "ants' eggs". Put out extra food in cold weather, and whatever you feed them do it regularly (early every morning is best) and make sure the feeding site is predator-proof. Insects will be attracted by water sweetened with jam, syrup or honey, but make sure they cannot drown.

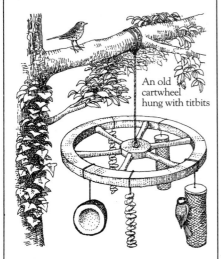

An old cartwheel hung with titbits

Nesting sites

Two sorts of nest boxes are shown below. An entrance hole of 30 mm is best for tree sparrows, small tits and nuthatches, while 40 mm will admit house sparrows and larger tits. The open-front type is used by pied flycatchers and robins. Twist a handful of hay and stuff it into a hedge or ivy wall; it will be taken as nesting material by birds or may be used as a nest by mice. Old flowerpots may be used by snails, frogs, robins or redstarts.

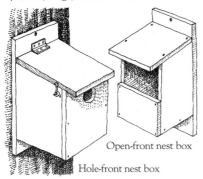

Open-front nest box

Hole-front nest box

thrushes and robins. Sunflowers will attract seed-eating birds—finches and sparrows and many others—in the autumn.

Remember that a number of butterflies and moths hibernate, so plant things that will provide them with nourishment when they wake up (you can even put out small pots of sugar solution, which they will welcome). As you attract more insects you will attract their predators with them, so that soon you can build up a complex community even in a small area.

Trees—providers for innumerable guests

If you have room in your garden do try to plant some trees. This is becoming increasingly important as trees in our forests and hedgerows are being chopped down to make way for agriculture. Trees are the most gracious hosts to wildlife, providing them with both shelter and food. Their importance can be understood when you realize that a single oak tree can support about 300 different species of insect and also provides food for over a hundred species of moth. All these, of course, are in their turn food for mammals and birds. So, although it is slow-growing, try to plant an oak tree. If you do, you will be planting the equivalent of a supermarket for your local bird population, from tits to woodpeckers, from warblers to shrikes.

Hazels are also good to grow, as are beech trees, for their nuts will attract squirrels and dormice, and when the nuts fall to the ground they provide sustenance for voles and field mice. Birds gorge themselves on the fruits of the berry-bearing trees, such as elder, holly, hawthorn and blackthorn, in preparation for their autumn migrations, and will come back to build their nests in spring among the dense protective foliage. Holly berries are available throughout the winter, and the resident thrushes, starlings and wood pigeons come to rely on this supply during a harsh season. Be sure to plant both male and female hollies.

Every sort of fruit- or nut-bearing tree you can plant will help your wildlife, for not only will trees provide food and nesting sites for mammals and birds, but also roosting places and perfect perches from which birds can sing. So never underestimate the importance of a tree, because it can be as bustling with wildlife as a city is with human beings.

Attracting wildlife

So far, I have only mentioned natural attractions for wildlife in your garden—you can of course add all sorts of other things. You can make feeders and special nesting places for anything from earwigs to owls. With very little trouble you can build a garden pond and, apart from the joy of watching the birds bathing, you can expect to see a whole new spectrum of animal life develop, quite different from that of a typical back garden. You may also add to it by putting dragonfly larvae or newts or sticklebacks in it. My sister has a small pond some two metres by three in her garden and she introduced some frog spawn into it. This hatched out and, naturally, the tiny frogs thought this was their real home, and they have stayed in the garden ever since (although they could easily escape), living around the pond and breeding successfully each year.

It is amazing how nature responds if you try to help it. Even in a great city, in a landscape of bricks and mortar, a handful of soil and a few plants can grow into a habitat for the creatures in this alien terrain that need a sanctuary.

SIMPLE GARDEN EXPERIMENTS

Your garden is an excellent place in which to investigate the habits and behaviour of various small animals, since it encompasses a great number of microhabitats and, after all, you are living in the middle of it. You can make daily, hourly or even minute-by-minute observations, and check your traps often (you should *always* regularly service any trap you set, and remove it as soon as you finish your trapping project). The projects described here are really just starting points; you will be able to go on from these and devise all sorts of fascinating experiments of your own.

How to make a pitfall trap
A simple pitfall trap (below) to catch insects and other small animals can be made as follows: Obtain a fairly large jam-jar about 75 mm across and 150 mm deep. Dig a hole, and sink the jar so that its rim is level with the soil surface. Place a piece of tile, slate or wood on four stones so that it is supported 25–50 mm above the mouth of the jar. Small surface-dwelling animals will fall into the jar and be unable to climb out because of the steep smooth sides (the same principle by which spiders appear in the bath), and the cover protects the trapped creatures for a while from rain, sun and predators such as hedgehogs and shrews.

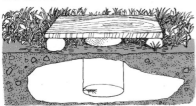

What attracts bees to flowers?
In the summer when bees, wasps and other insects are visiting flowers for nectar and pollen, you can learn what attracts them to the blooms. Is it scent or colour? Obtain some glass or plastic filters from a photographic supplier that filter out some colours of the spectrum but let others pass through. Be sure to get a clear one and one that lets through ultraviolet light but nothing else. Set up your experimental table near a flower bed active with insects. Flatten various flowers between sheets of white card below and various filters on top, and see which insects are attracted to which parts of the spectrum. They will probably be attracted to flowers under the clear filter and also to those under the ultraviolet one, even though to us the ultraviolet filter cuts out the flower's colours. This is a clue to the fact that many insects can see ultraviolet rays and use them as a guide to which flowers to visit. Those insects who do not visit the table at all probably use scent as a guide—they cannot "smell" the flowers under the filters.

Experiments with pitfall traps
1 Position several traps in one part of the garden—along the hedge, say. Check them in the early morning, at midday, and in the early and late evening. When are the animals most active? And does their activity relate to the weather conditions?
2 Sink traps in various garden microhabitats such as the lawn, flower bed, compost heap and vegetable garden. Compare the types and numbers of animals caught in the different areas.
3 Mark some trapped beetles and other creatures with small dabs of paint (see right), and then release them where you trapped them. Do they get trapped again, and if so, is it in the same trap? This should tell you whether the creatures stick to a home range.
4 Bait some traps with meat, cheese or fruit. Do different baits catch different kinds of animals?

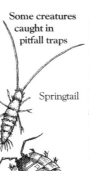

Some creatures caught in pitfall traps

Springtail

Woodlouse

Earwig

Marking small creatures
Many experiments require you to mark the animals in some way. The usual method for small things is simply to paint on a tiny dab of quick-drying typist's correcting fluid (right). Place the paint somewhere inconspicuous (so as not to attract predators) and on a hard flat part of the body. Very active animals can be slowed down for marking by an hour in the fridge or a few seconds in an anaesthetizing jar.

Snail habits
Snails tend to be creatures of habit. They often have a "roost" to which they return after the night's feeding travels. Put broken flowerpots in damp areas of the garden (right). After a few days some snails will probably be using the pots as daytime shelters. Mark them with dabs of paint (try not to disturb them too much); then check your shelters each day. Do the marked animals come back? What happens if you move the pot at night while it is empty, or put several other flowerpots nearby?

THE NATURALIST IN
MEADOWS AND HEDGEROWS

In a man-dominated landscape the naturalist can still find plenty of material in the meadow—a loose term meaning any land that was once cleared for farming, but which now is left for haymaking or pasturage. Meadows are sometimes called (appropriately enough) "tame" grasslands, for the activities of the farmer and his grazing livestock prevent the area from reverting to the original woodland vegetation. Hay meadows, of course, are sown by the farmer with plants that he knows his animals will like—rye, cocksfoot and legumes like clover and medick, which "fix" nitrogen. These sorts of meadows are mown every spring and ploughed under every few years which enriches the soil with nitrogen.

Some hay meadows that have not been sown or ploughed for a few years show definite changes. These undisturbed meadows are the best from the naturalist's point of view. Grasses still predominate, but you can see many other herbaceous plants and perhaps the tentative return of a few woody shrubs. Wild flowers establish themselves among the grasses, like the lovely yellow early-flowering lesser celandine, cow parsley and the speedwells with their beautiful blue flowers like tiny fragments of summer sky. There are buttercups, ragworts, plantains, docks and the bladder campion, so named from the curious bladder-like base of the flower which is all you can see when the petals close up in bright sunlight, and they look like little Chinese lanterns. This tapestry of wild flowers is rich in insect life, which in turn attracts swallows, martins and swifts and at night the bats. Voles and woodmice find these areas a paradise, even though they are hunted by the silent owls. In the dawn mist you may see deer grazing or a fox, bright as an autumn leaf, hunting for voles.

A rich meadow such as this sets up complex food webs. The multitude of plants provide roots, leaves, nectar and pollen for a host of creatures who, in their turn, attract the carnivores which prey on them. You can imagine, therefore, how disastrous the habit of early mowing (and ploughing) can be to this diverse world of living things. The flowers are ruthlessly decapitated and as the plants fall they expose the nests of birds and small mammals. Not only the adults but the babies are open to the attacks of predators, like hawks and foxes. The whole web that nature had so painstakingly built up can be shattered in a day and the lives of hundreds of creatures affected. Ideally—but how many farmers are idealists?—hay meadows should be sown side by side so that if one is mown at least some of the creatures have a chance to escape into the neighbouring field, but alas this happens all too infrequently.

Exploring the meadow
The wealth of plant and insect life means that a hay meadow is a wonderful place to collect. Furthermore, the insect life varies according to the time of the day. I remember once finding a lush Dorset meadow,

The reluctant sunbather
Most flowers love bright sunlight, but the bladder campion is not so enthusiastic. This fragrant flower grows in meadows, roadside verges and other dry grassy places. On cloudy days or in the shade the petals are fully open (as shown on the right). In bright sunlight, however, they retreat into the bladder-shaped base of the flower.

A successful partnership
A bumble bee (opposite) searches with its antennae for the nectar in a thistle flower, and receives a dusting of pollen in return. Both partners benefit—the thistle provides the bee's food, and the bee will carry the flower's pollen to other thistles, ensuring cross-pollination.

Straws and spoons for drinking nectar
The mouthparts of various insects have become adapted for obtaining nectar from different flowers. The butterfly's long hollow proboscis is normally coiled up like a watch-spring below its head. When it feeds, the proboscis extends like a drinking straw (left) so that the butterfly can reach nectar at the base of deep trumpet-shaped blooms. The bee, on the other hand, has a shorter jointed proboscis and uses it more like a spoon, drinking with a lapping motion from shallower flowers.

aglow with sweet-scented flowers. I spent the whole morning collecting and by lunchtime I had filled all my available jars and tins and boxes, yet there were still hundreds of specimens I had not acquired. I hastily cycled all the way home, unloaded my treasures, had a quick sandwich and made my way back to this wonderful meadow. By four o'clock my collecting equipment was full again with a completely new set of creatures that I had not seen in the morning. Once more I cycled home, deposited my catch and returned to the meadow where, to my astonishment, I found yet another new set of creatures called out by the sinking sun and the cooling air. It taught me what a treasure trove a field like this can be.

When you get to a meadow and make a reconnaissance the first things that are liable to attract your attention are the larger nectar feeders. Because of their colouring the butterflies will be the most obvious as they flit from flower to flower, and you can note how each species has its own characteristic flight. Then there are the numerous species of bee, as well as the "domestic" variety, and with luck you may see some of the day-flying hawk moths or the very curious and handsome burnet moth which, in some miraculous way, is immune to the cyanide in a killing bottle. Butterflies are attracted to most of the colourful blooms—the reds, blues, violets and so on—and feeding on such flowers you will find clouded yellows, skippers and blues. It is very curious that all the blue butterfly larvae (except the holly blue) have a gland on their backs that exudes a sugary substance which ants love and collect without hurting the larvae. So the larvae act as a sort of ant's milk-bar.

Not all butterflies you see in a meadow are necessarily looking for food; some, like the browns, graylings, ringlets and marbled whites, will be searching for the appropriate grasses on which to lay their eggs.

In addition to the butterflies you will see a number of different species of bees "working" the meadow flowers. There are nearly 250 species of bee in Britain alone, ranging from the honey bee to the well-known bumble bee. An enemy of the bumble bee is the cuckoo bee which, as its name implies, has the same bad habits as the bird. The female cuckoo bee finds a bumble bee's nest in early summer. At first the workers attack her but she employs some method of "charm" and is at last accepted by them. Her next step is to kill the queen bumble bee and become queen

BUTTERFLY FLIGHT PATTERNS

Each family of butterflies has its own characteristic flight pattern, though this is modified depending on what the insect is doing—looking for food, or searching for a mate or a place to lay eggs. Three commonplace examples are shown below; from your studies, you may be able to add extra details for different species.

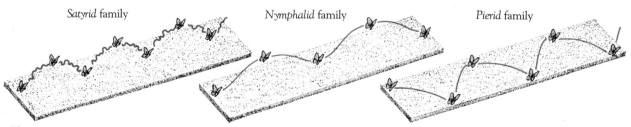

Satyrid family

Nymphalid family

Pierid family

The graylings, ringlets, browns and heaths fly in a straight line but flutter up and down like a roller coaster.

Long low swoops in a straight line indicate a tortoiseshell, fritillary, red admiral, comma, purple emperor or peacock.

A low looping zig-zag flight is characteristic of whites, orange tips, clouded yellows and brimstones.

COLLECTING BUTTERFLIES

Butterflies and other flying insects are best caught with a net designed specifically for the purpose, so that you harm them as little as possible. When I first started collecting butterflies I used to gallop about the fields and hillsides, getting hot, sweaty and exhausted. I soon found out that this was not the way to collect, as the most dim-witted butterfly could outsmart me. The answer is to use stealth and caution. By far the best way of making a butterfly collection is to catch and keep the insects alive and breed them (see page 278), keeping one of the offspring for your permanent collection and releasing the others at the place where you caught the parents. (Always follow the collector's code—see page 320—and never keep a specimen unnecessarily.)

Types of net
The two main types are the round-ended net and the kite net; the latter can be used without the handle. The bag must be of soft black cotton netting, not nylon (which will damage the butterfly's wings). The scissor net is closed quietly over the resting insect.

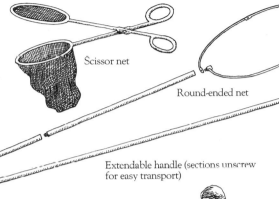
Scissor net

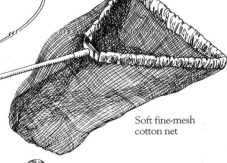
Kite net

Round-ended net

Soft fine-mesh cotton net

Extendable handle (sections unscrew for easy transport)

Using the net
For a butterfly resting on a flower, approach it slowly, making sure that your shadow does not fall on it. A quick sideways sweep (right) and the insect is in the bag. At the end of the sweep, twist your wrist

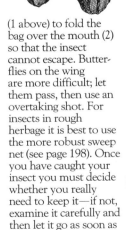

(1 above) to fold the bag over the mouth (2) so that the insect cannot escape. Butterflies on the wing are more difficult; let them pass, then use an overtaking shot. For insects in rough herbage it is best to use the more robust sweep net (see page 198). Once you have caught your insect you must decide whether you really need to keep it—if not, examine it carefully and then let it go as soon as possible, at the place where you caught it.

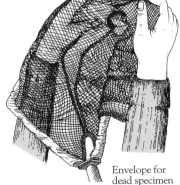

Envelope for dead specimen

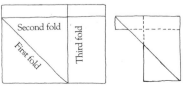
Second fold

First fold

Third fold

Removing the butterfly
A suitable carrying container for live insects can be made from a cardboard tube cut from the centre of a toilet roll. Tape thick black paper over one end, and have black cloth and an elastic band ready to cover the other end. Hold the container inside the net, perch the butterfly on your finger (see left), then manoeuvre it in. To preserve the specimen, load it into the killing jar until dead, then place it in an envelope (left).

Smoking out moths
Small moths and other insects lie deep in clumps of undergrowth during the day. A bee-keeper's smoke machine, or even tobacco smoke blown down a tube, can be used to smoke them out so that you can catch them with a net.

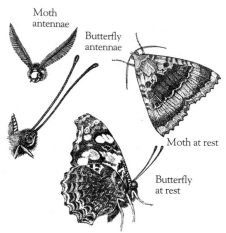

Moth
antennae

Butterfly
antennae

Moth at rest

Butterfly
at rest

Butterfly or moth?
There are three main differences between
butterflies and moths. Butterflies are out by
day while moths usually fly at night, but this
is not an infallible guide since some moths fly
by day. Second, moths spread their wings
sideways at rest whereas butterflies hold them
together over their backs, though again there
are exceptions. Third, the butterfly's antennae
are long and slender with clubbed ends,
whereas a moth's are shorter and feathery.

Colourful inhabitants of the meadow
The top picture on the opposite page shows
an Essex skipper probing into a thistle with its
proboscis. This little butterfly lives all over
southern Europe but only reaches the south-
east coast of England, hence its name. Bottom
left is the six-spot burnet moth, a day-flyer,
resting on field scabious. Its vivid colours (red
is a common warning colour in the animal
world) tell birds and other potential predators
that it is not to be tampered with; it can exude
an extremely distasteful liquid when
molested. The two common blues (right
centre) are mating on horseshoe vetch; the
greenish eggs are usually laid on rest harrow
or birds-foot trefoil. Bottom right is a
marbled white, which is in fact a member of
the *Satyridae* (browns and heaths). The female
scatters her white eggs in the grass as she flies
over meadows and rough pasture.

herself. The workers then tenderly look after her eggs and young, which
hatch out into queens and drones. Once the queens are fertilized, the
drones die and the queens hibernate. When spring comes they fly off and
start other palace revolutions among the poor bumble bees.

Now watching the various bees busy about the multicoloured meadow
flowers you know it is the beautiful colours that attract them, with the
exception of red which they see only as drab grey. However, bees have
an advantage over humans for they can see further up the light spectrum
into the ultraviolet. So what we see as pale yellow or white flowers may,
to the lucky bee, look like a rainbow and be beautifully patterned in these
ultraviolet colours. It is always as well to remember that most living
creatures have abilities which we would certainly consider ourselves
lucky to possess.

Flowers, leaves and stems

In your exploration of the meadow, do not concentrate only on the
colourful flowers. Among the less obvious but sweet-smelling ones you
will find a host of insects—hoverflies, plump furry beeflies and several
species of small dark solitary-living bees will all be gathering nectar there.
Buttercups are interesting; there are two sorts, a "nodding" kind and an
upright kind. The nodding one droops its head in rain and the insects use
it as a sort of yellow umbrella and thus help pollinate it. The upright kind
is more independent, being self-pollinating, so has no need of turning
itself into an umbrella. Another very good source for insects is the
umbellifers—hogweed, cow parsley and so on. Their tiny flowers stand
shoulder to shoulder on stalks as though on the spokes of a bicycle wheel.
Feeding on these you will find all the creatures just mentioned, as well as
many other species of small fly and many beetles feeding on the nectar or
the pollen. There are handsome longhorned beetles, for example, and the
carpet beetles like little clockwork jewels, who look so pretty despite the
fact that their larvae feed on dead and rotting animal matter. The lovely
flower clusters of cow parsley form little worlds of their own: shieldbugs
siesta in them, while they provide rich hunting preserves for their
relatives the assassin bugs, beautiful but deadly and in competition with
the equally lovely and fierce tiger beetles. Robber flies "dive-bomb" these
flower heads after their insect prey, and among the flowers the crab
spiders (who can change colour to match their surroundings) crouch in
ambush for unwary insects. Here you may also see the larvae of the oil
beetle, a curious black insect that looks as if it had sent its wing cases to
the laundry and they had come back badly shrunk. The larva, however, is
an unattractive yellow bristly louse-like creature. This climbs up into the
flowers, waiting for one of the honey-making bees. When one alights the
audacious larva hitch-hikes a lift on the bee which, apparently un-
knowingly, carries it back to its nest. Here the larva gets off the bee and
on to one of the bee's eggs floating in its cell of honey. It sucks the egg
dry and then moults into a naked white grub which gorges itself on the
bee's supply of pollen and honey. It seems to me that most species of bee
lead very troubled home lives with so many unwanted guests, like the oil
beetle and the cuckoo bee.

Do not forget to examine the leaves and stems of the meadow plants,
for here you will find various grasshoppers, leafhoppers, leaf beetles and
weevils, as well as the caterpillars of many different species of butterfly

and moth. If you sweep with your linen bag you will be surprised at the quantity of insects and larvae you obtain. You will probably catch a lot of the adult microlepidoptera, the tiny moths that you see flying at night. Many of their miniscule larvae "mine" the leaf surface of meadow plants leaving little "roads", while other cause blisters and yet others roll the leaves up into tubes.

Down at ground level you get all sorts of organisms. You might find the shaggy ink cap (also called "lawyer's wig"), an elongated mushroom that looks as though it has tiles on its roof. This forms part of the diet of the large black slug which, curiously enough, when full grown is brick red, brown or black with a red "rim" along its foot. The other slug you are liable to see is the field slug. Small compared to the black slug, it is white to brown with irregular dark spots and streaks and with a sharp keel along the end of its back which the black slug does not have. Here on the ground is the domain of the wolf spiders that stalk any insect they can. Under stones you may find resting millipedes, waiting for darkness before emerging, looking like little clockwork trains. Under flat thinnish stones (ones that can be easily warmed through by the sun) you will find ants' nests. Search particularly for red ants, for these make suitable pets if you want to set up a home ant observatory or formicarium (page 281).

The meadow at night
Meadows at night are fascinating for they take on a whole new aspect. Most of the moths, for example, have spent the day resting deep in the cool herbage of the meadow and they now emerge, attracted to the sweet scent of certain pale flowers like evening primrose or bladder campion. This plant, as I have said, closes its flowers in sunlight and so it relies on the moths to pollinate it at night. The nectar-feeding moths have long proboscises like butterflies and so they take over where butterflies leave off; they are on the night shift, so to speak. However, the moths hover like helicopters in front of the flowers as they feed, not perching on the

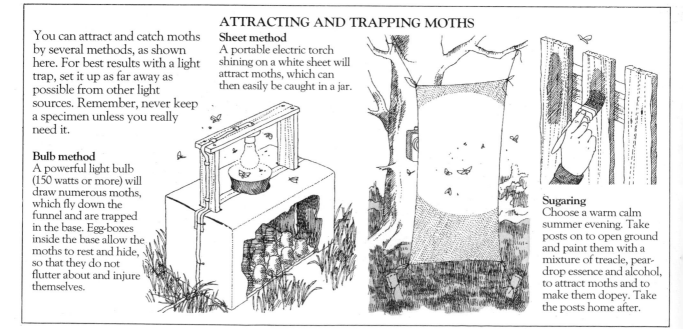

ATTRACTING AND TRAPPING MOTHS

You can attract and catch moths by several methods, as shown here. For best results with a light trap, set it up as far away as possible from other light sources. Remember, never keep a specimen unless you really need it.

Bulb method
A powerful light bulb (150 watts or more) will draw numerous moths, which fly down the funnel and are trapped in the base. Egg-boxes inside the base allow the moths to rest and hide, so that they do not flutter about and injure themselves.

Sheet method
A portable electric torch shining on a white sheet will attract moths, which can then easily be caught in a jar.

Sugaring
Choose a warm calm summer evening. Take posts on to open ground and paint them with a mixture of treacle, pear-drop essence and alcohol, to attract moths and to make them dopey. Take the posts home after.

blooms as butterflies do; and unlike butterflies, who plunder the flowers throughout the day, some moths have only a short feeding flight. That of some hawk moths lasts only about half an hour at twilight.

The largest family of the lepidoptera is the noctuid or owlet family. There are nearly 400 British species, and over 6,000 worldwide. Some of the owlets are extraordinary, for their hearing is so acute they can even dodge bats. As the bat swoops, uttering its supersonic cries that act like radar to tell it where its prey is in the dark, the moth picks up these sounds (inaudible to human ears) and swerves out of the way—a really remarkable feat and, one feels, very depressing for the bat to have such wonderful built-in radar only to be defeated by a moth. Some moths, of course, do not feed at all in the adult stage. Among these are the handsome tiger moths searching the meadow for plants to lay their eggs on, and the ghostly pale swift moths, the females of which simply scatter their eggs over the grasses as they fly, like someone sowing seeds.

If you want to observe moth and other insect behaviour at night in a meadow, use a strong torch with a red light. While you are looking for moths you will find plenty of other activities going on. At this time the larvae of the satyrid butterflies, the noctuid moths and the sawflies, for example, climb high up the grass stems to feed, as do the bush crickets. You might be lucky enough to see the noisy low-flying mole cricket, an extraordinary beast with front legs just like those of a mole.

Down among the bases of the plant stems, the surface of the soil becomes a veritable jungle during the hours of darkness. Webless hunting spiders stalk in the gloom, feeling for any insect prey. The fast voracious *Lithobius* and other centipedes tackle any prey indiscriminately—worms, beetles and other insects, and even the large field slugs. Earthworms are also hunted by the *Carabus* ground beetles. Worms are very shy of light and only at night or on warm damp evenings will you see or hear them nosing around the grass roots, half extended from their burrows as they search for a mate or for a dead leaf to pull down underground. Very early in the morning, just as it is getting light, birds search for the foraging worms, but they do not always succeed in catching them since the worm has tiny bristles along its length and these firmly anchor it in its hole so it can only be tugged free by a fairly hefty bird, such as a blackbird.

Wet meadows

Low-lying areas of farmland, especially where a stream or river meanders through, are kept as permanent pastureland. In times past—and sometimes even today—the fields were flooded regularly in winter with slowly moving water, to prevent frosts and so encourage continued plant growth; these flooded areas formed the water meadows. Here the soil is rich and oozing with moisture and the plants are different from those in the drier hay meadow. Here can be found dainty flowers with such pretty names—the delicate lilac of the lady's smock, one of the plants on which various species of froghopper form their balls of "cuckoo spit", and the great rich golden-yellow kingcups. I have seen a sizeable water meadow so thickly covered with these lovely flowers you could hardly put a finger between them. It looked like a sheet of gold and almost blinded the eye. These blooms are wells of nectar for various insects like the soldier fly, so called because it has two spikes on its back that resemble spears. In water meadows you get the meadowsweet, delicate,

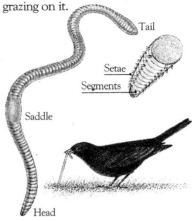

NATURE'S GARDENERS

Earthworms are incredibly numerous. In a meadow there might be over 500 in each square metre, which means the worms in a pasture can weigh more than the livestock grazing on it.

Tail

Setae

Segments

Saddle

Head

Body hairs
On each segment of the worm's body are 4 pairs of tiny hairs (*setae*). Using these it can anchor itself surprisingly firmly in its burrow.

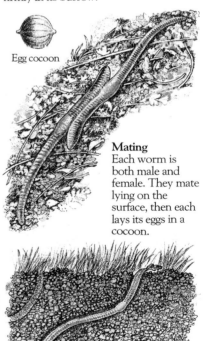

Egg cocoon

Mating
Each worm is both male and female. They mate lying on the surface, then each lays its eggs in a cocoon.

Tunnelling
Worms usually eat their way through the soil, but there is the odd permanent burrow which can be safely closed by pulling in a dead leaf.

creamy and with a beautiful scent, but it has no nectar. However, it is host to many flies and beetles that come to feed on its pollen. You can also find the curious sulphur-throated oxlip, all the flowers pointing in the same direction like a crowd waiting for a procession.

In some areas of water meadows you will find certain plants growing in clumps—bracken, ragwort and the creeping buttercup, all plants unpalatable to the grazing cattle or horses. You may also come across tussocks of tufted hair grass. Now each of these conglomerations of plants represents a sort of island in the meadow and forms a sanctuary for resting moths. You might want to try your hand at smoking out some of these clumps for you are sure to get any number of the microlepidoptera and you might be lucky enough to find a slumbering drinker moth, so called because its caterpillar drinks dew, an odd habit for a caterpillar. In the smoking-out process, of course, numerous other creatures of all kinds will emerge from the vegetation.

Now water meadows are meant for grazing and from a naturalist's point of view this is a gift. Consider every horse dropping and every cowpat as a bounty, and this does not only apply to water meadows. A fresh cowpat, moist and warm, is a most desirable source of nourishment to dung and other beetles and several butterflies; and it forms the warm rich bed for many fly larvae. Some fungi love manure; the dung round-head, with its slimy yellow bell-shaped cap and long delicate stalk, is often seen sprouting from horse droppings. Do not be prissy: dung to a naturalist is a natural treasure provided for him by domestic stock (one of the few intelligent things they do, in fact).

Wet meadows are very lucrative hunting grounds for amphibians. I remember one such large flat expanse of meadow in Corfu which, during the spring rains, would be ankle-deep in water. This, of course, provided ideal conditions for the tree frogs, grass frogs, toads and newts to spawn in, and this is where I went to collect spawn and adult specimens. I very soon found I had a number of rather extraordinary helpers in my collecting. The gipsies from mainland Greece—a black-eyed, brown-skinned, wild-looking people—would bring their herds of cattle (huge black beasts with gigantic curved white horns) to graze on the water meadows. As these great ebony-coloured monsters splashed very slowly along, grazing on the lush grass, they drove the amphibians before them. I found that if I kept just ahead of the herds as they grazed they literally drove the creatures into my net. In fact I was acting like the cattle egrets in different parts of the world; these birds do the same thing to collect grasshoppers and other insect food disturbed by the buffalo or other large animals grazing slowly across grassland.

In the wettest parts of the meadow you will see rushes and sedges and amongst them, if you are lucky, you might come across an amber snail. I was never lucky enough to catch one, to my annoyance, but I knew the story of it which I found bizarre and fascinating. The amber snail is the intermediate host for a kind of worm-like fluke. This means that the snail acts as a home at one period of the fluke's life until the fluke passes on to its final host, a bird. What happens is this: a fluke enters the snail and its larvae take up residence in the snail's horns. Here they pulsate like a lighthouse in green and brown rings, sending up to 70 flashes a minute. This catches the eye of a bird, who would otherwise probably not notice the snail. The bird eats the snail, and out of this complicated business the

The cowpat society
Despite our view of dung, it is a wonderful source of food to small creatures such as flies, beetles and worms. The dung beetle and the yellow dung fly lay their eggs in any crack or crevice they can find in the pat, and the larvae hatch and eat the dung. The dor beetle digs a burrow under the pat and puts several little balls of dung at the end of it. Eggs are then laid in the burrow, so that when the larvae hatch they are well protected and have their food waiting for them.

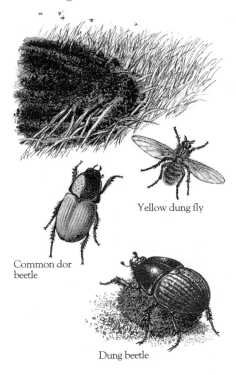

Yellow dung fly

Common dor beetle

Dung beetle

Insect armour
The *Picromerus* shieldbug (opposite) is a broad flat insect whose body is strengthened and shaped like a shield. Most shieldbugs are plant-eaters and need their armour for protection. One species, the parent shieldbug, actually guards her young and puts herself between them and any danger—this is a rare example of an insect caring for its offspring.

Old meadow

For this expedition we found an old meadow in Cambridgeshire near the famous Monkswood experimental station. It was a warm day with occasional sun and on this sort of day meadows smell magnificent. The old "ridge and furrow" area we explored had once been agricultural land as we could tell from the legacy of many different grasses we discovered. Some plants have splendid old names like the rest harrow, so called because its tough stems literally arrested the harrow. Grasslands like this are rich in insect life as you can soon discover when you start to explore this miniature green jungle. A host of small mammals like mice and shrews and ground-nesting birds like meadow pipits and larks all feed on the insects or seeds.

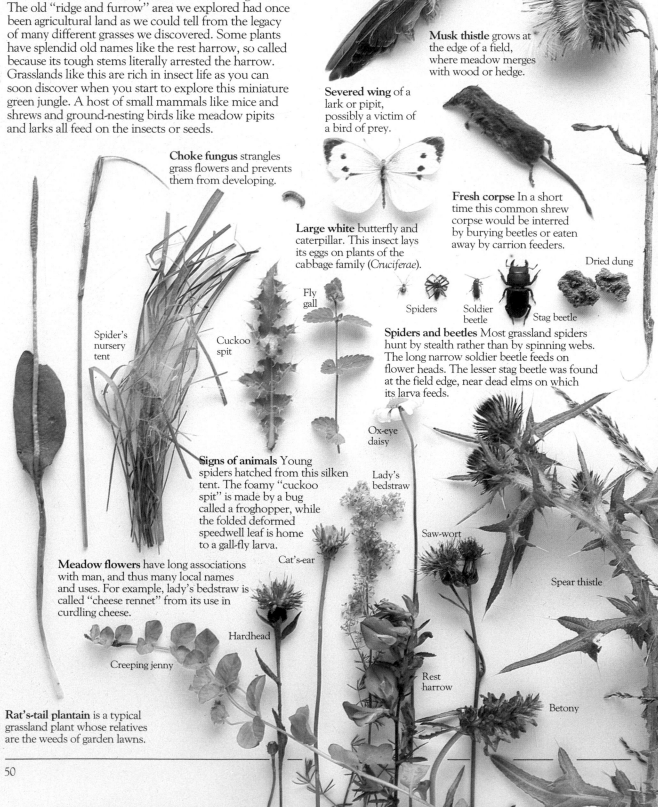

Musk thistle grows at the edge of a field, where meadow merges with wood or hedge.

Severed wing of a lark or pipit, possibly a victim of a bird of prey.

Choke fungus strangles grass flowers and prevents them from developing.

Large white butterfly and caterpillar. This insect lays its eggs on plants of the cabbage family (*Cruciferae*).

Fresh corpse In a short time this common shrew corpse would be interred by burying beetles or eaten away by carrion feeders.

Dried dung

Spiders

Soldier beetle

Stag beetle

Fly gall

Spider's nursery tent

Cuckoo spit

Spiders and beetles Most grassland spiders hunt by stealth rather than by spinning webs. The long narrow soldier beetle feeds on flower heads. The lesser stag beetle was found at the field edge, near dead elms on which its larva feeds.

Ox-eye daisy

Signs of animals Young spiders hatched from this silken tent. The foamy "cuckoo spit" is made by a bug called a froghopper, while the folded deformed speedwell leaf is home to a gall-fly larva.

Lady's bedstraw

Saw-wort

Cat's-ear

Meadow flowers have long associations with man, and thus many local names and uses. For example, lady's bedstraw is called "cheese rennet" from its use in curdling cheese.

Spear thistle

Hardhead

Creeping jenny

Rest harrow

Betony

Rat's-tail plantain is a typical grassland plant whose relatives are the weeds of garden lawns.

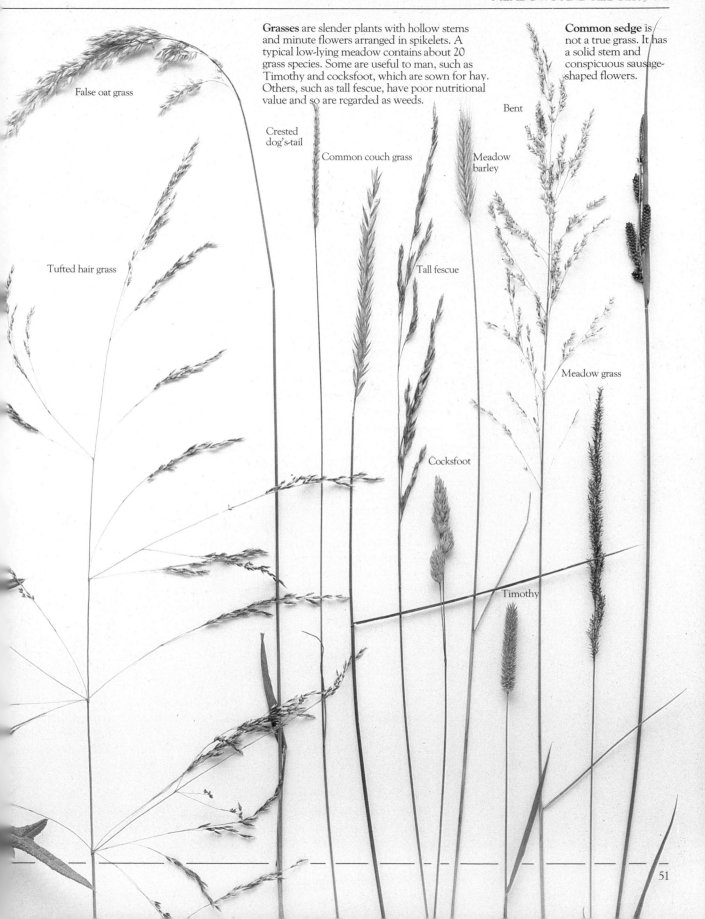

False oat grass

Grasses are slender plants with hollow stems and minute flowers arranged in spikelets. A typical low-lying meadow contains about 20 grass species. Some are useful to man, such as Timothy and cocksfoot, which are sown for hay. Others, such as tall fescue, have poor nutritional value and so are regarded as weeds.

Common sedge is not a true grass. It has a solid stem and conspicuous sausage-shaped flowers.

Crested dog's-tail

Common couch grass

Bent

Meadow barley

Tall fescue

Tufted hair grass

Meadow grass

Cocksfoot

Timothy

only ones who are satisfied are the flukes. The snail is dead, the bird now suffers from internal parasites and only the flukes are delighted at the outcome of the whole affair. But I have always wanted to find an amber snail with flashing horns.

Hedgerows

The guardians of the fields and meadows in Britain, and in parts of Europe and eastern North America, are the hedgerows which map out the countryside in chequerboard fashion. Many hedgerows are hundreds of years old—some in Britain are a thousand years old—and these are rich communities that have flourished under man's control. The hedgerows bordering the fields are of great ecological importance, for they provide food and shelter to a hundred different forms of life, from butterflies and other insects to wild flowers, birds, voles and mice, foxes, and weasels and stoats.

I remember, in Hampshire, finding a hedgerow that divided two fields, one of wheat and one of potatoes. The band that formed the demarcation line between these two large open areas consisted of about 500 metres of undergrowth, principally made up of brambles, hawthorn, hazel and elder, and interspersed with a few small oak and beech trees, each wearing a thick overcoat of ivy. This line of undergrowth between two big fields really constituted an island of wildlife. Quite apart from the wealth of insects under rotting logs or on the low vegetation of violets, primroses and other plants, there were ten different species of bird nesting. In the hazel brakes there were dormice and in the clutter of dead and rotting undergrowth, overgrown with ferns and weeds, there lived voles and woodmice, hotly pursued by a family of weasels. At one end of this island a fox was rearing its cubs, and hares who inhabited the fields on either side used to come into the strip of hedgerow to gain sanctuary and coolness during the heat of the day. It was a miniature world of its own wedged between those two great fields.

The disgraceful way that these ancient hedgerows are bulldozed to death to make larger and even larger fields is criminal folly, for it is not only changing the whole face of the countryside and destroying the creatures that live there but it is creating vast open areas which, as has often happened in other parts of the world, are then ripe for erosion. I have seen drifts of rich soil piled uselessly along the roads beside fields deprived of their hedgerows.

Examining an ancient hedgerow in detail can be a fascinating task, for they generally have a verge (which borders on the road), a ditch, then the bank on top of which the hedge itself grows. The hedge consists of closely-planted bushes, often woody hawthorns with their branches intertwining and the whole being lashed together, as it were, by climbing vines such as the honeysuckle, with its sweet scent that strengthens towards nightfall and attracts the hawk moths. Here and there in the hedge you will find an emergent tree like an oak or an elder. So in the typical hedgerow you can find many micro-environments.

Mammals of the hedgerow

There are some very charming mammals that make their homes in the hedges, such as the enchanting tiny harvest mice with their toast-brown fur and their big eyes. They are sweet little things and look very fairy-like

Harvest mouse outside its nest

as they dance through the hedgerows or the corn stalks in the adjacent field, using their prehensile tails to cling on as some monkeys do. Their beautifully-made grass nests, about the size of a tennis ball, can be found just above the ground in the hedgerows and in the corn and wheat fields alongside. In the nests the female gives birth to five or six young, so small that the whole litter weighs no more than a small coin. In the hedgerows too you will find the summer homes of dormice, round balls of grass, moss and twigs. Sometimes the bank vole builds a nest aloft as well, and seems to favour feathers as nesting material. The dense leaf litter at the base of the hedge is patrolled day and night by the voracious common and pigmy shrews in search of worms, spiders and insects. These tiny creatures' lives are divided up between short periods of calm followed by long periods of frantic activity. The pigmy shrew, Britain's smallest mammal, uses up so much energy that it has to eat its own body weight in food each day to survive. Using the hedges as hunting grounds to forage for ground-nesting birds and small mammals are those fierce little carnivores, the stoats and weasels. Stoats are much bigger than weasels and are a paler colour with black tips to their tails. Only hedgehogs can defeat these bloodthirsty predators, simply by rolling up in a spiny ball.

Hedgerow birds

Hedges, particularly old ones, are beloved by many species of bird. The hedge not only supplies abundant food but also provides protective cover and different situations for nesting. Small insects, spiders and worms are found in large numbers in the litter under the hedge, and this litter is gleaned by such birds as robins, wrens, dunnocks and that summer visitor to Britain, the whitethroat. The robins make rather untidy nests, whereas the cock wren will make several beautiful domed nests out of moss, so his female can have a choice. Dunnocks—wrongly called hedge sparrows—build fairly low in the hedge and their eggs are the most wonderful sky blue. The whitethroat is so secretive that probably all you will see is the flash of the outer white tail feathers as it dives into the undergrowth. They, like dunnocks, make dainty grass cups for nests, slung low in the hedge. The larger birds such as thrushes and blackbirds can take bigger prey, like snails and grasshoppers. These birds build very substantial nests, the blackbirds' being thickly coated inside with mud. Linnets, chaffinches, tree sparrows and yellowhammers (the bird whose call is supposed to sound like "a little bit of bread and no che-e-e-z" and never does as far as I am concerned) feed on the seeds of the field and hedge-bank weeds and weave small well-constructed cup-shaped nests.

Hedge-nesting birds make fresh nests each spring, so when the autumn comes and the birds have finished breeding you may collect a few nests. This will not only provide you with some interesting comparisons of nests for your museum, but is also a cunning way to collect many larvae and pupae of various micromoths and beetles. Chaffinch and sparrow nests are particularly rich in infesting insects. In these old nests you may even find a dormouse hibernating.

Inhabitants of the hedgerow bank

Now the bank that supports the hedge is generally dry, since water runs off it. This particularly applies to banks facing south, and here you will find the early annuals which flower and set seed before the summer sun

COLLECTING NESTS

Collect nests only in the autumn or winter, when their makers no longer have any use for them (see page 270). Wear gloves to protect your hands from the bites of fleas and other insects that probably infest the nest.

Freeing the nest
Cut each twig in turn, above and then below, using secateurs.

Extracting the nest
Work the nest free slowly, draw it gently from the hedge, and place it in a plastic bag to carry back to your workroom.

THE HEDGEROW— A VANISHING SANCTUARY

A hedgerow grows and matures like any plant or animal, and a well-established hedge provides an array of microhabitats for numerous resident species. In addition, it is a cool sanctuary for creatures from the surrounding fields. The oldest British hedges date back to Norman times, nearly a thousand years ago. You can estimate the rough age of a hedge by the number of woody plant species present, using the following method: Pace out a 30 metre length of hedge (30 large steps), and count the number of well-established woody plants—things such as hawthorn, hazel, elder, ash and oak—in that stretch. For each established species count 100 years; so, five species indicate a hedge planted in Tudor times. Unfortunately, what has taken many centuries to grow into a small but intricate and teeming habitat can be destroyed in one day by a man with a bulldozer.

Verge and ditch

A weasel returning from its night's hunting noses through the profusion of wild flowers and grasses that spills from the undisturbed hedgerow verge. The plants grade from those that grow well in dry soil near the edge to the damp-loving ferns in the ditch at the back. Nowadays many roadside verges are cut regularly with a flail, or sprayed with chemicals, and this reduces the diversity of plant species growing there.

Hedge-nesting birds
The robin (top left) sits on its eggs, and the linnet (above left) is feeding its nestlings. The yellowhammer (right) sits on its perch, looking for food.

Hogweed

Hedge mustard

Couch

Perennial rye grass

Primrose

Weasel

Chickweed

Common frog

Shrubs and nests
A thickly-growing shrub layer gives excellent cover for nesting birds, three of which are shown on the left. Most birds have two or three broods each year, and they build their later nests higher in the hedge as the shrubs, climbers and other vegetation become more rampant. It is estimated that about 40 species of birds both breed and feed in the hedgerow, and many more use it as a temporary shelter or feeding ground.

Bullfinch

Oak

Hawthorn

Holly

Hawfinch

Hedgerow trees
Standing alone and unhindered above the hedge, a tree can attain its full shape—in woodland, the surrounding trees usually affect its growth. Young oaks, like the one shown here, are often rather straggly but thicken and fill out with the passing years.

Bramble

Wren

Bank
The dry light bank soil is ideal for all kinds of burrowers, such as this bank vole hiding from the passing weasel. The vole has made its grass nest at the centre of its network of tunnels and runs. In quiet areas bank voles are out and about by day, but in most places they are nocturnal. Principally a vegetarian, the vole will gnaw along the spiral of a snail's shell to get at the creature inside.

Bank vole

gets too hot, like the thale cress and the hedge mustard, chickweed and shepherd's purse. But the bank can be moist if it faces north and is shaded by thick hedge, and here you will find the ferns, violets, primroses and the lovely lords-and-ladies in bloom. Indeed, the various micro-environments created by a hedgerow produce such a diversity of wild flowers that this is an excellent place to learn to identify them. There are few better or more relaxing pastimes than investigating a shady bank with the aid of your field guide book. Once you look closely you will be astonished at the different species you will discover. Some of the commonest ones are beautiful, like the tiny white mouse-ear and the gold lesser celandine. Not only are the colours of the flowers so lovely but their faces are so different; compare, for example, the shy, rather sad-looking face of a violet to the flamboyant trumpet-like honeysuckle. Now you may want to take some of these flowers home for closer examination or to add to your permanent collection, but before you do so remember that it is illegal in Britain to uproot any plant without the

MAKING A PLANT PROFILE

A wet meadow is an excellent place to learn to make plant profiles. The increasing moisture from a dry area to the wettest part, combined with the grazing patterns of livestock, provide a widely varied series of microhabitats for different plants. (Before you start, get permission from the landowner.) Drive a stake into a wet area and another into a dry part, and stretch a string between them about one metre high. Moving along the string, make a sort of straight-line map and note the plants that grow along the moisture gradient and their heights.

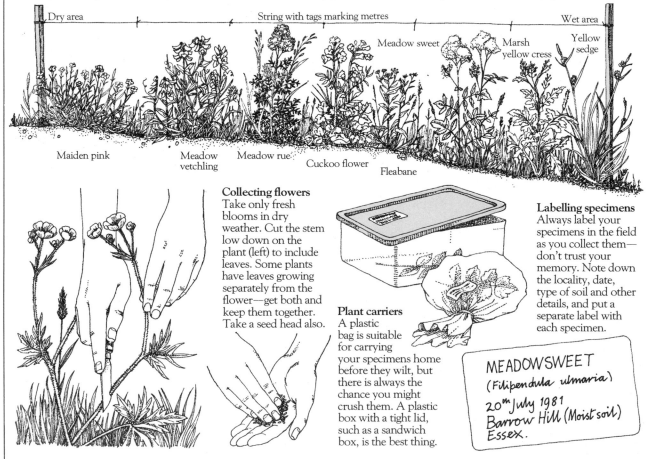

Dry area · String with tags marking metres · Wet area · Meadow sweet · Marsh yellow cress · Yellow sedge · Maiden pink · Meadow vetchling · Meadow rue · Cuckoo flower · Fleabane

Collecting flowers
Take only fresh blooms in dry weather. Cut the stem low down on the plant (left) to include leaves. Some plants have leaves growing separately from the flower—get both and keep them together. Take a seed head also.

Plant carriers
A plastic bag is suitable for carrying your specimens home before they wilt, but there is always the chance you might crush them. A plastic box with a tight lid, such as a sandwich box, is the best thing.

Labelling specimens
Always label your specimens in the field as you collect them—don't trust your memory. Note down the locality, date, type of soil and other details, and put a separate label with each specimen.

MEADOWSWEET
(Filipendula ulmaria)
20th July 1981
Barrow Hill (Moist soil)
Essex.

permission of the landowner. There are over 20 species of wild flower in danger of extinction in Britain (page 320) and it is against the law to disturb these plants in any way. Make sure that your collecting does not help edge another species towards extinction.

On some banks the soil is reasonably loose and crumbly and thus forms an ideal habitat for a variety of burrowing creatures. There are the various digger wasps, for example, and the spider-hunting wasps who fill their tunnels with spiders they have paralyzed with their stings. The spiders lie there, a macabre immobile larder, until the wasp grub hatches out and feasts on them. On these banks you will find the burrows of wood mice and—since they are making a comeback—the homes of innumerable rabbits, who emerge from their burrows in the early morning and evening to feed in the adjacent fields. Unfortunately, rabbits do great damage to hedgerows, not only by honeycombing the banks and disturbing the soil but also by undermining the hedge plants and trees growing above.

One of the most fascinating burrowing creatures I have come across is the trapdoor spider. These were very common on the banks in Corfu but you had to really search for them, for each trapdoor, round as a small coin, was so carefully camouflaged with moss it merged perfectly with its surroundings. Beneath the trapdoor—made of silk, with a bevelled edge and little recessed "handles" by which the spider held the door closed—there was a long silken tunnel where the spider lurked, a fat shiny chocolate-coloured beast with heavy fangs. I would dig out these tunnels and establish them in biscuit tins, and the spiders would soon get so tame they would lift up their trapdoors to be fed. When the babies first hatch in the bottom of the burrow they are bright green, and as they cluster all over the mother she looks as if she is wearing a green fur coat.

Of course, everybody uses everybody else's homes in a bank community. In spring an old rabbit hole might be a home for a female stoat and her young or a hedgehog and her babies. Hedgehogs may choose a nice cosy rabbit hole to hibernate in when winter comes. Deserted wood mice holes make ideal homes for pigmy shrews or the blundering noisy bumble bees.

There is an interesting story about bumble bees connected with Charles Darwin. He was investigating the red clover and discovered that bumble bees, because of their long tongues, were the only insects that could pollinate the clover; he further observed that there were many more bumble bees' nests in banks and hedges which were sited near towns and villages. He realized that this was because in towns and villages there were more cats, which ate the wood mice and kept their population down, which in turn helped the bees because the wood mice raided their nests and ate the bees' honey and grubs. It is really a classic piece of ecological detective work to link clover with cats. Nowadays we know quite a lot about these food chains, and it is astonishing the links you find that can connect, say, a tawny owl with a primrose. Hedgerows are the perfect places to find such links, being so rich in plant and animal life.

Ditches and verges

Below the bank you generally get the ditch and a good ditch acts like a magnet to the naturalist. Ditches can be always dry, always wet, or wet and dry at different times, depending on whether they are used for

The hidden hunter
The trapdoor spider is common in hotter Mediterranean areas. Instead of weaving a web, it fashions a small silk trapdoor at the entrance to a deep tunnel in the loose soil of the hedgerow bank. Trapdoor spiders eat virtually anything they can drag into their tunnels—even other spiders.

Hedgerow

To the naturalist, a mature hedge makes a wonderful hunting ground. On our May trip to a hedgerow flanking the Pilgrim's Way, in Kent, it rained for most of the day, but we still found many interesting things in the ecologically rich habitat. The large number of woody plant species packed into a small area, covered with creepers and fringed by grasses and wild flowers, attracts an enormous variety of insects, birds and mammals. Most butterflies, bees and other flying insects were sheltering from the downpour, but slugs and snails revelled in the dampness as they chewed their way through the vegetation. Along the bottom of the hedgerow were numerous bones and other relics of nocturnal hunting.

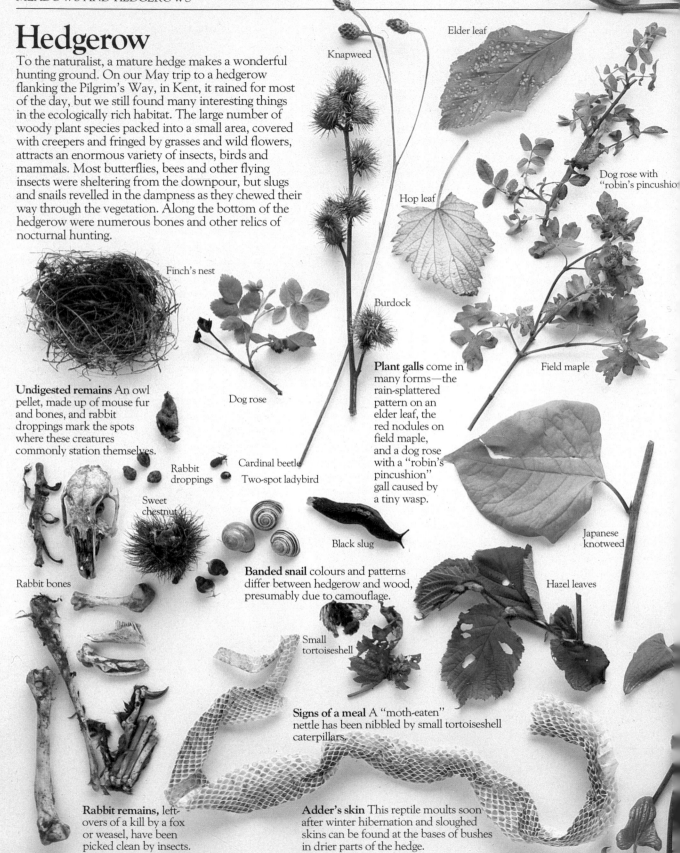

Knapweed

Elder leaf

Hop leaf

Dog rose with "robin's pincushion"

Burdock

Field maple

Finch's nest

Dog rose

Undigested remains An owl pellet, made up of mouse fur and bones, and rabbit droppings mark the spots where these creatures commonly station themselves.

Plant galls come in many forms—the rain-splattered pattern on an elder leaf, the red nodules on field maple, and a dog rose with a "robin's pincushion" gall caused by a tiny wasp.

Rabbit droppings

Cardinal beetle

Two-spot ladybird

Sweet chestnut

Black slug

Japanese knotweed

Rabbit bones

Banded snail colours and patterns differ between hedgerow and wood, presumably due to camouflage.

Hazel leaves

Small tortoiseshell

Signs of a meal A "moth-eaten" nettle has been nibbled by small tortoiseshell caterpillars.

Rabbit remains, left-overs of a kill by a fox or weasel, have been picked clean by insects.

Adder's skin This reptile moults soon after winter hibernation and sloughed skins can be found at the bases of bushes in drier parts of the hedge.

Dog's mercury (male)

Dog's mercury (female)

Barren brome

Wood melick

Cocksfoot

White deadnettle

Ground ivy

Dog's mercury is a member of the spurge family. Its milky sap is poisonous and was once used by poachers to poison salmon and trout water.

Hedge parsley

Yellow archangel

Comfrey

Bracket fungus

Vetch

Speedwell

Nettles and relatives (family *Labiatae*) have flowers like snapdragons and square hairy stems.

Common flowers of a hedgerow bottom include vetches, which climb up the rampant vegetation by means of tendrils, and sweet-scented crosswort which trails over other herbs and grasses.

Black bryony has small yellow-green flowers and red berries; its name comes from its black roots.

An old log and stump invaded by fungi. The wavy brown bracket fungus can be found on wood throughout the year.

drainage. Some ditches are cleaned out periodically and you will then find them invaded by annuals which colonize quickly. In damp ditches look out for the mosses and liverworts, and these are ideal places to find the common toad or the curious natterjack toad, with the single stripe down his back and his habit of running instead of hopping. In the drier ditches you get the tough-rooted perennials such as the stinging nettle which, in spite of its bad habits, can be cooked and used in salads and as a vegetable, and is also the caterpillar food plant for some of our most beautiful butterflies, like the red admiral and the peacock. The red admiral folds over the nettle leaf, lays an egg in this "envelope" and seals it with silk. The caterpillar, dark and prickly when it hatches, is as unappetizing-looking as its food plant. Peacock larvae are also dark and spiny-looking, and like the red admiral caterpillars they feed within a silken tent from May to July. Then they go their separate ways, burrowing underground to change into browny-green pupae.

Often there is a verge which lies between the ditch and the road or farm track and this is the area that gets the harshest treatment. It is frequently trampled on, mown and subjected to chemicals in herbicides and exhaust fumes. Only the toughest plants survive here—things like thistles, cow parsley and wild parsnip.

Studying small mammals

Meadows, hedges and fields abandoned by farmers are excellent areas to study the ranging behaviour of small mammals, using live traps. There are two sorts of trap. One catches only one animal at a time, like the excellent Longworth and Havahart traps. They are not too expensive, but you can make something yourself using the same principle. The other sort catches any number of animals. In Corfu as a boy I "invented" (at least I thought it up for myself) a particularly useful flip-top trap, which is described opposite.

There are, however, problems with the multiple-capture traps. I remember once I watched a large bank vole over a period of days and when he came out in the evening he always used one special path across the meadow to the hedge. He had traversed this so often that he had worn away the grass with his little feet and his pathway looked like a miniature road. I decided that I wanted to trap and keep this vole for a while in order to study him, so I set a flip-top trap alongside his run and baited it with a piece of apple. The first thing I caught was a shrew and the day after that a wood mouse. For some reason my vole was avoiding the trap. So I reset it and this time I baited it with a piece of bread soaked in aniseed, a scent and taste that a lot of small mammals find irresistible. When I visited the trap after a few hours I found a disastrous thing had happened. My poor vole, attracted by the aniseed, had gone into the trap. But once he was caught he was quickly followed by a weasel, which promptly killed him and had eaten most of him when I arrived. Although I was sorry that my fat friendly vole had met such a sticky end I was delighted to have the weasel, for I had tried unsuccessfully to trap one of these little predators for many months. This story shows that *traps must be checked regularly*. I do not recommend leaving traps out all night without checking them, but if you must, use a Longworth-type trap and plenty of bait and nesting straw or grass so that an early-evening captive will not go hungry or cold.

Hedgerow small mammals

Most hedgerows have a sizeable population of small mammals who scurry about in the leaves and undergrowth, usually during the hours of darkness, searching for food. The voles, with their short tails, blunt noses and small ears, are herbivores. Mice have long tails and large ears and are more versatile as regards their diet, eating seeds, nuts, berries, succulent plants and small creatures such as insects. The shrews, who have long twitching pointed snouts and ears virtually hidden in their fur, are voracious insectivores.

Bank vole

Long-tailed field mouse

Pigmy shrew

TRAPPING AND STUDYING SMALL MAMMALS

Hedgerows, meadows and the edges of woods are excellent places to investigate the behaviour of the small mammal population and how far individual animals range in their search for food. The basic equipment you will need is one or several live traps such as the ones shown here. Before you do any trapping you must be familiar with the law, as some animals are partially protected while others are completely protected. Your local natural history society should be able to give you details, and also perhaps arrange for the loan of traps. If you intend to keep small mammals at home see page 283; remember that the welfare of your "captive" (for this is what it is) always comes first. For the collector's and trapper's codes see page 320.

Types of trap
Two small mammal traps in common use are the Longworth trap, which seems to be preferred in Britain, and the Havahart, which is American. Both are single-capture types. They work on the fact that most small mammals are nearly always both hungry and curious. The creature smells the bait and enters the trap; when it is a good way inside, the trapdoor clicks shut and a locking bar falls to prevent any other animals from going in. Useful baits for voles and mice are fresh greenstuffs or grain, whereas shrews and small predators such as weasels like raw meat. Don't forget straw or grass as nesting material to keep the captive warm, and put the Longworth on a slight downward slope so rain cannot enter. If you have several traps, you can lay them in a grid pattern as shown below. This allows you to follow the wanderings of individual animals if you mark them by the fur clipping technique shown below.

Havahart trap

Longworth trap

Dyed bait in container

Grid system for bait positions

Field
Hedge
Bank
Ditch
Verge

How to study ranging behaviour
A small plastic tray can be used as a bait container to hold bran, oats or similar mouse or vole food. If you have several trays and some coloured edible dye (such as is used for coloured icing) you can carry out an experiment to discover how far small mammals range in search of food. Lay out your bait trays about 2 or 3 metres apart on a grid system, as shown above. Use undyed bait at first, then after a few days substitute dyed bait at one of the stations. Small mammals often leave droppings as they feed, and often prefer to leave them on bare surfaces, so the flat bait tray makes an ideal latrine. Coloured droppings appearing at other stations show how far the animals who ate dyed bait have travelled. Using other dyes, and placing bait at different stations, you should be able to build up a picture of their movements.

Flip-top trap
Cut the lid of a large biscuit tin to fit just inside the top rim. Hinge this with a central wire at the point of balance so that if anything walks on the lid it will be tipped into the tin. Bury the tin and cover the lid with loose soil, twigs and leaves.

Fur clipping for individual identification
Hold the creature gently, wearing soft gloves to avoid being bitten, and trim a small patch of fur. With the positions shown left you can mark up to nine individuals and identify them when they are trapped again, and thus see how far each mammal travels.

THE NATURALIST IN
SHRUBLANDS

Near my house in southern France is a shrubby country called the "garrigue". Here the holm oak and other trees that are dotted about rarely grow more than three or four metres in height and the dry land is covered with wild herbs and heath. My property runs alongside a huge tract of garrigue belonging to the French Army. Here, several times a week, the soldiers assemble to learn how to kill each other and you can hear the thump of mortars, the crack of rifles and the cackle of machine guns. Yet, in spite of this—or maybe because of it, since the Army is too interested in itself to worry about wildlife—an astonishing variety of creatures inhabit this area. There are herds of wild boar and numerous forms of southern European reptiles such as the southern smooth snake and the asp viper. Many species of rodent thrive here also, and because of this you see a number of birds of prey, including quite rare ones like Bonelli's eagle, nesting in the stunted growth.

Shrubby landscapes like the garrigue are found in other places bordering the Mediterranean, and in the western half of North America you get the sage brush and the chaparral. In the southern tips of Africa and Australia there are similar types of country and even an area of it in central Chile. To produce this sort of shrubland the climate has to be right—hot dry summers (usually with drought) followed by moist cool winters. Most of the shrubs are perennial and have complex and very extensive root systems, so as soon as spring arrives they are the first of the plants to be able to start utilizing the moisture and the nutrients that have soaked deep into the soil with the winter rains. Some of these shrubs even secrete poisons from their roots and stems which successfully prevent grasses and other plants from growing beneath them and thus depleting their water and food supplies.

Generally the leaves of shrubs are hard and leathery, like those of the holm oak, or else slender and silvery, like those of the olive. These sorts of leaves greatly reduce moisture loss—very necessary in such a dry terrain. The stems tend to be slim but tough and woody, and most shrubland is fairly low growth, rarely reaching five metres high. The stems put out a mass of small branches which intertwine and end in tight crowns. They give admirable cover for the animals that live among them while at the same time providing coolness and shade for the ground-living creatures during the baking-hot summers, and their autumn fruits are a rich harvest for seed-eaters.

The origins of shrubland

The garrigue and the other shrublands bordering the Mediterranean are not nearly as thick as they were originally. In former times the areas were blanketed with forests of holm oaks, strawberry trees, pines, cork trees and the spiky-branched tiny-fruited wild olives. In these forests lived

Wild boar

Asp viper

A handsome reptile
The sun-loving sand lizard (opposite) crawls over heather in search of insect prey. It lives in open shrubland in southern Britain and spreads onto coastal dunes in the north. Since 1981 it has received protection from the Wildlife and Countryside Act, but disturbance of its habitat continues.

Mediterranean shrubs

Brushing through the short shrubs of the garrigue, the odour rises warm and pungent from the plants' volatile oils, which help reduce water loss. The resinous scent from cistus is particularly pungent after rain. Rosemary's blue flowers are visited by bees, while its leaves are familiar as a culinary herb.

Cistus

Rosemary

wild sheep and goats and at one time even lions. But man has been despoiling these regions for over 8,000 years and the forests had no chance against the herdsmen and their domestic animals, the encroaching farmers and the timber-hungry people who wanted to build boats, cities and palaces. Even Plato, 2,500 years ago, noticed how the landscape was changing and wrote: "What now remains, compared with what then existed, is like the skeleton of a sick man, all the fat and soft earth having wasted away and only the bare framework of the land being left."

Because of the droughts and, in consequence, fires in these regions, true forest would take many centuries to regenerate even if the area were left alone. But in the meantime, the shrubland—which is an early stage in the slow regeneration of this forest—has become a natural habitat in its own right, and the vegetation is not only resistant to savage droughts but is fire-resistant as well. After a swift and terrible fire has scorched and blackened an area you will see that the shrubs quickly sprout fresh growth from the roots. Indeed, in some species the seeds actually need the heat of fire to be able to germinate. Another fire-survival device used by shrubs is the production of numerous seeds, and these provide a rich source of food for both birds and mammals.

There are many annual plants which spring up between the shrub thickets in the early summer sun and flower, seed and die before the fires of autumn can kill them. These produce a prodigious revenue of insects for the naturalist, so it is not surprising that here in the garrigue was the place that Jean-Henri Fabre, the great entomologist, spent nearly 40 years studying the many fascinating species of insect that live their lives in this dry and desiccated terrain.

One of the interesting things Fabre found out during his researches was that some male moths discover the whereabouts of the newly-hatched female by scent, sometimes over very long distances. He used the giant peacock moth in his experiments and so you can imagine my delight when—quite inadvertently—Lee and I made the same observation as Fabre's, only 50 kilometres from his old house but nearly 100 years later. It happened like this.

We were having Lee's birthday party when there came wandering up to our patio a huge fat roly-poly caterpillar longer than my finger and covered with knobs on which were bunches of spikes. Its colouring was greenish yellow and the knobs were blue. It was really a handsome creature and Lee said it was the best birthday present she had received. She put it in a wooden box until she could attend to it after her party, but when she went and looked in the box she found it had spun an elongated egg-shaped cocoon on the side of the box and had retired into it to turn into a pupa. It remained in this state, showing no signs of life, all through the winter. Then, on our wedding anniversary, it hatched out into the largest and most beautiful moth you have ever seen. Its wings were striped with various shades of grey and fawn and had cream-coloured rims to them. In the centre of each upper wing was a large eye marking consisting of a black rim, then a cream-coloured ring and in the centre a blue spot with touches of white and red. These lovely "eyes" were repeated on the underwing, and their similarity to the markings on a peacock's tail give the moth its name. Lee, of course, said this was the best anniversary present she had! She decided to release it, so that evening the box was left out on the patio with no lid on. Imagine our astonishment in the morning to find two giant peacocks in the box, the male mating with our female who must have attracted him with her scent. They stayed coupled for some time and then the female laid a battery of eggs on the side of the box. When she had done this we released her and the male. Two weeks later the 50 eggs hatched successfully, producing tiny black caterpillars with reddish hairs. Although they seemed to feed well on apple leaves, unfortunately none of them survived.

The fact that male moths are attracted to the females in this way is extremely useful to the naturalist, for you can use virgin females (in other words, newly-emerged and unmated specimens) as "bait" to catch yourself good males for breeding. The very best way is to collect wild larvae and allow them to pupate. Then separate the sexes—yes, you *can* tell the difference in some species, believe it or not, by examining the pupa closely (the male has two tiny bumps at the end). When a female emerges, put her in a "male trap" early in the evening. By the next morning you will be astonished at the choice of husbands she has attracted. Pick out the finest male to make up your breeding pair and let the others go.

Another interesting thing to do is to study the moths' behaviour. It's best to try this with a day-flying species like the oak eggar or the emperor moth, both beautiful insects. The females of these species send out their scent signals in the afternoon, so it is easier for you to see the whole process. As soon as your female has hatched put her in a large muslin bag and hang it in the place where you collected her as a larva. Here you can sit and watch the males flying in to her "call", and you can observe whether she responds to them and if they respond to each other. You can even catch some of the males, mark them with tiny spots of quick-drying typist's corrector fluid on the top of the thorax and then transport them downwind and see from what distance they can scent the female and how long it takes them to return. It's by doing this sort of experiment (and you

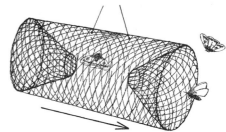

Trapping a male by scent
Place the freshly-hatched female moth (see page 278) in a trap made out of fine-meshed chicken wire, folded in at one end to form a funnel with a hole about 3 centimetres in diameter. Within hours, male moths will be attracted to her airborne scent and will enter the trap and flutter frantically around the female. Move the female and a male to a large shaded cage and leave undisturbed for at least 24 hours to allow mating. Then release the male, keep your female in a cage with her food plant, and from the eggs you should be able to rear caterpillars.

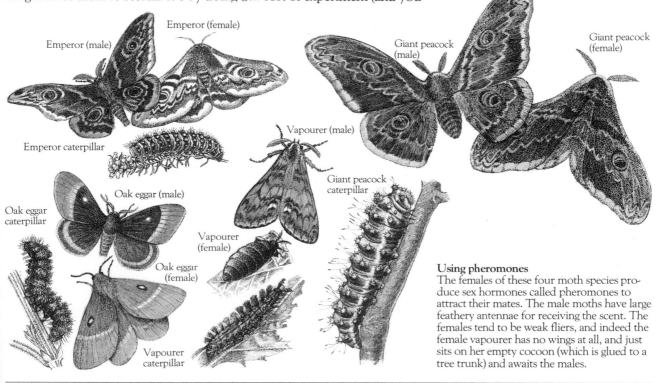

Emperor (female)

Emperor (male)

Giant peacock (male)

Giant peacock (female)

Emperor caterpillar

Vapourer (male)

Oak eggar (male)

Giant peacock caterpillar

Oak eggar caterpillar

Vapourer (female)

Oak eggar (female)

Vapourer caterpillar

Using pheromones
The females of these four moth species produce sex hormones called pheromones to attract their mates. The male moths have large feathery antennae for receiving the scent. The females tend to be weak fliers, and indeed the female vapourer has no wings at all, and just sits on her empty cocoon (which is glued to a tree trunk) and awaits the males.

Garrigue

Take a short journey inland from the holiday resorts lining the Mediterranean and you find a dry scrubby habitat where low scattered bushes are intersected with patches of wiry grasses, rock and sand. Such enchanting shrubland habitat goes under the name of "garrigue". The garrigue that surrounds our villa in southern France was the site for our August collection. The hot air was heavy with aromatic scents from wild herbs and busy with a constant humming and droning from the insect hordes. One oddly-shaped twig suddenly snatched at a passing insect and turned out to be that deadly hunter, the mantis; a loud rustle in the undergrowth signalled a large ocellated lizard, which in turn tackled the mantis with impunity.

Southern tor grass

Longhorn Its wood-boring larvae do much damage to mature trees and have progressed to the wood in buildings.

The owner of these feathers, a red-legged partridge, is common in dry shrub and arable land. In summer it gorges itself on grasshoppers, locusts and crickets.

Wasps' comb In the summer heat, workers fan the nest with their wings to cool the larvae.

Robber fly

Juvenile plant bug

Adult plant bug

A powerful bristly predator, the robber fly is found in arid areas. It crouches on a leaf or twig, dive-bombs a passing insect, injects it with poison and then sucks it dry.

Worker bumble bee

Locust

Look like your food This bug strikingly re-sembles a seed head of its food.

Noisy cicada Cast skin of this incessantly chirping insect, which spends several years underground as a larva.

Leuzea th

Giant peacock moth pupa

Swallowtail hindwing

Meadow brown

Common blue

Male oak eggar

Butterfly pupa

Looper moth caterpillar

Mantis skin

Mantis egg case

Thickly-woven cocoon of the peacock moth is similar to that of its commercially-valuable relative the silk moth. Can you make out the delicate "looper" caterpillar camouflaged against the grass stem?

Almond eaten by squirrel

Almond eaten by rat

Lower jaw of rat

Deadly lover The female mantis usually eats her partner (all except for his wings, that is) a few hours after mating.

Solitary wasp's nest

Snails in summer are usually aestivating—lying dormant, their shell mouths sealed with dried mucus.

White snail

Decollated snail

Banded snails

Ram horn snail

False oat grass

Spanish oyster plant

Shrubland flowers By late summer the blooms have withered and died, leaving only various strange-looking seed heads. Most of these plants have dry prickly tough leaves both to deter grazers and to help conserve moisture during the torrid Mediterranean summer.

Poppy seed head

Timothy

Kermes oak has a stunted bushy shape and remarkably holly-like leaves, but its acorns are more familiar and provide valuable food for rodents and other herbivores.

Star clover

Brown-berried juniper is distinguished from other juniper species by its berries, which take on a pinky rust colour in their second year instead of the blue, purple or black of the other types.

Bugloss

Lichens
Aspicilia (white)
Caloplaca (orange)
Verrucaria (black)

Cladonia lichen

can work out many more for yourself) that you can, as Fabre did, unravel some fascinating mysteries of insect behaviour, and might indeed discover something quite new and unknown.

Heathland and moorland

There are two special kinds of shrubland that are well known in Britain, the heathland and the moorland, where vast stretches turn a rich royal purple with the heather that blooms in late summer. Heathland occurs mainly in southern England where the soil is light, dry and sandy, whereas moorland is found in the north and west of Britain on wet shallow soil covering ancient bedrock. Moorland is generally higher, colder and there is a greater rainfall than in heathland.

Since the soils of both heath and moor are generally poor in mineral nutrients, especially nitrogen, many of the plants that grow there are leguminous, such as the golden gorse. The dominant heath and moor plants, the heathers, have developed *symbiotic* relationships with certain species of fungi which live among their roots and "fix" atmospheric nitrogen. The plants of heath and moor are also adapted to combat water loss, as are plants in other shrubby areas in the world. The root systems of heather are deep and the leaves are tough, with thick cuticles and sunken stomata, and in some species they are "hairy". The leaves of the wiry shrubland mat-grass roll up like tiny cylinders; the stems of gorse

LOWLAND, HIGHLAND, HEATH AND MOOR

Heath and moor owe their existence to man's activities. Where woodland was felled on low-lying sand or gravel, heath has developed; moor is typical of deforested upland areas. The main difference between them is the amount of peat the soil contains. Heaths have a thin peaty covering, whereas on moor the peat can be up to ten metres deep. The flora and fauna of these two habitats have similar elements, the differences being mainly due to their temperature and rainfall.

Common heather

Wavy hair-grass

Broom

Heath grass

Spotted orchid

Natterjack toad

Heath rush   Cranberry

Heath speedwell

Heathland soil

You don't need to take a spade with you to examine heathland soil, for sunken paths cut through it will expose the layers. Just beneath the thin top soil (1) is a shallow layer of dry peat (2) which acts like a lid, preventing oxygen from circulating. An orange band of iron salts forms a "pan" (3) which hardens and prevents plant roots from penetrating into the subsoil (4). Only plants with shallow roots can grow over the pan.

are furrowed, and young gorse stems are hairy. All these features are designed to retain valuable moisture.

Why should mineral nutrients be in such short supply in these areas? And how can one talk about a place being "waterless" when you are referring to rainy boggy moors? The answer to the problem of mineral shortage concerns the rate of decay of plant and animal remains to form a rich soil. In heathland the decay is slow because the ground is so dry. Water drains away quickly in the sandy soil and any nutrients that have accumulated are washed away by it. In the northern moors, on the other hand, the shallow soil becomes waterlogged and cold, and this too inhibits decay. On the question of water loss, rapid drainage of rain in heathland explains the heath plants' adaptations to *lack* of water, but one must remember that in moorland, too, plants can be "thirsty", although there appears to be water everywhere. This is because the thin soil layer is cold and has not much free oxygen simply because it is nearly always waterlogged. This adversely affects the plants' ability to absorb water and, of course, the ever-present winds on the moors increase the evaporation of water from the leaves. So plants in moorlands suffer from what is called "physiological" drought—they simply cannot take up enough water and it is not just a matter of lack of rainfall or excessive drainage.

There are differences in detail, however, between the two habitats. In moorland you may easily find the cross-leaved heath with its clusters of

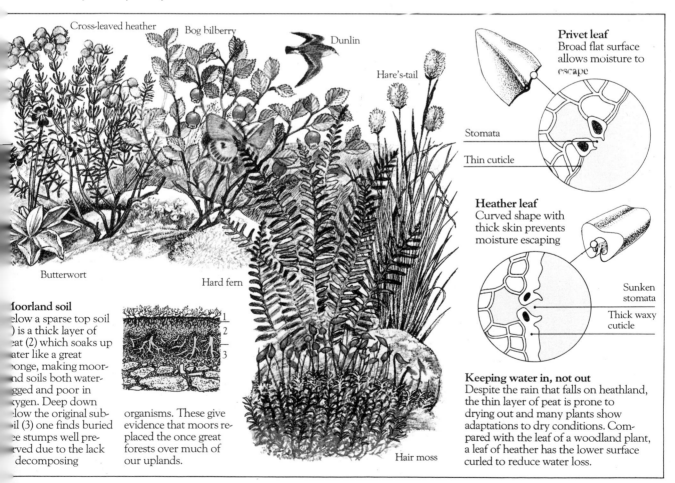

Cross-leaved heather

Bog bilberry

Dunlin

Hare's-tail

Butterwort

Hard fern

Hair moss

Privet leaf
Broad flat surface allows moisture to escape

Stomata

Thin cuticle

Heather leaf
Curved shape with thick skin prevents moisture escaping

Sunken stomata

Thick waxy cuticle

Moorland soil
Below a sparse top soil (1) is a thick layer of peat (2) which soaks up water like a great sponge, making moorland soils both waterlogged and poor in oxygen. Deep down below the original subsoil (3) one finds buried tree stumps well preserved due to the lack of decomposing organisms. These give evidence that moors replaced the once great forests over much of our uplands.

Keeping water in, not out
Despite the rain that falls on heathland, the thin layer of peat is prone to drying out and many plants show adaptations to dry conditions. Compared with the leaf of a woodland plant, a leaf of heather has the lower surface curled to reduce water loss.

Heathland

On this collecting trip we went to Dorset, one of my favourite parts of England. It was a really awful day—cold and cloudy, with rain later—so we did not see many flying insects. The flat landscape consisted mainly of heather and gorse, with fox-coloured bracken and rabbit droppings everywhere, broken only by boggy depressions. Down in the jungle of heather stems were beetles and spiders but the ants stayed underground on this cold day. In nutrient-poor boggy areas sundews fed on insects for their nourishment. The layout of our specimens reflects the make-up of a heath—the dry soil with heathers, bracken, gorse and grasses, and then a clump of plants from the bog.

Cladonia lichen

Sharp-flowered rush

Bog plants grow in depressions where water drains slowly and small semi-permanent ponds develop. Rushes, sedges and grasses grow in these oasis-like areas in the sandy "desert" of heather.

Common heather

Lichens, plant-fungus partnerships, can grow almost anywhere—even bare rock—but cannot cope with polluted air.

Common milkwort

Sundew copes with poor nutrient availability by "eating" insects. We saw this one trap a cranefly, clasping it like some vegetable octopus.

Dwarf gorse

Bell heather

Dodder on heather

Heathers are tough wiry plants, adapted to poor dry soil but in their turn a rich nectar source for numerous insects.

Yellow sedge

Moss

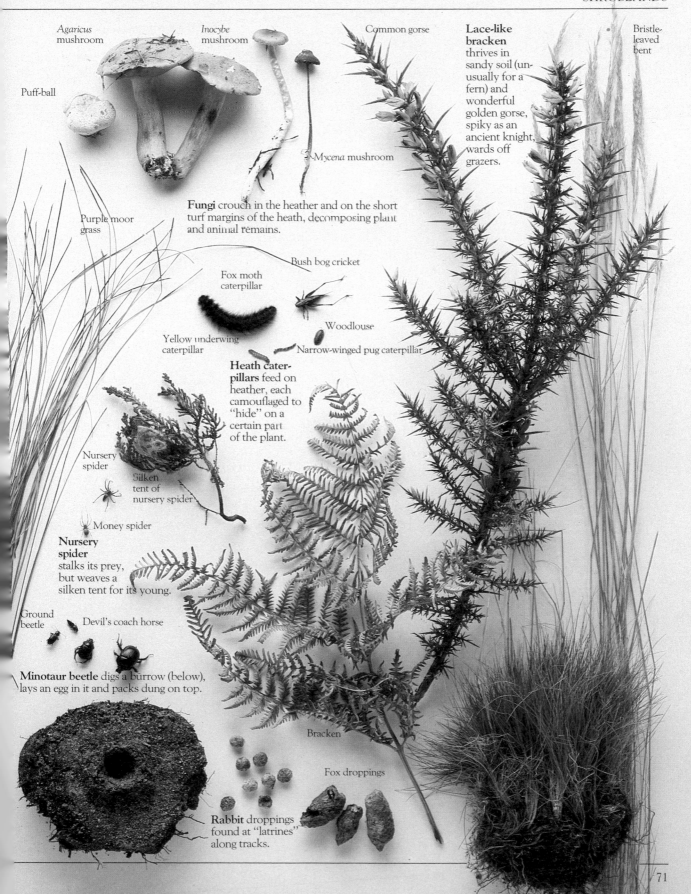

Agaricus mushroom

Inocybe mushroom

Common gorse

Lace-like bracken thrives in sandy soil (unusually for a fern) and wonderful golden gorse, spiky as an ancient knight, wards off grazers.

Bristle-leaved bent

Puff-ball

Mycena mushroom

Purple moor grass

Fungi crouch in the heather and on the short turf margins of the heath, decomposing plant and animal remains.

Bush bog cricket

Fox moth caterpillar

Woodlouse

Yellow underwing caterpillar

Narrow-winged pug caterpillar

Heath caterpillars feed on heather, each camouflaged to "hide" on a certain part of the plant.

Nursery spider

Silken tent of nursery spider

Money spider

Nursery spider stalks its prey, but weaves a silken tent for its young.

Ground beetle

Devil's coach horse

Minotaur beetle digs a burrow (below), lays an egg in it and packs dung on top.

Bracken

Fox droppings

Rabbit droppings found at "latrines" along tracks.

small rosy flowers on the ends of the stems, like bunches of balloons. There are various stiff and upright sedges and rushes, and mosses and liverworts grow between the tight clumps of taller vegetation. On heathland the legumes—broom and birdsfoot-trefoil, the latter coloured like a plate of eggs and bacon (which it is sometimes called)—are found as well as gorse, and flax and marsh violets peep in white and lilac amongst the heather. In the heath communities of the south you can find the innocent-looking common dodder. This has no leaves but a hard red stem and clusters of delicately-shaped pink flowers. But don't be misled by its appearance, for it is a sort of plant Dracula. It climbs up its victim anticlockwise and the roots from its wiry stem pierce the flesh of its host and suck out the nutrients. In some cases the dodder, like Dracula, leaves its victim dead, drained dry.

Where I once lived on the south coast of England there was a lot of heathland around us and it supported an interesting selection of plants and animals. In low-lying areas which were boggy I used to collect and then keep on my windowsill the tiny sundews, the minute insect-eating plants that look and act so like a sea anemone. I remember it was a sundew that was responsible for my obtaining a creature that I had long wanted to capture, the strange centipede *Geophilus electricus*, which is about as long as a finger, thinner than a matchstick and has the curious ability to glow in the dark. I had long hunted for this creature but with no success. Then one day I made a trip to a bog in the centre of a great stretch of heath, where I knew I should find on the heather the cocoons of the beautiful emperor moth. Having obtained several cocoons I made my way further into the bog to collect some sundews which grew there in great profusion. To my astonishment the first one I came across was clasping in its strange lentil-like leaves one of the luminous centipedes that I had searched so long for. With some difficulty I rescued him from the sticky embrace of the sundew and he seemed none the worse for his adventure. That evening in the special cage I had fitted out for him he gave a wonderful display, zooming over the leaves and moss and bark and glowing eerily in the dark, looking for all the world like a lit-up train seen from the air at night.

Sundews are found on boggy moorland as well, of course, as is another insect-eater, the common butterwort. In this plant the rather fleshy leaves are curved inwards at the edges and covered with a sticky fluid. When an insect lands on this it acts like an old-fashioned fly paper and traps it. Then the leaf curls round to engulf the prey, the leaf glands exude a digestive fluid and the insect is dissolved and absorbed. Thus sundews and butterworts extract the all-important nitrogen that enables them to live successfully on acid peat bogs which are lacking in this element. The leaves can also digest pollen or stray leaves of heather that happen to fall on them, so providing more nutrients.

Creatures of the sandy heath

A careful search of heathland is almost certain to reveal dung beetles and the bloody-nosed beetle (so called because it exudes a blob of red moisture from its mouth when you pick it up). In sandy areas between the heather you may find the conical pits of the ant lion larvae. Woe betide any ant who wanders into one of these cones. Immediately, the larva—which lies buried in the sand at the point of the cone—starts to

It's only self-defence
In response to being prodded, a bloody-nosed beetle expels red fluid from its mouth. This habit (known as "reflex bleeding") is common throughout the insect world—as anyone who has ever handled bugs, caterpillars or grasshoppers well knows. These creatures warn off a predator with this drop of liquid which lingers with its foul smell.

bombard the ant with sand grains. Eventually the ant slides down to the bottom of the cone, to be devoured by the triumphant larva.

On the heathland, too, you may catch sight of the bright-green sand lizards, gleaming bronze slow-worms, and smooth snakes and adders. I remember once in the Purbecks I found a female adder sunning herself in the heather. She was the biggest I had ever seen and when after an exciting chase I finally caught her she measured the length of my arm and was very beautifully marked. She was so fat that I felt she must recently have eaten something like a small grass snake or a nest of young voles. So you can imagine my surprise when three days later I found she had given birth to 17 young, all as beautiful as their mother. I kept them for a few weeks but the catering difficulties of feeding my huge addery soon forced me to pack up the whole family except two and release them back on the Purbecks. (Remember, adders are poisonous and you should never handle them—leave that to an expert.)

Wildlife of the high moors

The great moorlands in the north of England and in Scotland can provide you with some of the most spectacular forms of animal life to be found anywhere in the world. Among the birds of cliffs and rocky outcrops you can see the majestic golden eagle, the heavy buzzard or the tiny merlin, one of the daintiest and smallest of British falcons that nests in the heather. Here too you can see the great black sombre ravens, biggest of the crow family, perhaps rolling and swooping and teasing a golden eagle as it circles through the sky, for ravens are no respecters of other birds. Ravens, of course, eat virtually anything they manage to overcome, and will steal the eggs of gulls and even of the golden eagle if they can.

In the streams that tumble through the moorland you might be lucky enough to see the ring ouzel, unmistakable as the male is a rich black, the female duller, but both have a white crescent like a half moon across the top of the breast. Then amongst the heather in the evening you may see that most beautiful and curious of birds, the nightjar, hawking for moths and other insects while uttering its strange churring cry. Its beautiful plumage, a subtle combination of browns and greys and creams, provides this bird with the most perfect camouflage in the day when it rests on the ground with its eyes closed. I once sat down and had a picnic almost within arm's length of a nightjar and for an hour I did not realize it was there. It was only when I got up to go that I almost stepped on it. They have short beaks but such enormous gapes that, in their swift silent flight, they can even engulf hawk moths or large flying beetles. The male nightjar has a lovely mating flight in the twilight when he courts his lady love with a wonderful display of aerobatics, twists and turns, rolls and somersaults, clapping his wings over his back. At one time these birds earned themselves the name of goatsuckers, because they were seen to hawk around goats and other cattle in the evening and people got the idea that they would suck the milk from the udders of goats and cows. But they were, of course, preying on the insects attracted by the animal herds.

Up in the high moorlands probably the most spectacular courtship displays belong to the black grouse, a very beautiful bird that has a black body with a bluish sheen, brown wings, a white rim to its tail and great red "eyebrows" about the eyes. The cock birds perform singly or in groups of from five to over 50 individuals in their display grounds or

Buzzard

Nightjar

Merlin

Moorland

Dartmoor on a chilly cloudy day in the middle of July may, at first sight, look similar to its related habitat, heath. But delve under the spreading heather and in amongst the plant roots and the first difference is immediately obvious: moorland soil is thick, springy, damp and peaty, quite unlike the dry sandy heath. Clumps of moss and lichen abound, and most flowers are hardy low-growing individuals that can withstand the cold sweeping winds typical of upland moor. The dry tough-looking heather supports surprisingly numerous creatures. If you sit quietly you may hear the melancholy "coor-lee" of the curlew; move too noisily and you may catch sight of a red grouse whirring away over the heather.

Down feather

Drinker moth cocoon

Eggshell of wood pigeon

Empty cocoon of drinker moth pu

Nipped-off grouse feather indicates it fell victim to a fox. Large areas of moor are burned to encourage grouse, which feed mainly on young heather shoots.

Communal caterpillars Emperor moth larvae live in groups for the first few instars (moults), and are the only British representative of the silk moths.

Bilberry bumble bee Heath bumble bee

Dung fly

Meadow brown *Noctuid* caterpillar

Cladonia lichen

Dog lichen and bog moss

Stonecrop stores water in its succulent leaves; a few red fruits can be seen beneath the pink petals. It is common on acid soils and old dry-stone walls (hence its name).

Ants A few representatives and eggs of the probable 10,000 ants that make up a colony; the nest was situated under a large flat stone.

Lichens show striking differences in growth forms. The bright red tips are the fruiting bodies of the fungal partner; whether the algal partner is carried with the fungal spores to new sites is not known for sure.

Parmelia lichen

Sheep's rib

Ram's horn

Badger droppings
containing beetle
wing cases

Rabbit
droppings

Victims of winter Probably
caught in the cold weather, the
owners of horn and rib died
high up on the moors. The
remains were found in the
shallow headwaters of a
moorland stream.

Living on dung Ink-cap is one
of several fungi that grow
readily on dung, which provides a rich source of food in
an otherwise nutrient-poor
habitat. Other dung-feeders
include species of dung fly and
dung beetles; nature wastes
nothing and within a few days
all traces of these sheep
droppings would have
disappeared.

Early bilberries develop from
the drooping pink flowers.
The delicious berries have a
blue-grey bloom and are much
prized by both birds and mammals.

Dwarf gorse

Young bracken

Common heather

Tormentil

Bell heather

Heathers Common heather is
dominant but bell heather takes over
on slightly drier ground—a fact used
by knowledgeable travellers to
navigate through dangerous bogs.

Black grouse stamping grounds
In the early spring dawn, well before the Highland mists have cleared, grouse cocks congregate on their leks—patches of ground worn bare by the activities of generations of birds. Within this arena a fascinating yet bloodless battle ensues. The aim of their display is territory, the centre area of the lek being the choicest site where a male can show himself to the best advantage to the females, who watch from the perimeter. A dominant male with an optimum territory near the centre will eventually mate with up to 80 per cent of available females.

"leks" as they are called. First they look around, and then they start crowing. After this they stretch their necks out horizontally, beat their wings rapidly against their sides, fan their tails out and give their gobbling cries which they might keep up for half an hour at a time. Sometimes they will stop gobbling and give a crow, accompanying it by jumping into the air and fluttering their wings. The displays of the black grouse are so spectacular that they are well worth making a special trip to see.

It is in the moorland also that you get the chance to see some of Britain's larger mammals. The fox, of course, is always to be found there, but in addition you may be lucky enough to see one of the fiercest mammals, the wild cat, a beautiful creature like a very large domestic tabby cat with a huge fluffy tail. When I worked at Whipsnade Zoo I looked after a pair of these lovely animals for a time but I could never get them tame, in spite of bringing them titbits in the shape of rabbits or mice. They would snarl and spit at me and would be quite prepared—if I got too close—to jump at me and try to rip me up with their dagger-sharp claws and needle teeth.

The king of the moorland is the red deer, the largest and most spectacular wild mammal in Britain. Standing nearly a metre and a half at the shoulder, and with magnificent branched antlers like trees, they are wonderful-looking animals. The best time for seeing red deer is in the rutting (mating) season from mid-September to the end of October. It is then that the stags collect a group of hinds around them and start their loud roaring to guard their territories and fight other males with their massive antlers.

I remember once in Scotland I was high up on the moors early one morning. The mist was very thick since the sun had only just risen and had not gained sufficient strength to disperse it. Shivering in the dew-drenched heather, I could hear the stags around me roaring their fierce belligerent cries. As I waited for the mist to lessen, some six or seven metres away a fox trotted past, so intent on his own business that he never saw me. He glowed like a flame in the mist and I could see the pearls of dew clinging to his flanks and brush. It was a momentary glimpse and then he vanished into the mist. All around me the stags were roaring and it was very frustrating to hear them but not see them. Then, very close at hand, I heard one stag's challenge answered by another and through the mist came the clatter and rattle of antlers as they started to fight. I started to creep forward and, as luck would have it, a slight breeze blew up at that moment and parted the veils of mist to reveal two magnificent stags locked in battle. In the background their respective groups of wives huddled, gazing at them apprehensively. The stags, heads down, antlers interlocked, strained to and fro like wrestlers, the steam spouting from their nostrils. You could see (as in many other animals that have contests like this) that it was not a fight to the death but a test of skill and strength. Their neck muscles stood out and their legs strained as they wove round and round in a circle. Sometimes they would release their antlers only to clash them back together again in what each hoped would be a more advantageous position. Unfortunately, I did not see who won, for after watching them for about two minutes the wretched mist came down again and blotted out the deer and moorland as well. But brief though the glimpse had been it was a wonderful sight and certainly well worth getting up at dawn to see.

THE FAMILY LIFE OF THE RED DEER

Though originally a forest animal, the red deer has adapted well to open country and herds live on moorland in Scotland and the West Country. Here they exist mainly on a diet of grasses supplemented with lichens, fungi and raids on agricultural crops. The hinds live in small groups, often just a mature hind with two or three calves, while the stags remain solitary or assemble in loose herds. Female groups show more cohesion—for example, when danger threatens, the highest ranking hind will utter a series of sharp warning barks to alert her companions, who immediately turn tail and follow her in single file. With males, no equivalent "fatherly" protection is shown.

The enigmatic antlers

Only red deer stags possess the antlers, which grow as paired bony structures well-supplied with blood vessels. The amazing fact is that the antlers are shed each year in the early spring, and start to regrow immediately. By the end of July the soft skin covering, known as velvet, flakes off and exposes the hard bone. After several years, antlers reach their maximum development with three branches at the top (known as "tops"). At this stage a stag is said to have "all his rights" Oddly, it seems that certain stags without antlers (known as "hummels") are more successful than antlered stags at fighting and breeding.

One year's growth

Early spring

Midsummer

Late July

Successive years' growth

Second year

Third year

Fourth year

Fifth year

The rut—a mating ritual

Autumn is the rutting season when stags in prime breeding condition answer each other's calls and come together to fight with great clashes of antlers. Once victorious, a stag secures a territory and a harem of hinds. But, although he remains in breeding condition for a month, he rarely stays master of his harem for more than a week before being ousted by a younger more vigorous male. The combat ensures survival of the fittest.

THE NATURALIST IN
GRASSLANDS

True grasslands exist in most parts of the world where the rainfall is not sufficient to produce thick forest, and yet sufficiently high to prevent the creation of a desert. The great grasslands once covered nearly half of the earth's land surface, from the rolling prairies of North America to the great savannahs of Africa and the vast steppes of Eurasia. For me one of my favourite places in the world is the Argentine pampa. Standing in the middle of this great grassland, you can turn slowly in a circle and the pampa stretches all around you to the horizon, as flat as a billiard table. The main vegetation is huge clumps of pampas grass with their tall white feathery seed heads and narrow razor-sharp leaves. Here there are very few trees, their place being taken by clumps of giant thistles, like huge spiky candelabra, which grow higher than a man on horseback.

Generally speaking, grasslands have few, if any, trees to break the drying winds and most of these areas go through periodic drought conditions. As a result, the major plants of these regions live more *in* the soil than above it. Just under the surface there is a tangled mat of roots and rhizomes (rhizomes are the underground stems that propagate the plant and store food for it). Some grass roots grow down to the depths of a metre into the soil, while the tap roots of other soft-stemmed plants may penetrate to five metres in their search for water and nourishment.

The grasses, the legumes and other soft-stemmed plants, particularly members of the *Compositae* (daisy) family, flower from spring to autumn and so the face of the grasslands is like an ever-changing painting in subtle colours as the different plants come into flower. The American prairie ranges from the rich deep purple of the pasque flowers to the brilliant burnished yellow of the swathes of goldenrod. The Russian steppes are starred with flowers, stamped with great patches and cushions of emerald green moss, flooded with the wonderful blue-violet flowers of the irises and then great sweeps and billows of steel-grey grasses. In the dry season, of course, the grasses change to a variety of browns and other sombre shades. It is at this time, when the vast expanses are tinder dry and heavy summer storms bring violent flashes of lightning, that there is a great danger of fire. Grasslands in many parts of the world have been "maintained" by fires long before the arrival of man, particularly in areas that have a short wet growing season and then a long hot rainless period which bakes all the plants crisp and dry. Once in West Africa I saw lightning strike a small stunted tree in the grasslands. Immediately the tree and the golden grass, stiff and brittle, took fire, and the tide of pink and yellow flames spread over the hillside at almost the speed of a running man. Rushing in panic before the fire were the fast movers like hyraxes, cane rats, ground squirrels, snakes and so on, but the slow movers like the tortoises were engulfed by the flames. It was interesting to note how all the hawks in the area gathered above the rim of the fire, swooping

Pampas grass
from Argentina

Subterranean grassland mammal
The mole (opposite) is a common inhabitant of grasslands, where it can burrow easily in the usually loose soil and has abundant food in the shape of worms, slugs and soil insects. Although it rarely surfaces and has poor or even non-existent eyesight, when the need arises the mole is an excellent swimmer.

Pastures new

The seed is the part of a plant's life cycle that enables it to spread to new ground. Some seeds, like the dandelion "clock" and fluffy white cotton-grass, are wind-borne and have a light feathery structure and parachute-like shape. Burdock is a hooked seed that hitches a ride on a passing bird or mammal.

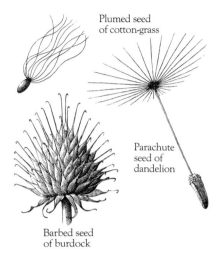

Plumed seed
of cotton-grass

Parachute
seed of
dandelion

Barbed seed
of burdock

Running repairs

Worker termites (opposite) use mud to seal a breach in the wall of their nest mound. These termites are among the smallest creatures of the African savannah, but are one of the most numerous—there are over one million individuals in a large colony. These insects have a primitive anatomy, like their near relatives the cockroaches, but their social organization is tremendously complex.

down on lizards or squirrels trying to escape, and all the insect-eating birds were doing the same for fleeing locusts and other insects.

After a grassland fire has spent itself the rain washes mineral nutrients into the soil from the black and charred remains of the burned plants. Almost at once there is a spurt of new growth from the quick-growing plants, especially the legumes. Any woody vegetation has been burnt off, thus freeing the grasses from competition for growing space and "maintaining" the grassland habitat. Almost in the twinkling of an eye the black ash-covered area becomes dotted with emerald spots and before long the scars of the fires have vanished under a green sea of new growth.

The pollen of grasses is carried by the wind whereas most of the other plants are pollinated by insects. The way the various seeds are dispersed is fascinating. Some take to the air and, like the grass pollen grains, have developed parachute-like growths to help them ride the wind and travel great distances. Other seeds have developed a formidable array of hooks, harpoons, anchors and spikelets which grimly grasp feathers or fur, to be carried away from the parent plant by a bird or mammal host. Too often the socks and trousers of an ardent naturalist resemble chain mail with the numerous seeds adhering to these garments, picked up during investigations in knee-high grass. Trying to de-seed yourself is one of the most frustrating tasks of a naturalist's life, but it certainly is an excellent way to collect such seeds.

Insect life in the grass

The herbaceous layer, knee- or even waist-high as in meadows, has the greatest abundance of insect life of the grassland, coming to a peak in the summer and to a lesser peak in the autumn. As well as the nectar-feeders there are many species of insects that chew on various parts of the plants—members of the grasshopper family, for example, and the various chafers and leaf beetles. Other insects are equipped to pierce the fragile "skin" of the plants and suck out the juices, things like the true bugs (the *hemipterans*) such as aphids and leafhoppers. When you sweep for insects in this sort of domain be sure to use your thick linen net, which is less likely to pick up burrs and sharp-spiked seeds.

The soil is the home of ants and earthworms who aerate it by their "ploughing" activities and who help incorporate the mulch into it. Mulch is the thick layer just at ground surface of the remains of plants in various stages of decay. Here you can find scavenging beetles and hunting spiders and the ground-hugging plants, like the dandelion and wild strawberry, all in a tiny world of their own. The rich carpet of mulch provides moisture and a mild even temperature, and the thick layer of plants growing above gives protection from the wind and sun.

Termites, the so-called "white ants" (which are not true ants but very primitive insects related to cockroaches), are an extraordinary group of creatures that can create their own microclimate even under the baking sun of the tropical grasslands. Termites are inconspicuous in temperate zones but become flamboyant and brilliant architects and engineers in the tropics. The grassland species build their gigantic rock-hard mounds and within each is a royal chamber for the tiny king and the massive queen, along with numerous galleries for workers and soldiers, nurseries, air-conditioning systems and sometimes fungus "gardens". These are carefully planted and tended by "gardener" termites who bring in mulch

to grow the fungus on, which provides food for the colony. Some of the mounds are so huge in relation to the tiny termites that it would be as though men, using mud bricks, had created an edifice several times higher than the world's tallest skyscraper.

Termite nests are not only used by termites. In mounds where the walls have been breached I have found in residence giant cane rats, porcupines, pangolins (scaly anteaters), spitting cobras and once a five-metre python coiled round and incubating a clutch of three dozen eggs.

Grassland birds

The small grassland birds eat insects or seeds, or a combination of both. In North America on ungrazed prairie (ungrazed by domestic stock, that is) you find such things as meadow larks and grasshopper sparrows feeding on the insects and their larvae that inhabit the tall herbaceous layer. On heavily-grazed prairie the seed-eating horned larks take over, presumably because the over-grazed grass does not provide the right environment for a wealth of insect life. This alteration of food supply may be directly due to the undisciplined grazing habits of man's domestic stock. On the savannahs of East Africa, for example, the zebra and antelope herds thickly cover the grasslands, and yet there is an abundance of insect life which is gleaned by bustards, cattle egrets and bee-eaters as the herds move through the grass.

Many of the small grassland birds have special song flights by which they delineate their territories and attract their mates. Some of the songs are long and complex, and many of them consist of trills or buzzy notes which seem to carry well in such open country. Others have extra-ordinary displays in flight like the whydahs of the African grasslands who shimmer their long tails, and the scissor-tail of the Paraguayan grasslands which, as it flies, crosses its two long tail feathers so that it looks as if it is cutting up the blue sky. Many of the *gallinaceous* (game) birds, like the partridges, quails and guinea fowl, have marvellously-marked plumage which forms perfect camouflage for their life on the ground. In the breeding season a number of these species, grouse in particular, create trampled arenas in the grass in which the males do their elaborate and beautiful strutting "dances" to stake out territories and attract mates.

The birds of prey that live in the open grassland areas are generally strong fliers like the Bateleur eagle, famous for its acrobatic flight, and the little kestrel, which searches for its prey by hanging immobile in the air like a black cross, as manoeuvrable as any helicopter. Then there are the specialists, like the African secretary bird whose incredible eyesight, hawk-like beak and scaly "armour-plated" legs ideally fit it for its role as snake hunter.

Herds, packs and townships

On the great grasslands of the world you find an enormous diversity of mammals. Many of the herbivorous ones are highly socialized. Living in herds is a sort of safeguard, for if one zebra gives the alarm on seeing a predator like a lion, all the zebra rush away together and so, in the general confusion, the lion stands less chance of catching one. In a large expanse of grassland where (in comparison to a forest) it is more difficult to hide you get interactions not only between individuals of the same species but between species as well. A giraffe, for example, because of its superior

Scissor-tail of the Paraguayan grasslands

Secretary bird of the African savannah

height, might become aware of danger before a zebra does, and its behaviour will warn the zebra that there is danger nearby. The hunters of the grasslands are sometimes social too. On the steppes you get the wolf packs working as a unit. In East Africa the Cape hunting dogs have developed their deadly routine into a fine art and once an antelope is singled out and pursued by these relentless canines it rarely escapes. Lions, too, cooperate when hunting and an accomplished pride is an efficient killing machine, with each animal playing his or her part in the murderous chess game of the hunt.

Many small grassland mammals live together underground in complex burrows. The prairie dogs of North America tunnel out underground coteries or towns that can cover many hectares in extent. On the Eurasian steppes their place is taken by the charming suslick, a type of ground squirrel, and various species of vole. On the pampa of South America I have captured the big black-masked rodents called viscacha, animals as large as the European badger. They live in large colonies and have a passion for collecting things with which to decorate the entrances to their nest burrows. It is said that if you lose anything on these grasslands, you must look for it at the nearest viscacha "town". A man once lost his gold watch while riding across the pampa from one estancia (ranch) to another. Three days later he decided to visit every viscacha town that lay along the route he had taken. Sure enough, at the second town he discovered his watch being used to decorate a burrow, together with twigs, flowers, some white stones and a couple of tin cans.

Most of these burrowers stay fairly close to a convenient hole leading to their underground city and do a rapid disappearing act when danger threatens. The prairie dog utters a shrill whistling alarm call before diving head-first into its burrow, thus alerting the entire colony. In Argentina you can hear the tucotucos underground giving their alarm call (from which they get their name) as they hear footsteps above their burrows.

The burrowing activities of the rodents and other mammals have a considerable effect on the soil in the same way that worms do. In parts of the Russian steppes, for example, the suslick burrows are of great importance because the animals mix the top soil with salt-rich lower layers, and there is a noticeable increase in plant growth on the suslick mounds. In areas where the winter is severe mammals like prairie dogs, suslicks, tucotucos, marmots and so on hibernate. But in the case of suslicks parts of their range have such severe winters and such desiccating summers that these little animals both hibernate *and* aestivate.

The small sociable grassland creatures can usually be seen at night or in the early mornings or evenings, but there are some beasts who spend their life underground (they are called *fossorial* when they do this). Those who rarely if ever see the light of day are creatures like the golden moles and the mole rats. On the other hand, there are great hoppers and leapers, like the jumping mice, the jackrabbits and the jerboas. They are not adapted to take prodigious leaps just for fun but to escape from predators and, by leaping above the grass, to see if enemies are around.

Some of the small grassland predators also use tunnels; they keep their defenceless newly-born young in holes they excavate for themselves. Animals that dig their own homes include ferrets and badgers of North America, the mongooses of Africa, the cosac fox of the steppeland and the enchanting bat-eared fox of Africa, which is said by the Africans to

SOCIABLE RODENTS

The basic social prairie dog "unit" is the coterie, which consists of one or two males, two or more females, a few juveniles and several babies. Adjacent coteries are always squabbling over territories, but the whole township will bark a warning if a predator is about.

Ever alert and watchful
The prairie dog keeps a continual look-out to protect himself and his near neighbours against danger in the shape of a coyote, ferret or badger.

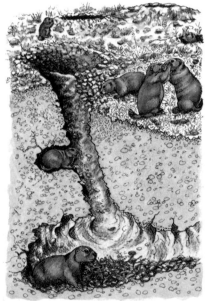

Burrow layout
The earthen rim around the entrance protects against flooding, and about a metre down is a small guard room where the occupants hide if threatened. Below is the grass-lined nest where they sleep at night and where the young are born.

Grassland felines
The sandy-coloured caracal (top), sometimes called the desert lynx, is built for speed with its long powerful legs rather like a cheetah's. It roams over the grasslands and semi-desert of North Africa and western Asia, searching for and running down all sorts of animals (including the fleet gazelles) with tremendous bursts of speed. The serval (above), with its leopard-like rows of spots, is also very fast. Its favourite food is guinea fowl and other birds—it can run as fast as some birds fly, and can leap nearly two metres to grab a low-flying victim.

roll ostrich eggs round with its paws, bumping them into stones till they break. Some predators take advantage of other species' digging activities, such as the gleaming black and white skunk I used to hunt through the thickets of giant thistle on the Argentine pampa. Skunks usually spend the day in rodents' burrows. Sometimes they kill and eat the rightful householder but on other occasions they simply make the poor creature move over so the skunk can fix itself a home alongside. If a skunk asks you to move over, it is unwise to argue. The small cats like the pallas cats, pampas cats and the servals, sleekest and most beautiful of the small predators, use the deserted dens of porcupines or wild pigs in which to rear their young.

Sometimes different species co-habit in underground dens. Deserted prairie dog towns may be taken over by rattlesnakes and burrowing owls. In South America, burrowing owls frequently dig nest burrows in the viscacha townships. On the pampa I have seen these owls sitting, stiff and upright like soldiers, on guard at the entrances.

Great herds of the plains
The most magnificent and impressive inhabitants of the grasslands are those large hooved grazing mammals, the ungulates. Herds of bison and prong-horns roam the North American prairies, and on the steppes lives the strange soulful-looking saiga antelope with its curious bulbous nose. But it is on the African savannah, of course, that you find more large ungulate species than anywhere in the world. A grassland community of immense complexity and great beauty has evolved, centred on this splendid wealth of herbivores. They are in perfect balance with their grassland home. They have high-crowned flat molar teeth for grinding grass, yet they feed selectively, each species having its own menu. For example, zebra graze on the crown of the herbaceous layer of plants, while such things as topi and wildebeest feed on the middle layer, and the Thomson's gazelle on the shorter grasses and fallen fruits. Giraffe, elephant and rhino each feed at their different levels in the trees and shrubs dotted about the savannah. This selective feeding combined with the herds' constant movement protect the environment from over-grazing and subsequent erosion.

Ungulates have hooves and long legs which allow for a springy, seemingly tireless gait over the long distances they have to travel in search of green pastures during the inevitable dry periods. The mass movements of the ungulate herds are one of the most spectacular sights on earth: in the Serengeti you may still see herds of hundreds and thousands of Thomson's gazelle and gigantic herds of wildebeest on the move. In the early 1800s in South Africa the migration of the springbok was an incredible sight. Herds of over 25,000 animals were common and swept along remorselessly trampling lions, hunting dogs or other predators caught in their midst, and even men have been known to be caught and engulfed in these monstrous marches. Now, because of the so-called march of civilization we can no longer see these magnificent natural spectacles—only pitiful remnants of springbok herds remain.

On the journeys in search of food it would not be right for the animals to fight and bicker, since there are enough dangers in the vast migration without adding to the problems by quarrelling among themselves. But in the breeding season it is different. Males fight each other over breeding territories or females. However, in the horned ungulates the fighting is ritualized—that is to say it is more like a judo contest, pushing forward, banging the base of the horns together, or entangling the horns as a test of strength. The sharp tips of the horns are kept out of the way and, like the moorland red deer and many other similar animals, the males rarely actually hurt each other.

The newborn young of the horned ungulates can be divided into those that within a very short time follow their mothers (like the wildebeest) and the kind that "stay put", waiting until their mother returns to their hiding place to feed and clean them (such as the Thomson's gazelle). But most baby ungulates must be up and about very soon after birth if they are to survive, and even the "stay-puts" quite competently follow their mothers if necessary.

Pageant of plains wildlife

The huge ungulate herds have their "friends", their enemies and even what might be macabrely described as their "undertakers". Friends to the gnu and wildebeest are the ostriches, because the birds' extra height and good eyesight will warn the herd of approaching predators; this is why you often see the ungulates grazing close by their large flightless lookouts. Also, as I have mentioned, zebra and others like to keep near giraffe. Other friends are the oxpeckers and tick birds, who are helpful to the ungulates in two ways. First, they perch on the animal and walk over its skin, eating the external parasites. In addition to this they are quick to spot danger and by giving the alarm call will alert their hosts. There have even been cases of tick birds waking a sleeping rhino when danger

The ungulate way of life

The teeth and the legs and hooves of grassland ungulates such as the North American bison (bottom) are interesting in the way they have become adapted to a grazing way of life. An ungulate's cheek teeth are large and have crescent-shaped crests of hard enamel which are excellent for mashing and grinding up the tough grasses and other plants. The long legs are well designed for running; this is very important in herbivores, which are often fairly defenceless and much preyed-upon. The foot bones are strong yet light, which reduces the momentum needed to move them as the animal gallops across the plain.

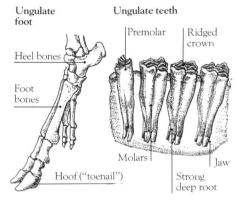

Ungulate foot

Heel bones

Foot bones

Hoof ("toenail")

Ungulate teeth

Premolar

Ridged crown

Molars

Jaw

Strong deep root

Yellow-billed oxpecker

Spotted hyaena

threatened, and on one occasion they were seen trying to rouse a rhino that had been shot and killed, even as the hunter approached.

The enemies of the grassland herd animals are, of course, numerous. Exceptions are the elephant and the rhino, who have few enemies (apart from man, of course) though a lion has been known to tackle a baby elephant or rhino with success. But in the main lions go after zebra and wildebeest, while the cheetahs and Cape hunting dogs go after the smaller species like gazelle or impala. However, the hunting dogs are not above tackling a wildebeest or a zebra if their pack is large enough. Servals, bat-eared foxes and eagles are always ready to pick off the young of the smaller ungulates, so all parents have to keep a constant close watch on their newly born babies.

The "undertakers" or scavengers are the jackals, the hyenas with their bone-crushing teeth and jaws and the ever-watchful and ever-present vultures. All these will kill and feed on unprotected baby ungulates if they can find them, as well as sick or wounded adults. But by and large their job is to clean up after the other predators, feasting on the remains of the kill. The jackals clean up the fragments of meat, the hyenas pulverize and gulp down everything—bones, hide and hooves—while the vultures plunge their naked heads and necks (no feathers to get matted and bloodstained) deep into the carcass.

If you are lucky enough to visit any of the grasslands of the world, binoculars or telescope are essential. You may be able to visit a hide near a waterhole and see the enormous diversity of wildlife—the many species of antelope coming down to drink, sometimes grazing quite close to where a pride of lions, having feasted well, lies in the shade. They seem to know the lions have fed and are therefore, for the moment, relatively harmless. You may see Cape buffalo herds shouldering massive and dark through the elegant antelopes, or watch a giraffe straddle its long legs wide so it can get its head down to water. You may see warthog, giant forest hog, elephant and rhino and many other exciting and interesting animals that have the grasslands as their home.

But if you are lucky enough to see scenes like these remember that they are only left-over fragments of the once huge miraculous pageant of African wildlife. Much of the natural grasslands with their innumerable species of plants and animals has been put under the plough, and a lot more of it is being destroyed through over-grazing by man's cattle, goats and sheep, and as a result the natural plains animals are dwindling in number. Man has destroyed so much and is still destroying by shooting, poaching and altering the habitat. Wildlife is being steadily pinned down into ever-shrinking areas. In 50 years there may be no room left for the wonderful ungulates of the world's great grasslands. One of the saddest journeys I have ever taken was when I travelled by train for days across endless desiccated eroded American prairie which once supported one of the biggest conglomerations of land mammals ever known—the bison herds. Yet today those once lush grasslands are dusty, dry and threadbare through our thoughtless over-cropping and over-grazing.

The downlands
Despite man's ruination of large tracts of natural habitat, there are some rare cases where our interference has actually promoted the diversity of life. Hedgerows are one example, and another is those "grasslands" of

southern England—the chalk downs. Of course, they are not to be compared to the great grasslands of the world, but in their own small way they are fascinating. Created originally by the felling of the beech forests by Stone Age man, these areas were then grazed by rabbits and sheep. This was not damaging for there was plenty of space for not very many sheep to roam and the rabbits were kept in balance by the fox population. Both these herbivores kept the downland free from encroaching forest growth. So, open thus to sunlight and warmth, and well drained because of the chalk beneath the shallow soil, there started to flourish a host of plants and insects rare elsewhere in Britain. The orchids, for example, with their tuberous roots, are well adapted to lack of water and nutrients encountered on the downlands. Of the three petals on the orchid flower the lip or *labellum* takes on the most peculiar shapes, and it is these shapes that give the orchids their strange and lovely names—monkey orchids, man orchids, ladies' tresses, spider orchids and even the lizard orchids, tall as a man. The lip of the bee orchid is shaped like a female bee and is so lifelike that the male bee attempts to mate with it. For the bee it is a futile attempt, however, but not so from the plant's point of view since the bee, during his "love-making", gets a mass of pollen attached to his head. When he attempts to mate with another orchid this is transferred, and the orchid is fertilized.

Many of the downland plants grow in shapes that protect them from drying up or being eaten. The thistles, for example, take the rosette habit, their leaves growing in flat wheels, while the rockrose and wild thyme creep along the ground. In fact grassy areas such as the downs, along with meadows and shrublands, are excellent places to carry out a survey of the plant life using what is called the quadrat. The best way to "randomize" your sample, so that you do not "accidentally" choose the most interesting-looking piece of terrain, is to throw the quadrat over your shoulder—wherever it lands is the chosen plot. By examining a number of sample areas you will soon build up a representative picture of the flora, and by doing another survey at a later date you can reveal how plant populations change with the seasons or with changing usage—from common land, for instance, to being grazed by sheep.

There are many species of downland butterfly—the chalkhill and adonis blues and the silver-spotted skipper are, indeed, virtually limited to this habitat. Until recently you could see the large blue butterfly there, but it is now probably extinct in Britain. Its caterpillar fed on wild thyme at first, then in mid autumn it would wander about aimlessly until found and befriended by an ant that would carry it off to the nest. Here the ants would cosset and love it for the sake of the "nectar" it produced and in return the caterpillar would grow fat on a diet of ant larvae. Having spent the winter in comfort it would then pupate and hatch in the spring.

Of the other butterflies there are the handsome marbled whites and the richly-coloured delicate dark-green fritillaries. You will find lots of snails in the downs since chalk is an important ingredient for shell-making, and the light soil is rich in worms and makes easy tunnelling, so there are plenty of moles. Many parts of the downs are dotted with copses ("hangers") of beech and yew, so you find most of the hedgerow animals here as well. But we must be ever-careful because the downlands, small but with their interesting fauna, are, like many other grasslands of the world, suffering from encroachment by the man with the plough.

PLANT SURVEYING

The quadrat is a device for taking sample areas when making a plant survey. It consists of a wooden frame one metre square divided into smaller squares with wires. Put the quadrat on a randomly-chosen place on the ground (the best way to do this is throw it over your shoulder), then identify and count all the plants within the frame. Take at least five areas for a representative result.

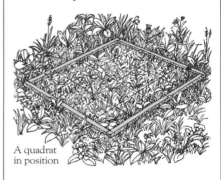

A quadrat in position

Ants and caterpillars
The caterpillars of some blue butterflies, notably the large blue (shown above), the long-tailed blue and the chalkhill blue, are "milked" by red ants just as ants milk aphids. The ant strokes the larva and drinks a sweet liquid exuded from a gland halfway along the caterpillar's back.

THE NATURALIST IN THE
DESERT

periox
hide
will f
"dro
increx

Prx
cacti,
bodix
cactux
surfa
courx
somx
carry
trap
vital
have
cacti
greex
and
to tex
the x
enox

The
If ar
the x
in p
opp
werx
ofte
grox
tern
For
ovex
exp
palx
brox
larx
and
arex
mo
ant
wir
grax
ind
eatx
thex
anx
wax
tex

Six

Deserts are not, as so many people imagine, dead landscapes. These great torrid desiccated areas of the earth are of exceptional interest to the naturalist since an amazing selection of plants and animals have adapted themselves to surviving in the harsh environment. The desert has a strange life of its own, lying dormant and dry until the rare occasions when it rains, and then it pulsates with activity and a pageant of plants and animals emerge from their dormancy to reproduce. The desert literally blossoms and a bleak lifeless landscape of red, yellow or white sand under a blistering sun can, overnight, become a shimmering carpet of flowers conjured up by the rain. The seeds of these plants, which have been lying dormant for perhaps ten years beneath the sand, suddenly awaken and come alive. With great speed, as if they know this season of moisture is quickly over, they germinate, push to the surface and flower, turning the bleak face of the desert into a multicoloured tapestry. Then they seed and die before the long wearisome waterless period starts again; by the time it does their seeds are safely locked asleep in the desert's hot depths, to ensure another season's growth. So, for that brief moment when rain washes the desert it changes the austere landscape to a riot of colours—floods of primroses and sunflowers, or pale gold sheets of desert dandelions.

This sudden blossoming of the desert naturally brings out a host of insects, for now they have food in abundance in the shape of nectar and pollen. To ensure pollination in a desert is of vital importance since the time in which conditions are right for seed production is short, and so almost all desert plants are animal-pollinated (as opposed to the chancier alternative of wind pollination). Butterflies, moths and bees all play their part, and in places birds and bats help as well. In North American deserts the giant saguaro cactus is pollinated by long-tongued bats who hover in front of its flowers and sip the nectar. But the prize for the most complex and extraordinary example of pollination must surely go to the yucca and the yucca moth. One can say that the strange spiky-leaved yucca (a huge succulent lily) and the yucca moth are so bound together that neither could exist without the other. It is a wonderful example of *symbiosis*. What happens is this: The female moth gathers pollen from a yucca flower in her mouthparts that are specially adapted for this task. She makes the pollen into a rather large ball which hangs down below her head. She then takes this ball of pollen to a nearby yucca flower and forces it into the style, which is the only way for the flower to be fertilized. After this she inserts one or two eggs into the yucca's ovary beneath its style. When the caterpillars hatch the plant seeds are partially developed. Some of these form food for the caterpillars but there are enough left over to make sure of the future of the yucca. So it is a sort of biological merry-go-round—the yucca cannot produce seed without the

Yucca moth ensures
pollination of the yucca plant

Stealing nests in an old saguaro
Safe nesting sites are at a premium in the exposed desert landscape. This elf owl (opposite) has commandeered a nest hole which was probably originally excavated by a woodpecker. In a similar way, the holes in trees of temperate woodlands are first dug out by woodpeckers and later taken over by owls, starlings or titmice.

Horned toad of Texas, as seen by its insect prey

the environment (take a snake which is cold to the touch and put it inside your shirt—you will soon feel it start to glow with the warmth of your body). The "warm-blooded" creatures (mammals and birds), on the other hand, automatically keep their bodies at a constant temperature regardless of whether it is hot or cold outside. But desert lizards are actually experts at minimizing the temperature changes of their own bodies, and they carry out this body temperature regulation by behavioural means. In the American desert lives one of the weirdest-looking lizards, called the horned toad because of its squat round toad-like body and the fact that it is covered with dozens of spines like giant rose thorns. Squatting in the sun, waiting for its insect prey, its apparently casual and haphazard movements are in fact very precise and controlled by the angle of the sun. These lizards have a very curious method of defence for they can squirt blood from their eyelids. I remember some years ago a friend in the United States sent me six of these strange creatures. They must have had a bumpy flight because they were very irritable when they arrived, and when I took the lid off their travelling box 12 jets of blood hit me, ruining my shirt.

For the naturalist, early in the morning and the evening are the best times for operating in a desert since it is cool. You may only see a few birds and lizards but there will be spread out before you a marvellous array of droppings, tracks and so on by which you can identify your animals. You can see daisy chains of footprints that will show you what lives there and by following these tracks I have managed to see and capture many desert creatures, from beetles to jumping mice. I once tracked a beetle for nearly a kilometre over the hot sand dunes before I caught him, and to my delight he proved to be quite a rare species.

The desert at night

It is really at night when the desert becomes alive, because during the day most creatures are sheltering from the terrible rays of the sun. When night falls and the temperature drops dramatically, out from holes or from under stones come the large hairy spiders, the centipedes and the lobster-like scorpions to pursue their insect prey. Feasting on the insects too come the bats, the swifts and the goatsuckers, who have spent the day in caves or cliff crevices in a state of torpor, reserving their energies for the vital and vigorous pursuit of their food after dark. This is the time, too, when the diminutive elf owls (tiniest of the owls, standing only 15 centimetres high) emerge from their homes which are, in many cases, holes in the giant saguaro cactus. These little birds hunt not only the desert rodents but spiders and insects as well. I once saw one catch, kill and gulp down in one mouthful a centipede as long as itself.

Aside from things like spiders and scorpions, the tiniest of the ground predators—and also among the fiercest—are some of the small mammals. When night falls in deserts all over the world a variety of strange creatures appear. Out come the dainty diminutive grasshopper mice and pigmy mice of the American deserts, the desert hedgehogs of Asia and Africa, and from Africa also the strange and brilliant golden mole with hair like gold spun glass. In Africa, too, live the extraordinary elephant shrews with huge eyes and long whiffly noses that really do look like elephant trunks. In parts of the United States and Mexico one even finds species of desert shrew—and shrews are animals that normally prefer to live in moist places. Then in the red Australian deserts you get all the strange little marsupial "mice" and jumping marsupials that resemble the jerboas. This host of little mammals eat arthropods, but many of the Australian marsupial mice will attack and eat the true mice that live there, some of which are as big as themselves.

All the tiny desert rodents have adapted in wonderful ways to their parched ecological niches. The kangaroo rats of North America, for example, rest during the day inside their burrows, the mouths of which they have carefully plugged up so that the air inside remains cool and moist. Even the seeds that they store in their burrows help conserve moisture by absorbing it from the kangaroo rat's breath. They lose little water in excreting waste products—their urine is four times as concentrated as a human being's. As their name implies, when they go foraging at night they escape predators by their prodigious jumping powers. Each hop is over half a metre long and they can cover six metres in one second, which is not bad going for an animal only as tall as your thumb. They can baffle an enemy, too, by changing direction in mid-air using the bushy-tipped tail as a sort of rudder. If these methods fail they resort to kicking sand in the enemy's face—not a very gentlemanly thing to do, but better than being eaten. They also chase and fight each other for they are charming-looking but quarrelsome little beasts.

Lots of other desert rodents, such as the gerbils, jerboas, hopping mice and some Australian marsupials, share the kangaroo rat's hopping abilities as it seems to be a good form of locomotion in desert terrain. Other rodents have other patterns of behaviour which are no less extraordinary. The Australian native mouse (a true mouse, not a marsupial) arranges little piles of pebbles outside the entrance to its burrow. These cool off during the night and in the early morning warmth

Hoppers and jumpers of the desert night
During the hours of darkness, the usually deserted desert is remarkably busy with nocturnal rodents out foraging for wind-blown seeds. In one hectare of Californian desert, scientists estimated they could have collected over 3,500 million seeds—enough to satisfy all the rodents, the seed-eating birds, the harvester ants and many other seed consumers, and still with a vast quantity left over to ensure the future of the plant species.

Hopping mouse
Canada and North America

Gerbil
Drier parts of Africa and Asia

Jerboa
Steppes and deserts of the Old World

Kangaroo rat
Arid areas of North America

Ears with a dual purpose

The photogenic fennec fox is a small nocturnal predator of the Saharan and Arabian regions. Its enormous ears serve as efficient radiators of heat, and they must also act like radar dishes in the detection of noises made by prey—the faint whisper of sand as a mouse trots along, or the trundling of a beetle. Other desert animals like the jackrabbit and the bat-eared fox have similarly oversized ears, and this follows a trend called Allen's rule. This "rule" states that warm-blooded animals living in hot regions tend to have relatively large ears, feet, tails and other extremities, to help with heat loss.

the water in the air condenses on them as dew; the native mouse gets his water supply by simply licking the damp pebbles.

All these little night creatures, including the small predators and the enormous variety of seed-eating rodents, themselves fall prey to hunters like snakes such as the African sand viper and the aptly-named American sidewinder. Both these snakes have evolved a curious sideways method of locomotion which gives their ribs purchase as they move across the soft sand surface. Higher up the scale are the larger mammalian predators such as the ferrets, badgers, skunks, small cats and—some of my favourites among the dog tribe—the various desert foxes with their huge ears, like the kit fox of the United States and the bat-eared foxes and tiny fennec foxes of Africa.

I vividly remember a meeting I had with a pair of young fennec foxes. Two friends of mine—wonderful amateur naturalists—worked in North Africa on various ornithological projects and whenever they returned to the zoo in Jersey they used to bring me a present in the shape of some lizards or a scorpion or something similar, the sort of lovely present one naturalist gives to another. On this particular occasion they returned from North Africa, bronzed and fit, came into my office and put on my desk a small cardboard box which I thought might contain perhaps a couple of interesting lizards or something of the sort. Gently I lifted the lid and inside were two baby fennec foxes. Huge ears unfolded like banners, black noses and black eyes surveyed me, quivering, and behind these wonderful masks were slender trembling bodies. I lifted them out of the box, and both fitted into my cupped hands like eggs in a nest. I put these captivating creatures on my desk and immediately one bit me and the other urinated copiously all over my unsigned mail. They were so wonderful to me as animals I did not care. You can always rewrite letters, but you can't always have the experience of two fragile and enchanting fennec foxes on your desk in the middle of a dull morning.

Birds of the desert

The creatures that seem to have the toughest time existing in desert conditions are the birds. Those who are not strictly nocturnal—like the owls—forage only at dawn and dusk when it is coolest, and spend the rest of the day in any shade they can find. The insect-eaters are very numerous and it is from their insect prey that they derive not only food but moisture. But the seed-eating birds must have extra water besides that in their food, and they often have to travel incredible distances to find it. Sand grouse living in the depths of the North African deserts fly several hundred kilometres to an oasis in order to drink and, more important still, to carry back water to their youngsters in the nest.

I remember in Central Australia when I went to see Ayers Rock, which is the biggest single rock in the world. It lay in the middle of the desert like a great jewel, changing colour with the movements of the sun. I found at its base several tiny pools of water, and at one of these I disturbed a flock of wild budgerigars drinking and bathing. The sun was just setting and Ayers Rock gleamed like a ruby in its dying rays. The little parakeets, scared by my sudden appearance, flew off together from the pool and then fastened for a brief second against the red rock face. It looked for one extraordinary moment as though somebody had thrown a green shawl that had attached itself to the rockwork. Then they wheeled round and flew off. This was probably the nearest source of water that these small birds had in 300 or more kilometres.

Many birds seen in the desert during the period of rains depart in the dry season. They court, nest and breed quickly, taking advantage of the host of insects with which they feed their young, and then, when conditions get dry, they migrate back to softer climates. The birds that live all the year round in deserts have adapted themselves wonderfully to this unfriendly environment, especially where their breeding is concerned. Most birds come into reproductive condition cued by what is called *photoperiod*—that is to say, the length of daylight tells them when to breed. But some desert dwellers are actually "instructed" to breed by rainfall. Several of the small insectivorous birds in Central Australia remain faithfully paired for years, but do not mate and lay eggs until a heavy rain falls and conditions are right for rearing a family. The Gambel's quail in the American South West goes even further; it doesn't even attempt to pair unless the rainfall has been heavy enough to bring out the vegetation, thus providing a food supply.

Gambel's quail

Creatures that cheat the drought

Some desert residents have evolved ingenious ways of evading the drought. There is an American bird called the whip-poor-will, a night-flying bird related to the nightjars, which "sleeps" during the cold dry winter and only comes out of its torpor in the spring when its insect food is abundant. Other small desert vertebrates, like the tiny jumping mice, spend the dry season in such a state of dormancy (hibernation when it is cold, aestivation when it is hot, remember). In the sandy desert around the city of Mendoza in the Argentine I have hunted the fairy armadillo, which is only found in this limited dry habitat. This, the smallest of the armadillos, is about 15 centimetres long, a beautiful shell-pink in colour, with long silvery fur scattered between its scales and along its tummy. It feeds principally on insects and their larvae, and constructs complex burrows beneath the sand. When the bitterly cold winter starts (for Mendoza lies at the foot of the Andes) the fairy armadillos burrow very deeply into the sand and hibernate.

Some of the most bizarre methods of avoiding inclement weather are practised by the desert amphibians. To begin with it is extraordinary to find frogs and toads, with their delicate skins and their reliance on water, in desert areas. But in the desert of North America the spade-foot toad spends eight or nine months sealed up in a burrow it has constructed, conserving the moisture in its body, and only emerges when a heavy rainfall soaks into the ground and arouses it.

The Australian water-holding frog is adapted in an extraordinary way. When the rare rains start, these frogs leave their underground caves and rush to the pools among the rocks where they gorge frantically on insects that have hatched out because of the rain. The frogs mate and spawn, and their tadpoles develop with remarkable rapidity. Then, as the pools start to dry up, the young and adult frogs absorb as much water into their bodies as possible and return to their caves, where they develop a skin like a layer of transparent cellophane. This retains all the moisture that they have absorbed, and so they lie there in a trance, encased in this cellophane skin until, in two or three years' time, the next rainfall brings them out of their coma.

Digging in against drying out
As soon as the desert soil begins to dry after the late summer rains, the spade-foot toad digs itself deep into the ground using horny projections on the soles of its hind feet (shown in detail below right). It remains underground, doubly protected inside a mucous envelope, for about nine months until the rains come again.

THE NATURALIST IN THE
TUNDRA

Travel northwards anywhere in the world and eventually the large forests dwindle into tiny stunted conifers, dwarf willows and birches. Soon even these cease and you find yourself on a vast treeless plain covered with a thin tattered blanket of dwarf heather, rocks shaggy with coloured lichens and tufts of blonde cotton-grass. This is the tundra. It is found on the roof of the world only, for at the other end of the planet the continents taper off and there are only the southern icy wastes of Antarctica, where even the most primitive vegetation cannot exist.

Like deserts or tropical forests the tundra is a world of its own, and its plants and animals are superbly adapted to its inhospitality. One of the most extraordinary things about tundra—and, you would think, one of the most daunting things from the point of view of *any* form of life—is the permafrost, the permanently frozen layer of soil up to a metre or more below the surface. This solid floor of ice will not allow water to drain through it nor will it allow roots to penetrate it. So the permafrost lies, like an impenetrable mirror, covered by a shallow waterlogged carpet of soil which freezes and thaws from winter to summer. This terrain is swept by driving winds carrying tiny sword-like bits of soil and ice which act like sandpaper on the plants, whipping and abrading them. In this refrigerator-like atmosphere dead things decompose slowly (as they do on moorland) and so nutrients are in short supply, even though the decomposers—bacteria and fungi—are just as numerous in this icy soil as in more temperate areas.

Not surprisingly, few plants can survive these sorts of conditions, but the more adventuresome take advantage of whatever they find. It might be a small difference in the lay of the land acting as a windbreak, or a tiny pocket of well-drained soil or even the tenuous advantage of a south-facing slope. On well-drained sites dwarf willows and birches flourish with various low-growing grasses and herbs. Where there are depressions like giant footprints, bogs form and in such havens you get sedges, rushes and lovely velvet-soft mosses. One of the most extraordinary things is the way that some plants take advantage of the snow as a protective blanket (much as Eskimos use it to build their igloos). Beneath the soft snowy winter eiderdown you will find mosses and herbs cosily resting, insulated from the icy weather above; and when the snow melts in the growing season it provides water.

It is only for two or three months of the year that the sun is high enough to melt the ice and snow and allow the plants to grow; but then, as if to compensate, the summer sun never sets and this allows the plants to work overtime at producing the life-giving sugars from its light energy. Some of them like the holy-grass even carry their leaves bolt upright, the better to catch the slanting rays of the sun. In the short arctic summer it seems that the whole process of life is more rapid as if the inhabitants

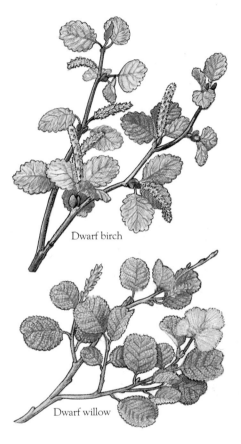

Dwarf birch

Dwarf willow

Following the sun
The Arctic tern (opposite) nests in colonies on the tundra coast. At the end of the northern summer it migrates south on the longest journey of any bird, over 14,000 kilometres, to the Antarctic seas where it spends the southern summer.

Brent goose nests in the arctic tundra but over-winters on British estuaries

Ringed plovers who nest in the tundra over-winter in Africa

were hurrying to take advantage of the precious few weeks allotted to them. The insect life is not as diverse as elsewhere in the world but there are millions of mosquitoes and flies, so many mosquitoes, indeed, that you wonder what they find to feed on. The marshes and ponds are suddenly alive with water plants and crustacean, insect and fish life. The water's surface becomes black with great flocks of ducks and geese, and its edges thick with droves of plovers and sandpipers. All these birds quickly raise their broods and then fly southwards to warmer, less hostile climes. Because it is essential that things are speeded up it means that the youngsters of the northern robins, for example, take only half the time to mature compared to their southern cousins, simply because they are—in perpetual sunshine—fed nearly around the clock. The ground squirrel babies reach adult weight and are ready for their long hibernation only three and a half months after they are born. Everywhere you sense the urgency, the hurry in the arctic summer.

When the long dark winter sets in the frozen wastes appear devoid of life of any kind. It seems a trifle peculiar that the only true hibernator in the tundra is the ground squirrel. But what happens to the other creatures during the savage winters? Most of the birds fly south, very wisely, but the ptarmigans brave it out, digging long burrows in the snow to find the twigs and leafbuds of sleeping plants. Ptarmigans, unlike other grouse, develop feathers on their toes in the winter and this gives them extra purchase on the snow, their feet being turned into miniature snowshoes. Except for a few black speckles their feathers turn pure white in winter and so if an enemy like a fox or a wolf is about the birds can crouch in the snow and, by virtue of their camouflage, become part of the white scenery and so escape detection. Another bird that goes white in winter is the great yellow-eyed flamboyant killer, the snowy owl. This huge and spectacular predator is patterned in a delicate tracery of black and white in summer, but when winter arrives it sheds this for a milk-white plumage in which only its eyes glow like fierce lamps. The snowy owl can hunt and catch ptarmigan in spite of their camouflage but it relies for a large part of its diet on the lemmings.

Fact and fable about the lemming

Lemmings are fascinating little animals, toast-brown and gold and very endearing to look at. It is curious that such a diminutive and charming little rodent could almost be described as the life blood of the tundra, since it plays a most important part in the whole food web of these vast grim areas of the world's surface.

The breeding ability of lemmings is astonishing. One female can have five litters a year, with eight or even ten babies in each, and the females from the first (and sometimes even the second) litter can start breeding when they are less than three weeks old—a real lemming production line. Even when the snow lies thick these little herbivores dig tunnels underneath so that they can feed on roots, mosses and grasses and *still* go on breeding. Small wonder that, over three or four years as their numbers swell, they eat off all the vegetation cover in an area and have to undergo one of their famous migrations simply in an attempt to survive.

The fertility of lemmings has given rise to an extraordinary mythology about their behaviour. Besides their normal migrations from winter to summer feeding grounds there are, every four or five years, vast movements of these little creatures, but the idea that they commit suicide is a myth. Like the springboks of the African grasslands, when the lemming population grows too vast it is necessary for them to move on, to find new areas with more food and less competition. And so they migrate hopefully, thousands starting to move in great carpets, whistling softly as they go to keep in touch with each other. When they reach an obstacle like a stream, a lake or even the sea they pause momentarily to gain courage and then, under pressure from the ranks behind, they plunge in. They are good swimmers but not many survive; as they swim out to sea (and if you are a lemming the sea looks like just another stream to be crossed) thousands are eaten by gulls and fish.

Although most of the tundra predators—owls, weasels and arctic foxes—hunt a variety of prey such as ptarmigan, ground squirrels and the arctic hare, they are very reliant on the lemming populations for their food. The small weasel can eat more than one a day and a snowy owl up to seven, whereas an arctic fox may feast on as many as a hundred a month. Because of this reliance on a staple diet of lemmings, the fluctuation in lemming numbers can affect all the other tundra animals. During the periods when there is not a glut of lemmings you can see how the other animals are affected. With little food the foxes do not breed and snowy owls either have smaller broods or else migrate south.

When I was a student keeper at Whipsnade Zoo I was extremely proud to be given my own section to look after. This consisted of some anoas (the dwarf buffalo from the Celebes in Indonesia), some racoon-like dogs—and arctic foxes, one of the lemmings' chief enemies. The

THE LIFE OF THE LEMMING

Compared to even the "busy" beaver, the lemming's life is one of continual frenetic activity. While many Arctic animals sleep, the lemmings are still active, feeding on roots and underground shoots in a maze of burrows and runs up to a metre below the surface. Breeding also continues, bringing the average annual number of litters to five, with six to eight young in each litter. When the snow melts in May or June the lemmings surge from their burrows and spread out over the marshy ground to feed on sedges and grasses. By the autumn, the young born early on that same year have already reached breeding condition.

Spring to summer | Autumn to winter

A bird's eye view
The lemming's coat changes in colour from mainly brown in winter to mainly black in summer. The two-tone effect must break up the body outline when seen from above.

2 metres

Surface run

Earth removed by digging

Main entrance

Section through a burrow system
Norwegian lemmings prefer to burrow in deep soil on south- or west-facing hill slopes. The surface burrows extend in all directions under the matted tundra vegetation. The large nests, about the size of footballs, are sometimes built above ground, hidden at the centre of a grass tussock or among clumps of heather.

Arctic predator and its prey
The large all-seeing eyes of the snowy owl (opposite) detect fine movements of prey, such as the lovely ptarmigan (right) and the varying hare (below right). The owl featured is a male, who retains his all-white plumage throughout the year; the female's feathers are barred with black which affords camouflage during the incubation period. Winter ptarmigan live in scooped-out depressions in the snow, feeding on shoots and grasses. In spring their feathers darken, starting at the head and spreading backwards until by the breeding season they are an overall russet-mottled brown which blends in well with the roots and dried branches of tundra vegetation. Varying hares of the extreme north retain their white coats throughout the year while those further south turn brown, russet or yellowish in summer. All three creatures illustrated have widely-spread furred or feathered feet, which serve as efficient snowshoes.

fox is much smaller and more delicate looking than the ordinary red fox. Its summer fur is a lovely ash grey and turns to snow white in winter. When I used to go into the paddock to feed my foxes they would flit round and round me like ghosts, uttering high-pitched bird-like cries. It was some weeks before I discovered that, out of misplaced kindness, I was overfeeding them. While cleaning up their little woodland I became aware of an unpleasant smell. Searching about I discovered, under the leaves, a large cache of meat. It was then I realized that my foxes were doing what they do in the wild state. Periodically, when there was a population explosion of lemmings, the foxes would catch as many as they could and store the carcasses away in a "larder" to serve them over the scarce months ahead.

Reindeer, wolves and musk oxen

Probably the most spectacular animals of the tundra are the wolves and the wild reindeer (called caribou in North America), which in their own way represent the great predators and ungulate herds of Africa. Most of the wild reindeer herds are migratory, spending the short summers on the tundra feeding on the lush vegetation while it lasts and then moving southwards as the bad weather sets in. They make tracks for the comparative shelter of the coniferous forests where they scrape and scrabble the snow away to uncover "reindeer moss" (which is in fact a

WOLVES—SOCIAL PREDATORS

The social life of the wolf is an example of the extended family working as an efficient hunting unit. The pack is led by a dominant male and female who are the only breeding pair, the remainder of the pack being made up of cubs, young individuals and related adults. Pack hierarchy ensures that when one or both of the dominant pair succumb to death or injury, those second in rank take over as pack leaders, though there may be some "in-fighting" from up-and-coming youngsters. The female wolf is an attentive mother and the cubs spend weeks playing and fighting among themselves, learning all the gestures and postures which as adults they will use to maintain the family traditions. Though packs of 20 are common, during the winter wolves often associate into large packs of around 50 individuals which can efficiently run down large quarry.

Body language
Apart from the howls and yelps which enable wolves to talk to each other at a distance, communication at close quarters is effected by a system of tail postures and facial expressions, used in conjunction with each other to reinforce the mood or message. Most of these signals are concerned with maintaining rank among members of the pack. Those illustrated show the sequences of signals from normal posture through to out-and-out aggression (below left) and to complete submission (below right).

Menacing threat   Attack   Normal posture   Submission   Complete submission

lichen) and dead grasses to feed on. There is even a report that reindeer are not averse to feeding on lemmings if they can catch them.

The wolf packs, of course, are the main enemies of the reindeer as they circle and harry the herds. But a healthy adult reindeer is more than a match for them and for the most part the canines feed on the old, the ailing or the weakest youngsters. This natural selection system of culling keeps the reindeer herds in trim, weeding out the weaklings and letting only the vigorous survive.

Reindeer

The wolf has always had very bad publicity, being depicted as an evil killer and even tied up with witchcraft, werewolves, wolf men and the like. But we now know more about wolves and we have discovered they are not the merciless cold-blooded killers they are made out to be. They, like nearly all predators, kill to live— and no more. Their social life is complex and fascinating. The pack of about 20 individuals is usually an extended family and it is by and large a very peaceful one. Order among the wolves is kept by an elaborate sign language of the tail—wolves "talk" with their tails, signalling each other and maintaining the hierarchy. When they hunt, of course, they communicate with melodious howls and, periodically, they have grand choruses to keep in touch—or to delineate territory—with other wolf packs in the vicinity.

When the female she-wolf is ready to give birth (they have from five to seven young) she digs herself a hole. In this den she has her cubs and while she is suckling them the rest of the pack bring her food. As the cubs are being weaned the pack brings food for them as well. After they are weaned the mother can leave the cubs to hunt with the pack, but a young female stays behind to act as nursemaid to the puppies and guard them from marauding predators.

It is reported that wolves prey on the strongest and most formidable mammal of the tundra, the musk ox, though how they manage to break through the defences of these ungulates is a mystery to me. When musk oxen are threatened they form a circle, with the strong adults facing outwards and all the young in the centre. This presents to the enemy a solid wall of massive heads and sharp horns. Musk oxen are small stocky animals shaped rather like a brick with hair—and what hair! They have the longest hair of any wild animal, almost a metre of it, as protection against the cold.

I remember the very first musk ox I ever saw was a baby in the zoo at Copenhagen. I could hardly believe that it was not a toy, it was so enchanting. It was almost square, covered with a thick coat of hair that looked as if it had been clipped. It had shiny stubby little hooves, fat legs, a gleaming patent leather nose and huge eyes with a blue tint to them. It stamped its hoof petulantly at me when I tried to talk to it, but eventually condescended to come over and suck my finger, loudly and wetly, gazing at me earnestly out of its big blue eyes. From that moment onwards I fell deeply in love with the musk oxen and I hope that they never perish from this earth.

Musk ox

Sad to say, like so many other animals the numbers of musk oxen have dwindled, principally due to persecution by man. The brave circle they form to protect their young against wolves is no defence against a hunter with a gun. At last, however, the musk ox is protected and one hopes its numbers will increase. It is a brave wild-haired personality of the tundra and, like wildlife everywhere, it deserves our mercy.

THE NATURALIST IN
DECIDUOUS
WOODLANDS

In large areas of the world the climate is what is called "temperate"—that is, it is neither very cold nor terribly hot, neither very dry nor very wet. It is in such areas that you get the great deciduous woodlands and forests. The face of the deciduous forest is constantly changing; the broad leaves in many shades of green clothe the trees in spring, and when autumn comes the woods become a pageant of russets, browns, golds and reds as the leaves die and fall, carpeting the ground with their many colours. The word *deciduous* (from the Latin) actually means "to fall down", a rather ugly word to describe such a beautiful process. Also, some naturalists use the words temperate or broad-leaved to describe deciduous woodland; in practical terms they are all much the same.

Deciduous forests are found mostly on the top half of the world, to the south of the drier colder belt of evergreen pine forests. In fact the staid dark-green pine forests are like a rather severe dress, below which appears the paler gaily-coloured deciduous woodland like a frilly petticoat at the hem of the dress. There are few of these forests in the southern half of the world—only some at the tip of South America and in New Zealand. The broad-leaf forests of temperate Australia consist, in the main, of many different species of eucalyptus tree, which are evergreen. The three main stretches of deciduous forests are in Europe, with a thin easterly wedge tapering off into Russia; in the eastern middle of Asia, across China and Korea to Japan; and in the eastern middle of North America. Nearly the whole of the British Isles was once cloaked in dense woodlands of oak or beech and it has been said that in olden times a squirrel, should it wish, could travel from the Severn to the Wash through the branches without having to descend to the ground. This squirrel's American counterpart could have travelled as far as 1,500 kilometres in the same way. Unfortunately those days are long gone because man has cleared his forests for timber and to make way for agriculture. By the 19th century in Britain only five per cent of the land remained wooded, and in the eastern United States the giant cities and sprawling agriculture ate into the forests and destroyed them. There are a few sad pockets left, of course, on the sides of mountains where it is awkward for timbering or farming. The rolling hills in the Cevennes near our house in France form one such area; another is the mountain "coves" in the American Appalachians where as a child Lee spent her summers fishing in the lakes and chasing snakes in the woods. There is the great Bialowieza forest in Poland, where the last European bison roam, and in England the New Forest with its ancient and gigantic oaks and beeches, a forest steeped in history. Here I used to go collecting specimens on horseback, a splendid way of moving and observing. The New Forest is not really the best name for it, as it is very old; a much better name would be "wildwoods". The woods are remnants of primeval forests that covered England when the glaciers

Winter and summer clothes
Shrouded by leaves in summer, the coming of winter for a deciduous tree reveals the true shape of its woody skeleton. The oak (top) has a wide irregular dome with massive twisting lower branches; the foliage is set in bunches and the bark is deeply fissured. The sycamore (above) has a more regular shape with a crowned dome and billowing branches, and light grey-brown scaly bark.

A youngster's inquisitive look
This fox cub (opposite) is at last branching out on its own after several weeks spent playing with its brothers and sisters outside the family earth. The vixen tutors her cubs in hunting by taking them on nightly expeditions but eventually, like all youngsters, they wander off and learn to fend for themselves.

MAKING A SOIL PROFILE

A soil profile will tell you which soil conditions suit which creatures, and is a useful "framework" on which to base your studies of litter-dwelling and burrowing organisms from different habitats.

Woodland soil profile
Dig a small pit (remember to fill it in afterwards), keeping one side smooth so you can see the soil in cross-section. Note the depths and colours of the various layers (called "horizons") and take samples. A typical deciduous wood profile (right) has deep litter (1) grading into thick rich humus (2). Beneath is a layer of subsoil stained with organic matter (3) which overlies the parent material, such as clay (4).

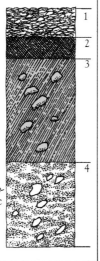

receded some 10,000 years ago, and which in those days were the haunt of large mammals such as wolves and wild boars.

Judged as a whole, a woodland is a giant organism made up of many component parts, in the same way that a human body is made up of numerous cells. A forest is a climate-creator—the millions of trees breathe out moisture and oxygen, creating rainfall and cleaning the atmosphere, and are thus of importance to the existence of all life. Inside a thick wood the air becomes warm more slowly and cools down more gently, so that extremes of temperature and moisture are smoothed out; and the crowd of trees turns and melts winds, so that in the tangle of leaves and branches a fierce gale becomes a breeze. Thus the woodland dwellers live in a relatively even environment.

Within the body of the wood you will find areas which provide everchanging habitats for a host of different plants and animals. As a dying tree falls, its bulk rips out a clearing in the fabric of the woodland, letting in more sunlight and thus providing a sort of brilliantly-lit windowbox in which light-loving plants and creatures flourish until the canopy grows over and the wood has healed its wound. Natural clearings are also found because of small differences in the mineral content of the soil, or a slight change in slope which alters the drainage and the angle of sunlight.

The tree — ruler of the woodland

From the point of view of the animals and plants of the woodland the most important, the most dominating organism in a wood is, of course, the tree itself. A tree is, when you come to think of it, a sort of lovely fountain of earth, and its shape and size determine what life will shelter beneath it and in its branches. To an enormous range of animal life a single tree is not only a home but also a larder—the tree starts food chains with its leaves (both living and dead), flowers, fruits and seeds; for a vast range of plant life the trees provide the right conditions for growth—a rich soil, shade and support.

The whole character of a wood, its whole personality, is created by attributes of the individual tree. The way its roots grow, either plunging deeply into the earth or spreading below the surface like a web; the way the canopy grows—upwards in a spike shape or spreading out like a green ceiling; the kind of bark it has, from smooth as silk to rough as sandpaper to carunculated like the hide of a rhinoceros. Take the sessile oak, for example. It must keep its root system shallow in order to catch and drink the rain before it drains through the thin soil that this tree lives in, and so the field layer loses out on both its water and nutrient supplies. Under these oaks you only see bluebells in the spring, and later on bracken—both plants that can cope with the lack of water and nutrients. On the other hand the tall beeches form a dense glittering canopy of leaves that burst from their buds early in the spring, thus preventing the growth of anything except shade-lovers like bramble and wood sorrel. In autumn the millions of beech leaves fall and form a thick blanket that makes it difficult for the tiny ground flora like mosses to establish themselves. But fungi do well in this blanket, especially the infamous and poisonous death cap. There are even two species of flowering plant growing here that need no light—they get fungi to provide them with energy. These are the yellow bird's-nest (a sort of wintergreen) and the bird's-nest orchid. Neither of these plants attempts photosynthesis (they don't even have any chlorophyll, the pigment that is essential for the process of photosynthesis) but they rely on the fungi that live on their roots to provide them with a supply of food.

There are endless examples of how trees dominate in the woodland. You will find that oaks can be muffled up in great cloaks of lichens and mosses which cling to their rough bark, whereas these organisms cannot find a foothold, as it were, on the smooth bark of beech or ash nor the slippery peeling bark of birch.

Woods can differ from one another, too, if man has had a hand in their history. Coppicing and pollarding, wood-pasturing and planting, clearcut logging, and, of course, fire all have their particular effects on the trees and therefore on plant life sheltering beneath them.

The characteristics of the lower storeys of the wood greatly affect animal life. Millipedes and slugs need moist ground; earthworms don't like eating oak leaves; lizards enjoy a broken canopy that lets sunlight through to bask in; pheasants must have thick cover to nest in; deer also use cover to hide in; and so on. You can see that the various forest trees—their shape, size, age, bark, leaves and the way they grow—all dictate the plants and animals that live on, in and around them. In a woodland the tree is king.

There are many sorts of temperate deciduous forests in the world and each is named after the one or two dominant species of tree that make up the forest. Thus in most parts of England and Europe the common oak is dominant, and on drier soils the sessile oak and (in chalky areas) the beech take over. The way the deciduous forest is distributed in North America is interesting. From southern Canada to nearly halfway down the American continent there are forests composed of beeches and maples (from one of which the delicious syrup is made). Farther south the forests are ruled by the yellow poplar, although they are mixed with

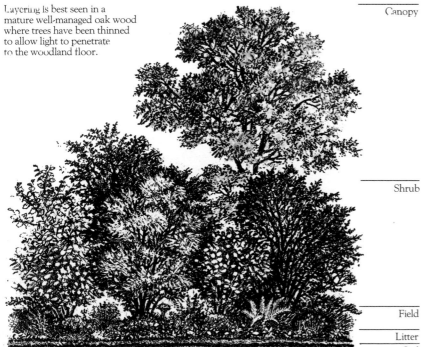

Layering is best seen in a mature well-managed oak wood where trees have been thinned to allow light to penetrate to the woodland floor.

Canopy

Shrub

Field

Litter

Soil

The layers of the woodland
You can think of the three-dimensional woodland as being divided into vertical layers or storeys, each layer with its characteristic plants and animals. The basic blueprint for a typical wood has been well worked out in terms of five levels, and using this the naturalist should make a map of his own local wood. Besides this "side-view" map you can also make the more usual plan view, using different symbols for the different tree species; this gives you a basis for further studies on the movements of various animals.

Canopy
The highest level is formed by the mass of intertwined branches, twigs, leaves and fruits of the mature trees plus the vast array of canopy-dwelling creatures.

Shrub layer
This is made up of bushes, shrubs, taller woody plants and young trees. Many creatures, especially insects and birds, take refuge in the thick shrub growth that springs up in gaps in the canopy.

Field layer
This is the floral and herb carpet, plus of course ferns, mosses and many other low-growing plants and the myriad of creatures, especially insects and other invertebrates, that live on them.

Litter
The "floor" of the wood is composed of dead and decaying leaves, other plant debris, various fungi and swarms of tiny litter-dwelling creatures, plus occasional visitors like surfacing earthworms.

Soil
The wood's foundation is the soil in which the trees are rooted and to which they constantly contribute gifts of leaves, branches, seeds, fallen fruit and, eventually, their own bodies as they die and decompose.

Deciduous woodland in spring

A really mature beech wood in the spring, before the millions of jade-green leaves have uncurled and formed a shimmering roof, is a light and airy place. Unfortunately, on the day we visited this lovely wood it was cool and damp, but nevertheless the wood smelt rich and moist like a plum cake. Underfoot were carpets of wild flowers and dark green shawls of wild garlic scented the air. The thick leaf litter under the enormous grey columns of the trees is home to numerous species of small invertebrate animals. Careful removal of this litter will let you watch a host of activities of the creatures which live on the forest floor.

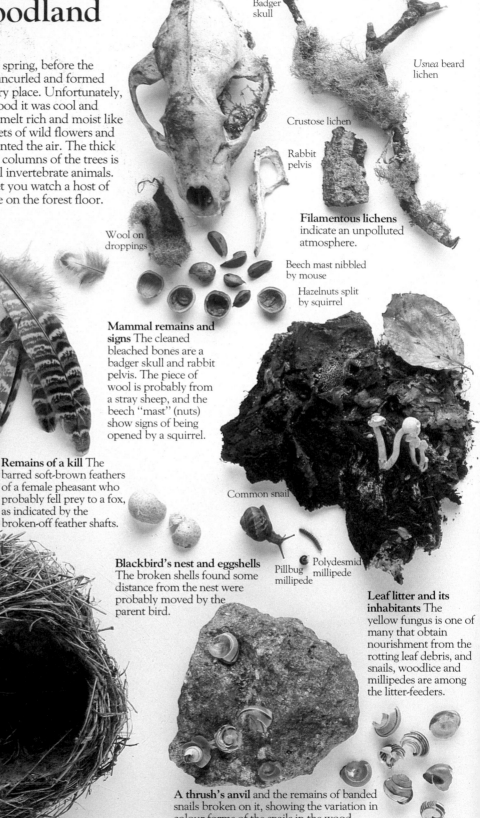

Badger skull

Usnea beard lichen

Crustose lichen

Rabbit pelvis

Filamentous lichens indicate an unpolluted atmosphere.

Wool on droppings

Beech mast nibbled by mouse

Hazelnuts split by squirrel

Mammal remains and signs The cleaned bleached bones are a badger skull and rabbit pelvis. The piece of wool is probably from a stray sheep, and the beech "mast" (nuts) show signs of being opened by a squirrel.

Remains of a kill The barred soft-brown feathers of a female pheasant who probably fell prey to a fox, as indicated by the broken-off feather shafts.

Common snail

Blackbird's nest and eggshells The broken shells found some distance from the nest were probably moved by the parent bird.

Pillbug millipede

Polydesmid millipede

Leaf litter and its inhabitants The yellow fungus is one of many that obtain nourishment from the rotting leaf debris, and snails, woodlice and millipedes are among the litter-feeders.

A thrush's anvil and the remains of banded snails broken on it, showing the variation in colour forms of the snails in the wood.

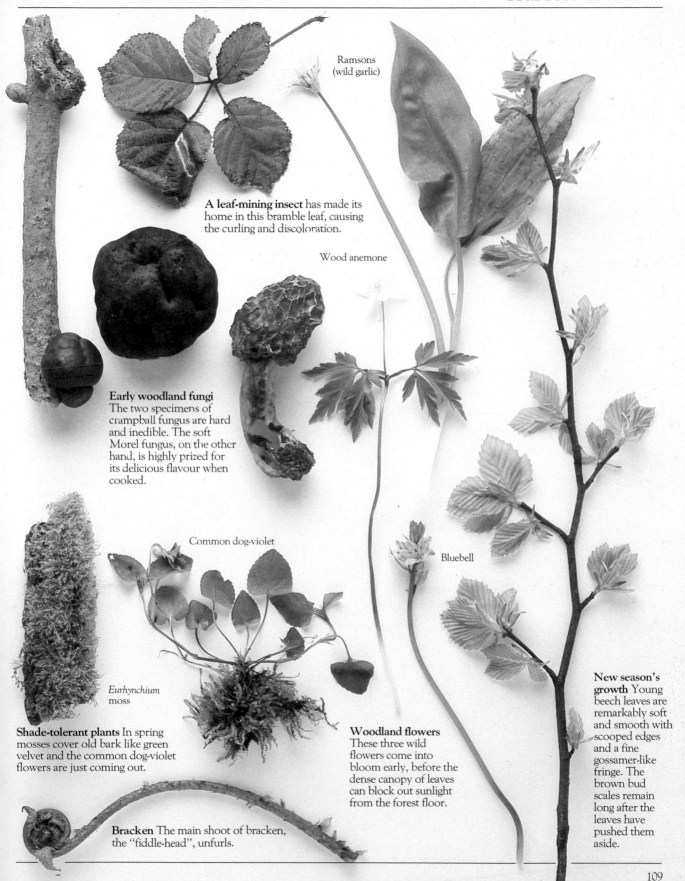

Ramsons
(wild garlic)

A leaf-mining insect has made its home in this bramble leaf, causing the curling and discoloration.

Wood anemone

Early woodland fungi
The two specimens of crampball fungus are hard and inedible. The soft Morel fungus, on the other hand, is highly prized for its delicious flavour when cooked.

Common dog-violet

Bluebell

Eurhynchium moss

Shade-tolerant plants In spring mosses cover old bark like green velvet and the common dog-violet flowers are just coming out.

Woodland flowers
These three wild flowers come into bloom early, before the dense canopy of leaves can block out sunlight from the forest floor.

New season's growth Young beech leaves are remarkably soft and smooth with scooped edges and a fine gossamer-like fringe. The brown bud scales remain long after the leaves have pushed them aside.

Bracken The main shoot of bracken, the "fiddle-head", unfurls.

several other species. As you travel into the deep south of the United States you see a combination of oak, hickory and magnolia.

Since in the woodland the trees are dominant, you should try to get to know the trees in your local wood. Build up a picture of the various species and the characteristics of the individual trees. For example, you can measure their heights and girths, map out the intricacies of their canopies, and estimate their ages. Do not forget to include the members of the shrub layer whose crowns are overshadowed by the dominant trees—holly, yew, cherry and youngsters of the bigger trees. They form important places for nesting birds and supports for the webs of spiders, strategically placed to catch insects flying their missions for mates or food.

Forest fungi

The most important inhabitants of the woodland soil are the ones whose activities are at first sight the least apparent. These are the many species of fungi. The colourful mushrooms and toadstools you see pushing up through the soil to decorate the forest floor are only the "fruit" of the fungi, that is to say the structures that contain the spores for propagation. The real hard work of decomposition is done by the mass of microscopic fungal threads (called *hyphae*) and small round fungal bodies in the soil itself. The threads of fungi run like a complex network beneath the soil and it is these, helped by bacteria and other microbes, that account for the decomposition of around 80 to 90 per cent of the dead plant and animal matter in a wood. In their turn, of course, the buried fungi are food for mites, springtails, and also the miniscule nematodes, most of which are microscopic worm-shaped creatures. But

THE VITAL STATISTICS OF A TREE

It is a useful exercise for the naturalist to "adopt" a tree in a nearby wood and follow its fortunes through the years. Keep track of its height, girth and canopy area as shown here. As far as age goes, it is said that a full-crowned tree is as old as there are inches in its girth, though there are exceptions such as the fast-growing poplars and the slow-growing chestnuts. You might try adopting two young trees of the same species, one in a dense wood and the other in a fresh clearing, and comparing their progress.

Girth and age
Measure the girth at the standard height of one and a half metres from the ground (right). Roughly, the age in years is equal to the girth in inches. A more accurate way is to count annual growth rings on a stump of the same size and species as your tree.

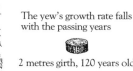

The yew's growth rate falls with the passing years

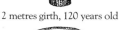

2 metres girth, 120 years old

5 metres girth, 300 years old

6 metres girth, 500 years old

Canopy area
Pace or measure the distance from the trunk to the outermost branches in eight directions and draw a scale map, as below. Trees in exposed places often have a wind-blown one-sided canopy.

Open woodland

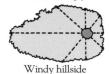

Windy hillside

Tree height
Walk 27 paces from the trunk and get a friend to hold a stick upright on the ground. Take three more paces, get down to the ground, and tell your friend to mark the stick where it crosses your sight line to the tree's top. The tree is ten times the height of the mark.

in the about-turns common in the natural world there is a form of fungus that preys on these fungus-eating nematodes, and it does so in the most extraordinary way. There are sundews and similar plants that use curious methods to entrap insects, but here we have a fungus growing beneath the soil that uses a hangman's noose for securing its nematode prey. The fungus has a series of little nooses which dangle from its filaments. The cells—there are only three—of which each noose is made are unusual in that they are highly sensitive to being touched. When a nematode, pushing its way through the soil, unwittingly wriggles through a noose the friction of its body signals the cells and the noose quickly inflates, thus strangling the nematode. This process takes only one-tenth of a second!

As I write this it is autumn in Provence and the mushroom season has just started. Our colourful local market has huge baskets groaning with a great variety of fungi, enough to make a gourmet's mouth water. The mushrooms have all been collected locally in the forests of the Cevennes and so are beautifully fresh. There are the fat rusty-brown ceps and the chanterelles, the latter looking like tiny yellow parasols that have been blown inside out by a miniature storm on the forest floor and tasting of champagne when you cook them. There are the young orange agarics, scarlet and egg-shaped and surrounded by a white fleshy "egg-cup". These are not to be confused with the orange white-flecked fly agaric, which is very poisonous. It was a baby fly agaric, incidentally, that Walt Disney got to dance so charmingly in "Fantasia".

Fungi have such wonderful colours and shapes that they make splendid subjects for drawings, paintings and photographs, all of which can be done in the forest if you wish not to disturb them. But you can carefully collect a few of the young ones—those whose caps are just open—and take them home for drying or making spore prints (page 254) without endangering their numbers. In fact the wealth of woodland fungi so impressed and fascinated the great French naturalist Fabre that he produced the most delicate, accurate and beautiful watercolour paintings of over 600 species. There are so many with lovely and descriptive names: parasol, Horn of Plenty, wood woolly-foot and stinkhorn. Blewits sometimes have a bluish tinge—their name comes from an old English word for blue. Puff-balls, white and round, explode to spread their spores and the skin of the related earth star peels away in sections to release its spores. The threads of the honey tuft, a rather nondescript toadstool itself, cause the glow you can see at night on bits of rotting wood. Dead men's fingers and black bulgar grow on rotting stumps and logs, and low on the trunks of old trees grow oyster, beefsteak and bracket fungi, the razor-strop and the gelatinous ear fungus, which looks like brown semi-transparent human ears. Not all of these are edible, of course, and some fungi are really poisonous, like the panther cap and death cap, the destroying angel and (a warning) the fool's mushroom.

Of course, when you say a mushroom is inedible it depends on your point of view (and generally we humans tend to have only one point of view—our own). But many fungi that we would turn up our nose at are considered delicious by slugs, snails, fly maggots and beetle larvae. The smelly fluid produced by the stinkhorn is attractive to slugs and is eagerly lapped up by flies who, in doing so, help spore dispersal.

The sudden appearance of so many different sorts of fungal fruiting bodies in the autumn is probably due to a combination of rain, warmth

Animal-eating fungi

To supplement their diet of plant matter, certain fungi have evolved ingenious ways of catching small invertebrates in soil and leaf litter. The sequence shows the "strangulation" of a nematode worm. Once caught and killed it is digested by the fungus, providing extra nutrients for growth. Many other fungi (which never appear above ground as mushrooms or toadstools) trap and digest animals. Some use sticky threads, others produce enormous numbers of sticky spores. Once a spore is swallowed or sticks to a passing amoeba or nematode it germinates, sending out a mass of filaments which eventually kill the host.

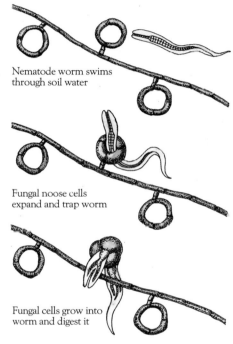

Nematode worm swims through soil water

Fungal noose cells expand and trap worm

Fungal cells grow into worm and digest it

Down among the dead leaves

A close-up of the spider *Drassodes* (opposite, above) shows how fearsome it must appear to the small litter inhabitants. It subdues its victims by wrapping them in bands of silk. The sculptured-looking *Polydesmus* millipedes (opposite, below left) also crawl through the litter and under damp bark, feeding on plant matter. Woodlice (opposite, below right) appear to be social creatures as they congregate under moist bark, but they are only brought together by their search for suitably dark damp conditions.

WINTER WORK

Many moths over-winter as pupae. You can usually locate them a few centimetres down and about 20 centimetres away from a tree trunk, in the angle of the roots; the north-facing sides of old oaks, poplars and limes are particularly profitable. Leaf litter also yields many small but interesting creatures, which are best sorted out at home (see page 264).

After locating pupae, replace earth carefully

Collect litter in a plastic bag to maintain humidity

and the showers of nutritious dead leaves that start to fall in late summer. Under these favourable conditions the fungi decorate the floor of the woods and quickly produce their spores; then, with the first frosts, they wither and die and merge back into the soil from which they sprang.

Life on the forest floor

When you walk on the soft carpet of the woodland floor, breathe in the rich and redolent smell and remember the teeming thousands of different creatures beneath your feet that have worked on the manufacture of the leaf litter. The "litter" in any wood is made up of innumerable kinds of dead plants and animals and, of course, animal droppings, but in a deciduous wood the bulk of the litter is made up of fallen leaves. The litter lies on the moist soil and is thus kept damp, and in its turn it acts like an overcoat to the soil and protects it from wide swings of temperature. There is really no fine line between where litter ends and humus, the end product of decay, begins. The millions upon millions of leaves found in a spring wood will, by autumn, have changed colour, died and fluttered down to the woodland floor; by the following summer these crisp dead autumn leaves will have been converted to a soft brown fragmented mat by the magical processes of winter rains, spring warmth and fungi and bacteria. Then a host of busy "gardeners" pass the leaf mould through themselves, thus breaking it apart still further—gardeners such as earthworms, potworms, snails and slugs, millipedes like little tubes with legs, woodlice and various insect larvae. Finally the decomposing mould is broken down even more by microbes and becomes part of the dark nutrient-rich humus underlying the litter.

The worms are, of course, burrowers, as are some millipedes and insect larvae. Underground they are protected from drying out, and in spells of drought they simply burrow deeper. The other litter-feeders—snails, for example, and pill millipedes and woodlice—do not burrow; they creep through the woodland floor mainly at night, for they cannot risk drying out and dying in the warmer daytime air. During the day they take refuge under litter, or beneath rotting logs or stones. Of all the soil invertebrates the most important litter-feeders are the various species of land snail and their close relations the slugs. This is because their body chemistry is best at digesting tough fibrous plant material. I have always kept large collections of snails for I think they are exceedingly interesting creatures, both from the great variety of shell shapes they assume and from their habits. There are many variations of shell structure. Some are flat and coiled like the neatly-stacked ropes on the deck of a sailing boat; this shape is constructed by the yellow and brown common round disc snail. The chrysalis snail's shell is fat and blunt with little "teeth" on the opening of the shell. A sharply conical shell probably belongs to the door snail, which is frequently seen on tree trunks and has a close-fitting door at the opening of its left-handed shell. Then you can find the glass snail whose shell, transparent as mica, allows you to observe the pulsating body inside as though you had the animal under an X-ray machine. Then, of course, you get the beautifully-marked banded snails which thrushes love to smash open on their "anvils". These handsome snails are not always available to the birds because in winter they hibernate underground with the mouth of the shell firmly sealed by a little door, called the *epiphragm*, that is made out of hardened mucus. In

Love in a damp climate
In late spring after rain, a pair of banded snails come together to mate among the mosses on the woodland floor. Each snail has both male and female reproductive organs, and after fertilizing each other they both lay batches of pearly opalescent eggs in the soil or litter. Land-dwelling molluscs such as slugs and snails go to great lengths to mate success-fully—most molluscs, being aquatic, simply cast their sperm and eggs into the water and leave fertilization to chance.

the spring their doors dissolve and they emerge. Spring is the time for snail courtship and although this is a somewhat slow-motion affair, as you can imagine, it is intriguing to watch. Snails do not have the complications that human beings have when choosing a partner. As far as a snail is concerned any other snail it meets is eminently eligible, for each snail is an *hermaphrodite* (both male and female). So if you want to breed snails you don't have to worry about getting a male and a female. Once you have a pair you have, in effect, two males and two females. The courtship dance is slow and gentle and consists of much pressing together of their feet and caressing each other with their "tentacles"—their eye stalks. In the banded snails, after these preliminaries each partner proceeds to "harpoon" the other with a tiny barbed "love dart" made from calcium. The love dart stimulates the exchange of sperm and when they have finished mating both the Mr-and-Mrs snails wander off to lay their round white pearl-like eggs in clusters under stones or logs, or buried in the soil.

The thousands of litter-feeders naturally fall prey to other creatures, the litter-hunters. There are wolf spiders, harvestmen, various ground and carrion beetles and their larvae and predatory fly larvae that prowl through the litter. Some of these predators tend to specialize in their diets. The pseudoscorpion, for example, pursues mites while some beetles and their larvae (like the glow-worm) pursue the slow-moving snails. But there are, in turn, predatory snails and slugs. The shelled slug, looking like a cross between a slug and a snail with its soft pale body and tiny flattened shell at the rear end, is an active burrowing predator and will pursue and devour earthworms, centipedes and even other slugs.

Vertebrate hunters of the leaf litter

The pigmy "eat-and-be-eaten" world of the woodland floor forms the food base for a number of small vertebrates who roam the leaf litter searching for prey. In particularly damp British woods you will find the amphibious toads and newts (the latter are a type of salamander). The toads, like all amphibians, need ponds or streams in the breeding season, but unlike the thin-skinned frogs the thicker- and drier-skinned toads can be found deep in the wood, quite a long way from water. In continental Europe there is the very strange midwife toad whose eggs do not need water for their development. In this species the male acts as a "midwife", taking the string of eggs and wrapping it securely round his hind legs like bracelets. He carries around this rather ungainly ornament, looking like a pop-eyed and very wealthy woman wearing too much jewellery, until the eggs are ready to hatch. Then he hurries down to the water and immerses his hind legs so that as the tadpoles hatch they can swim away. No one has discovered quite how the midwife toad *knows* his tadpoles are about to hatch. If you are lucky enough to find a male carrying his burden of eggs do not try to catch him or take him home, for they are nervous little creatures and will soon discard the egg bracelets.

Most of the purely terrestrial salamanders, in common with toads, rely on a water supply during the breeding season. Outside the breeding season you will be lucky to come across them, so well do they hide. The beautiful tiger, marbled and spotted salamanders of eastern North America spend most of their time in earth burrows and only emerge during rainstorms. The various European newts shelter their delicate skins under damp logs or stones and only emerge at night to prowl around in search of food, as do the spotted-and-banded fire salamanders of continental Europe. Fire salamanders mate on land, and if you are in the vicinity you might be able to watch this. The male is a terribly over-anxious ladies' man and in his enthusiasm will pursue every other salamander in sight, of either sex, and will even approach dummies, though he quickly throws these over. The right female, when he finally finds her, must respond correctly to his over-athletic embraces, and thus encouraged he will deposit a packet of sperm on the ground. Then he moves aside, having manoeuvred the female into the correct position so that she may pick up his "love letter" with her cloaca. In this species you can recognise each individual adult specimen by its markings; like fingerprints, no two fire salamanders are the same, and as they live up to 20 years and always inhabit the same area you could get to know the salamander population in your wood quite well.

Curiously enough, you get very few lizards hunting in the litter of the deep forest. I presume it is because they like more open places where they can not only hunt but also bask in the sun. But you will find a few representatives of the ever-present common lizard in the wood, as well as that legless lizard, the slow-worm. The common lizards hunt for arthropods whereas the slow-worms prefer soft-bodied prey such as earthworms and slugs. Both these reptiles produce live young rather than laying eggs. The slow-worm gives birth to between 12 and 20 babies, slender and beautiful, looking as though they have been cast in gold, with a vivid black stripe down their backs. The female generally gives birth in a concealed place like a rotting stump.

The snakes of the woodland, depending on their size, prey on insects, lizards, toads, rodents and even rabbits. In Europe there are the mildly poisonous adders that lie in wait for their meal and the smooth snakes who vary their diet by pursuing and eating other snakes. In Europe and North America are various kinds of ratsnakes and racers, both expert climbers. The racers, as their name implies, are very fast-moving and have been accurately clocked sliding through the undergrowth at almost nine kilometres an hour—quite a speed for a medium-sized legless reptile.

Not all snakes are venomous; indeed, the bulk of the world's snakes are harmless. A naturalist working in an area where there are poisonous snakes should not wander about in a constant state of panic. By and large, most snakes are just as afraid of you as you may be of them and they will do their best to avoid humans. It is comforting for the naturalist to know that an American is three times more likely to die from being struck by lightning than from a poisonous snake bite, and of course in Europe the risk of a fatal snake bite is even lower. Nevertheless you can take sensible precautions: walk carefully (a snake that is trodden on by a heavy naturalist can hardly be blamed for biting back); do not go stuffing your bare hands into holes or hollow logs before you investigate the interior; and wear thickish trousers and wellington boots or stout shoes which offer some protection.

Of the tiny mammals hunting through the litter bed there are various shrews who scuttle swiftly about in search of prey. Curiously enough, you may sometimes see young moles hunting above ground rather than waiting—as their parents do—for the litter creatures to drop through the roofs of their tunnels. Among the shrews will be both the white-toothed

Lizard-like amphibian and snake-like lizard
The fire salamander's bright colours advertize that even for an amphibian it is particularly distasteful—glands in its skin exude a repellent liquid if it is tampered with. A shy nocturnal creature who hunts earthworms, slugs and various arthropods of the forest floor, the salamander has long had a mythical association with fire. It is said that the amphibian's habit of resting up by day in rotting logs brought this about; imagine sitting in front of a roaring wood fire, putting on another freshly-collected log, and then a little while later seeing the astonishing sight of a salamander emerging from the flames! No such mythical powers for the harmless slow-worm, a legless lizard whose slender shape allows it to burrow into litter and soil for worms and slither into crevices in bark for slugs. Both salamander and slow-worm are what is called *ovoviviparous*—that is, the females do not lay eggs but retain them in their bodies until they hatch, and then give birth to the live young.

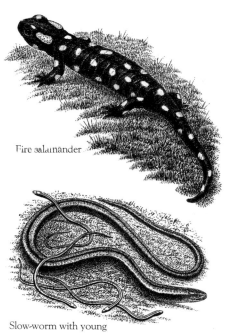

Fire salamander

Slow-worm with young

species and the red-toothed shrews who look as though they have got rusty teeth. They will very likely be the ones you find in hedgerows, because after all an old hedgerow is nothing but a wood in miniature. In south-west Europe, in the remnants of forest and the loosely-piled stone "hedges", lives the smallest mammal in the world. This is Savi's pigmy shrew, one of the white-toothed shrews, which is probably about as big as your thumb—it is less than five centimetres long.

Unfortunately, some people seem to dislike shrews. They are pugnacious, querulous and have prodigious appetites—they will even eat the distasteful salamander, and are not above a spot of cannibalism should it be necessary. In addition to these rather doubtful habits shrews are possessed of a musky smell which is rather unpleasant. You can often find dead shrews, presumably killed by one of the larger carnivores and then left because the body odour made them unpalatable. In spite of

DEATH, DECAY AND REBIRTH

A rotting log is a small but fascinating and almost self-contained society which shows clearly the process of ecological succession—one set of species being followed by another, and so on, until a stable end result (in this case the return of the pitted remnants of the log to the soil) is reached. The experienced naturalist can tell how long ago a log died by the particular species present at any one time.

Newly-fallen log
Lovers of damp—slugs and woodlice—are the first to find shelter, feeding at night on surrounding rotten vegetation. They are soon followed by carnivorous centipedes and spiders. The fearsome *Dysdera* captures woodlice with its huge jaws.

The following year
Still recognizable, the log is now scarred with beetle holes and covered with mosses and fruiting fungi. Bark bugs squeeze their flattened bodies under the tree's flaking skin; the larva of the cardinal beetle feeds on decaying wood, the adult beetle on nectar.

Another year past
The log is riddled with galleries of wood-boring insects, providing rich pickings for woodpeckers. The ruby-tailed wasp lays its eggs inside wood wasps' nests. The digger wasp stocks its cells with small flies. Stag beetles' larvae take several years to develop.

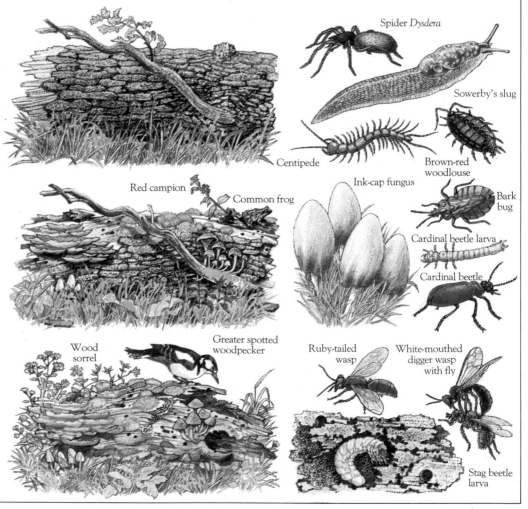

Spider *Dysdera*

Sowerby's slug

Centipede

Brown-red woodlouse

Red campion

Common frog

Ink-cap fungus

Bark bug

Cardinal beetle larva

Cardinal beetle

Wood sorrel

Greater spotted woodpecker

Ruby-tailed wasp

White-mouthed digger wasp with fly

Stag beetle larva

these defects, however, shrews are interesting animals, and one of the lovely things they do is to take their children out for a walk in what amounts to a school "crocodile".

The first time I saw this amazing shrew spectacle was in the New Forest. I was leaning against a bank, watching through my binoculars a woodpeckers' nest in a grove of gigantic oak trees. Suddenly I noticed something moving in the short undergrowth on the bank, and as it wove in and out of the low herbage I thought it was a snake. Then it came out into the open and I saw it was a mother shrew and her babies, six of them. The first baby had grasped his mother's fur near the root of her tail firmly in his mouth. The next baby had grasped the first one's fur, and the next baby his, and so on. The result was something that looked like a Conga line and, what made them appear even more as though they were dancing, they all kept in perfect step. The whole effect was at once charming and quite ridiculous. The reason why shrew families do this is apparently as a protective measure. They may be scattered over the leaf litter, learning to hunt, but at the slightest noise they form a Conga line which may help because it resembles a snake. Be that as it may, shrews are certainly very susceptible to noise. They will panic at the rustle of a leaf or the sound of a raindrop, and have even been known to die of shock on hearing a loud noise (such as a negligent naturalist sneezing).

Plants and animals of the field layer

In a wood the soil is nearly always damp. This is because it is shielded and shaded from above by the "storeys" of woodland; while below, most of the tree roots run very deep and so do not drain the top soil of moisture. In the damp soil the ferns and soft herbs flourish, and between them you find the mosses and liverworts, which lie on the surface sucking up the moisture like spongy carpets. Unlike most plants, mosses and liverworts have no water transport system so they can only grow just tall enough to emerge from the litter, or else they cloak soggy crumbling stumps or damp rocks. They are the plant equivalent of the amphibians in the animal kingdom, for fertilization must take place in water. However, even a very thin film of moisture is enough to allow the plant's sperm to swim and find the egg. The liverworts spread their "leaves" along the ground and get their name from the fact that the leaves are lobed, like a liver is. The mosses grow in green and gold cushions or in tiny spires, and the little stalked "turrets" that appear in spring carry the developing spores. With the dryness of summer the roof falls off the turret and then a ring of teeth beneath parts, letting any breeze gather up the spores and scatter them throughout the forest. You can collect the stalks and capsules and watch this intriguing process for yourself.

The ferns also need water for reproduction, and to make things easier their sperm and eggs are carried on the underside of a heart-shaped flattened "prothallus" that lies close to the ground. It is quite tiny, usually not more than a centimetre long, and not easily seen; however, if you collect spores from ferns in the autumn you can take them home and grow your own prothalli and watch the whole cycle (see page 256).

The autumnal leaf fall in deciduous woodland means that its inhabitants become much more exposed to the weather, and when winter really starts the forest settles to sleep. Among the great bare trees, stripped of their leaves and standing like dark skeletons, the only patches

A new start in life
A bracken frond seen from above has shadows of spore-bearing sporangia showing through on the underside of the leaf. The interesting development of a fern plant can be studied by taking home in plastic bags a few fronds with ripe sporangia. At home, shake out the spores on to small blocks of sterilized peat. Place the peat blocks in a shallow dish of water and cover with a sheet of glass or clear upturned bowl. A couple of weeks later, instead of a small fern developing, the peat will sprout roundish plants similar to liverworts which bear both male and female sex organs (see page 256). Providing you keep the plants moist, fertilization occurs and a young recognizable fern plant grows out from its fleshy-lobed precursor.

of green to lighten the gloom come from the glossy holly leaves and the delicately-veined ivy. On dead branches appear tiny explosions of colour from small fungi—the orange patches of witches' butter, the little white cups with scarlet insides that are pixie cups, and the gay canary-yellow winter fungus. At this season certain of the animals are safely curled up hibernating, and others are spending their winter in warmer resorts. But the ones who remain and are awake are very wary, for now the undergrowth they relied on as cover has died back and left them exposed.

Although it looks like it, wintry woodland is not dead but merely dormant. Most of the soft low plants of the forest floor, together called the field layer, have their food store-houses safely underground in the form of bulbs, corms, tubers or rhizomes. These are the plant larders in which energy is stored, untouched, until the spring when it gives the woodland-floor plants the headstart they need. They must grow and flower quickly before the trees deck themselves out in a million shimmering leaves. With the coming of late spring and summer, the leaves form a canopy which blocks sunlight and so restricts the growth of small plants on the now-shady woodland floor. The waves of early flowers include in their ranks the delicate wood anemone and the dog's mercury, the bunches of butter-yellow primroses and the blue haze of a thousand bluebells. Among them you will find some escapees from gardens, like the golden trumpet of the daffodil and the glistening white snowdrop. Later on the delicate mauve of the common violet and the stitchwort appear, and finally out come the summer flowers with their broad dark leaves and smaller, less colourful blooms—flowers like the bugle and lesser burdock, wood forget-me-not and sorrel.

The soft leaves and stems of woodland plants are ideal food for the vegetarians of the forest, and by examining them you will discover traces of nibbling by insect larvae, slugs, rabbits, mice, voles, and even some birds. You may find the woolly-coated red-footed oak eggar larva (which, incidentally, is misnamed since it never feeds on oaks but on bramble, hawthorn and heather) or the odd spiky-looking larvae of the beautiful silver-washed and pearl-bordered fritillaries, whose food plant is the violet. Other creatures feast on the early plants too. The plant bugs and capsid bugs can be found sucking the sap, and also in evidence are the strange hieroglyphics and patterns produced within the leaves themselves by the leaf miners, who are mostly the larvae of various moths and flies.

There is a great variety of tiny rodents in deciduous woodland. Some are vegetarian, like the yellow-necked field mouse of Europe. This lovely rodent, with its big eyes and yellow "collar", feasts on the many kinds of nuts that fall to the forest floor and stores in underground larders those it can't eat at once. Other woodland rodents are omnivorous, like the white-footed mice and the charming jumping mice of North America and the reddish bank voles of Europe. Though chiefly plant-eaters, they spend some of their time busily sorting through the leaf mould in search of insects and their larvae. The bank voles are rather unjustly accused of doing a lot of damage by "barking" trees, especially elder. These voles are good climbers and you can often see the marks where they have bitten off bark from low on the trunk to quite high up in the branches, but damage this extensive is generally done only in times of food shortage. The real culprit, the one who does harm trees by "ringing" them, is the field vole. He may chew off a ring of bark right round the base and thus kill the tree.

Wood forget-me-not

Bugle

Greater stitchwort

Spring and summer woodland flowers
During April the white-starred flowers of greater stitchwort are common in woodland clearings as well as along hedgerow banks. Bugle likes damp clayey conditions, and is often found carpeting large areas of oak woods in deep shade. From May to July its pagoda-like flowers, mainly blue, sometimes pink or white, are attractive to both bees and butterflies. Wood forget-me-not grows on black waterlogged soils. Its beautiful delicate-blue flowers are seen at their best towards the end of May.

One of the most attractive of all the forest rodents lives in Europe—the little birch mouse, with its distinctive black stripe down its back. A berry- and insect-feeder, this pretty little rodent is so gentle and placid by nature that it won't even attempt to bite if you try to pick it up. This nice disposition, one would think, would make the birch mouse an ideal pet, but alas you would rarely see it for of all the mammals its hibernation is the longest—eight months of the year.

Highways to the canopy

Despite the fact that some woodland shrubs act as stepping stones to the leaves and branches of the canopy above, most creatures commute between floor and ceiling of the wood along the main roads of the tree trunks. In building up your picture of a tree the trunk, and particularly the nature of the bark, is important. But never collect fresh bark—this is like skinning someone alive. Take bark rubbings, adopting the method used by those eccentric people who rush around old churches taking brass rubbings. (Come to think of it, they will probably think you are just as eccentric to take bark rubbings.) Loose or fallen bark may be delicately carved by the galleries of bark beetles and their larvae. These grooves, which may be on the exposed tree trunk or on the inside of the fallen bark, form patterns characteristic of the different species of beetle. You can, of course, take fallen bark home for your collection, or you can take a rubbing of the galleries from the bark or the tree trunk. It is one of the bark-tunnelling beetles that is responsible for spreading an infamous fungus which has destroyed so many elms—Dutch elm disease.

Before you invade the poor tree and start to take your rubbings, examine the bark carefully for both plants and tiny creatures that dwell there. On the shady side of a tree you may see a covering of light-green powder. This is the alga *Pleurococcus*, and it is worth collecting some of this in an envelope to examine under the microscope. In the fissures in the bark which are deep and moist enough to hold small quantities of soil you will find various mosses. You may also discover lichens, dry and crusty. Lichens are slow-growing and very tough so they can be found on any side of the trunk. They are true *epiphytes*, gaining all the nourishment they need from the dust and water in the air and using the tree merely as a convenient parking place.

Both the adults and larvae of many species of moth rest on bark and they are remarkably well camouflaged so as to escape detection in this exposed position. Some look like broken twigs or knots on the trunk, others are so beautifully coloured and patterned they simply fade into the bark and its covering of lichens. Various bugs and aphids sit on the trunks, busily sucking nourishment from the bark. You may also come across the coccid bugs, who either look like fluffy balls of wool made out of fine waxen threads or else have tiny shields to protect them. The capsid and deadly assassin bugs can be seen quartering the surface of the bark in search of the smaller creatures or under loose bark hunting for beetles and their larvae. At night some kinds of hunting spiders emerge from the leaf litter and climb aloft to hunt on the bark, while others live in the crevices of the bark itself. This wealth of insect life provides a rich food source for birds like treecreepers, nuthatches and woodpeckers.

Both Lee and I, when we were young—she in America and I in Corfu—constructed treehouses. Not only are these the greatest fun to

BARK RUBBING

One way of adding to your record of a tree is to take a bark rubbing, as shown below. Rub the lower branches as well as the trunk if you can reach; they often have different textures and patterns.

How to do the rubbing
You need a piece of waxed or grease-proof paper, some sticky tape and a large wax crayon. Tape the paper in position and rub slowly, all strokes in one direction, using the flattened side of the crayon. For an interesting result use crayon and paper of contrasting colours.

A classic case of camouflage
The light-coloured form of the peppered moth *Biston* shown resting on a lichen-encrusted branch is typical of open unpolluted country. Since the middle of the last century a dark form has developed to blend in with the bark of trees blackened by industrial pollution—an example of evolution in action.

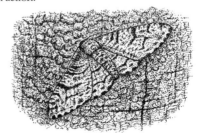

build (and remember to use ropes, not nails) but they are also extremely useful from a naturalist's point of view. They serve as hides *and* they allow you to climb aloft and study trunk and canopy life at its own level. From Lee's treehouse she could keep watch on a family of golden-eyed screech owls whose nest was in a hole in an adjacent tree, and she would frequently see opossums and racoons making their way up and down trunks or from tree to tree in search of insects and birds' eggs. She could also admire the wonderful acrobatics of the squirrels and striped chipmunks as they danced through the branches in search of nuts and other food such as soft shoots.

My treehouse was quite a large one built in a gigantic olive tree that must have been a thousand years old. In the warm summer I would spend the night there, amusing myself by "lighting" the interior with a host of fireflies whose flashing lit up the house with a weird greeny-white light. From my house, safely hidden from sight and with the aid of my binoculars, I could watch many birds rear their young. Kestrels nested almost within reach. Magpies built a nest directly above my house and I watched, fascinated, as three pairs of penduline tits wove their strange pear-shaped nests in the branches close to me. In the evening I could watch owls emerge from their daytime roosts in the olive boles and admire the ballet-like performances of the squirrel dormice, chasing each other and drifting through the trees like little puffs of grey smoke. One hot afternoon when I climbed up into my treehouse I discovered it had been taken over by a large Aesculapian snake. We were both equally surprised at the sight of each other, and in my attempt to catch him we fell out of the tree together. To my annoyance, he escaped.

Flowers, leaves and small creatures of the canopy

People do not often associate big forest trees with flowers, but of course trees are flowering plants and have a variety of attractive, but for the most part tiny, blooms. Every spring the woodland air is thick with billions of floating pollen grains from the male catkins, and if you look at slanting shafts of sunlight you can see the pollen in them, thick as smoke. Many kinds of tree release their clouds of pollen before or during "budburst"—the time the leaves appear—since it seems that the pollen can drift unimpeded on the free-flowing air currents and it is easier for the female flowers to catch it. After fertilization the development of the fruit begins, the leaves slowly start to enlarge, the canopy spreads and summer has finally arrived.

It is during summer that the fully decked-out trees in their shimmering leaves come under attack from a thousand kinds of insect. Not only are they beset by the leaf-eating beetles, but they are also nibbled at by the free-ranging larvae of many moths and even some butterflies. The leaves are scrawled upon by the activities of the leaf miners—larvae of wasps, weevils and sawflies, as well as of moths and flies. You can see how their passages increase in girth, widening as the larvae grow fatter. If you examine these tunnels carefully you can often identify the species by observing where it has deposited its droppings. The droppings can be seen as dark marks under the paper-thin green or brown leaf tissue that forms the roof of the tunnel. You might also find aphids, mites and certain weevils who create coloured spots on the leaves, or who feed inside protective rolls, folds or buckles that they fashion on the edges of

TRAY BEATING

Leaf and twig creatures may be collected by the technique of tray beating. You can make a cloth tray that fits on a folding wooden frame (below), or lay a sheet on the ground if it is a low branch. Some resourceful naturalists use an umbrella! Give the bough a sudden short sharp rap with a stick; don't disturb it beforehand, and don't tap it softly or shake it since this just mimics the wind and most creatures can hold on quite easily. Sort the ones you need, and replace the rest.

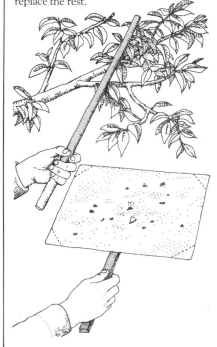

the leaves. But the trees' fresh green livery is really dominated by the gall-makers, some of the most extraordinary architects in nature, who decorate the forest leaves with their incredible structures.

The gall-makers are mainly gall-wasps, but there are also various beetles, flies and mites that cause galls. These creatures lay their eggs somewhere in the leaf—in the midrib, the stem, the petiole (leaf stalk), or in buds or even twigs. The larvae hatch out and their presence stimulates the surrounding tissues to swell and grow into a variety of weird shapes and colours, like trinkets and baubles from an Indian market. Their shapes and appearances are legion: they form patches, pits or protrusions; they can be pointed, round, spiky, hairy or smooth; they can be shaped like urns, kidneys or artichokes. Most of them are jade-green at first; some turn red or yellow, and others become striped like the old-fashioned bulls-eye sweets.

Each gall forms a home for a developing larva. In some the adult insect hatches out in the summer, in others the gall turns brown and the larva hibernates through the winter inside it. But the story of the galls does not end there, because within each gall you will almost certainly find other creatures which are either acting as parasites on the original owner-builder of the gall or who have just taken up residence as unpaying guests. The common oak apple, a very easy-to-find gall, has been known to give a home to 75 different species of insect as well as the rightful owner, the gall-wasp grub. So when you collect galls for hatching out at home you can never be sure what will emerge. Collecting galls is a most enjoyable occupation: not only are they such lovely shapes to draw and paint (and photograph) but you have all the excitement of wondering what is going to crawl out of them. The best time to collect them, of course, is when they are fully ripe in midsummer and autumn.

With the arrival of autumn the fruit of the tree is plump, ripe and ready to fall—acorns from the oak, beech mast, hickory nuts, the hairy seed of the poplar and the winged seeds of the sycamore that twirl towards the ground like little green helicopters. Now is the time for the rich pageant of autumn leaves. I have ridden on horseback through the autumn woodlands of the Canadian-American border and the multitudinous colours of the leaves—the reds, the bronzes, the yellows and scarlets—were so dazzling to the eye that it was like riding through a stained glass window.

Why do leaves change colour? The answer is prosaic. The brilliant canary-yellows of the leaves before they fall is due to pigments called *carotenids* showing through the normal green pigment chlorophyll, which is now inactive in the dying cells of the leaf. The vivid glowing reds are caused by the presence of *anthocyanin* pigments in the leaf, which are formed by the interaction of sunlight with the sugars trapped in the leaf during a cold snap. The forests of New England look as though they are on fire with their millions of scarlet leaves, painted this colour by the early frosts followed by sunny days that are so common in this region.

Why do deciduous trees lose all their leaves in the autumn while the darker giants farther north, the conifers, are perpetually green? The reason is that the broad soft leaves of the deciduous trees lose more water through evaporation than the narrow waxy "needles" of the conifers. Since tree roots do not easily take up water from cold or frozen soil the loss of a leaf (as in deciduous trees) or, alternatively, a water-tight leaf (as

Galls and miners

On virtually every plant there will be at least one leaf disfigured by a gall-maker or miner. The culprit is usually a larva of a wasp, weevil, mite or moth who makes a residence within the plant tissue. Gall-makers have taken advantage of the fact that when the plant finds itself invaded it tries to wall off the intruder with an extra growth of tissue, much as a firmly embedded splinter in your finger will cause inflammation and then a sealing-off tissue reaction.

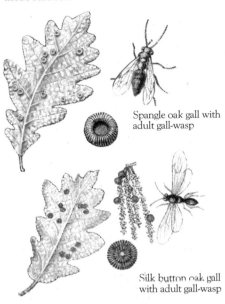

Spangle oak gall with adult gall-wasp

Silk button oak gall with adult gall-wasp

Ash leaf miner with adult miner-moth

Beech leaf miner with adult weevil

in evergreens) is a good policy for water conservation. Leaf loss is triggered by shortening day length rather than by lowering of the temperature, just as the change to white winter plumage or coat in some of the birds and mammals of the northern forests is initiated by less daylight.

Woodland birds

Temperate deciduous forests are fully used at every level by a vast concourse of birds. Your first observations of birds and their behaviour can be done by simply learning to drift gently through a wood: a naturalist in a hurry never learns anything of value. Details on how to go about bird-watching are on page 140, but remember that your most useful pieces of equipment for bird study lie on either side of your head—your ears. In deep thick forest you can track down your birds by the sounds they make: the crackling of seeds as finches feed in winter, the ring of woodpecker carpentry, the rustle of the leaves as robins or other small softbills winnow through the litter after insects or caterpillars. And, of course, there are the numerous sounds birds use to communicate with each other, from sudden piercing alarm calls to the soft gentle "talking" of a group of birds as they feed together.

Because branches and leaves hamper vision, sounds are a more efficient means of communication in a forest. Most of the smaller forest birds are well known as songsters, and by singing they tell each other the boundaries of their territories. Setting up a territory is a good strategy for a bird whose food is fairly stationary—such things as berries and seeds and, of course, tiny animal prey that does not move around much. The bird with a territory always gets enough to eat for himself and his family. You can, with patience, map out a bird's territory by noting the positions of his singing perches. The most versatile singers are usually the drabbest, even though you might expect them to have a brilliant plumage to match their delicious song.

Early spring is the best time for learning the various bird songs and calls. Not all the birds have started to sing and so there is less likelihood of your getting bewildered and muddled than there would be in summer, with a full dawn orchestra of a hundred different birds. Also, in early

Songsters in the dawn chorus
Even as the first glimmer of light creeps through the misty woodland, birds are up and about and preparing for the long day ahead. A few early morning visits to your local wood in early summer should enable you to make a "cast list" of the participants in the birds' dawn chorus and the order of their appearance. Quite why so many birds should sing so enthusiastically first thing in the morning has never been fully explained, but this should not stop you from sharing their joy. First to sing is usually the blackbird (who is also often the last to stop in the evening), followed by the song thrush, crow, pigeon and dove. The second relay consists of the robin, redstart, pheasant and blackcap, with mistle thrush, willow warbler and wren close on their heels. Finally the tits, woodpeckers and chaffinch join in as the early starters gradually cease their singing and turn to the chores of the day.

Turtle dove

Wren

Blackbird

Willow warbler

spring the trees are not yet in full leaf and this makes positive visual identification of bird species much easier than in summer.

Most naturalists work out their own private guides to songs and calls. For example, the blackbird, an early starter whom you may even hear on a fine February day, commonly sings the final motif of Beethoven's violin concerto. Since the blackbird could not have pinched the notes from Beethoven, he must have pinched them from the bird. After the blackbird comes the thrush and then the liquid, throbbing passionate song of the nightingale. A lot of birds are just tuning up in the spring, getting into practice for the real song carnival later on in the year. Some birds are very repetitive; in a North American forest you would have no difficulty in tracking down a red-eyed vireo, for example, as one patient naturalist recorded an equally patient vireo repeating its call an astonishing 22,000 times in a single day.

Once the feathered troubadours have finished with courtship and mating they settle down to the stern task of rearing a family. Woodland birds, like birds almost everywhere, are very particular about their nest site and the nesting material they use. Some nest in holes in trees, either hollowing out the hole themselves (like the woodpeckers do) or borrowing other people's holes (like the owls, redstarts and tits). Nuthatches, too, borrow holes and if they find the opening is too big they partly block it with mud, leaving the correct nuthatch-sized opening in the middle of the original hole.

When the babies are hatched the parent birds go to a lot of trouble to make sure the nest site is not obvious to predators. The eggshells are either eaten or carefully removed and deposited some distance away from the nest, sometimes with the smaller half of the shell carefully fitted into the larger half. You can tell the difference between an egg that has been moved by the parents and one that has been broken by a predator, for only in the former can you see a little roll of dried membrane along the inside of the broken edge and no trace of yolk or white. In the case of pheasants and grouse, the young are ready to follow the parents soon after hatching and so there is no need for the parents to get rid of the shells.

Feeding a nestful of clamouring baby birds is a full-time occupation. At one time or another I have hand-reared a number of baby birds, ranging from tiny fragile goldcrests to boisterous young magpies and solemn owls, and so I have a deep sympathy with bird parents and the hard work they have to do. I know that when you have just stuffed a baby bird full of food to such an extent you are sure it is going to burst, you walk past its box five minutes later to be greeted by a great gaping red mouth and a bird rolling its eyes and wheezing, trying its best to convince you it is on the borderline of starvation.

Most birds are devoted parents and, regardless of their own diet, choose the best for their young—protein-rich insects and soft nutritious caterpillars, for example. A pair of great tits has been observed paying 900 trips to the nest with food in one day. From a hide it is fascinating to watch and count the number of food trips the parent birds make. Does the number of trips vary with the number of young in the nest? And do nests that are parasitized by cuckoos or cowbirds need greater efforts on the part of the foster parents?

There are two exceptions to the frantic parental searching for live prey to feed the young, and these are the pigeons and the finches. In the early

Setting up house
A nuthatch puts the finishing touches to its new home—the deserted nest hole of a woodpecker. The new owner makes its front door smaller as protection against predators, and soon there will be between six and ten hungry nestlings to feed.

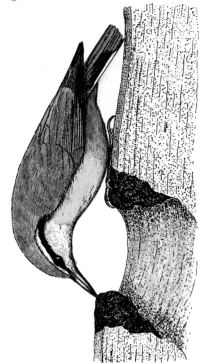

Deciduous woodland in autumn

Our autumnal woodland was typical of the damp clayey Surrey soil. Huge oaks predominated and between them were clumps of smaller silver birch, aspen and the odd maple. Many leaves had already fallen and the wood had an eerie stillness about it on this grey mid-October day. Among the trees there were so many fungi of all shapes and colours that you were virtually walking on a multicoloured carpet of them. The litter underfoot was thick and spongy, and searching carefully among the dead leaves and fallen wood, we found numerous small creatures preparing for the winter.

Aspen

Silver birch

Norway maple

Bark growths This lichen is common in areas of moderate pollution. Fungi attack the bark of particular trees—crampball is typical of ash.

Orange peel fungus

Stereum fungus on oak

Lecanora lichen on silver birch

Incrustosporia fungus on birch

Slime fungus

Stump puff-ball

White coral fungus

Fungi galore T bracket is so to big one can sup the weight of a human. Comp this with the sl fungus's strang consistency; it actually crawl s along a trunk i manner of a sl The spectacula agaric lives in association with the roots of birch trees and is poison- ous. Puff-balls have discharge their spores an resemble half- emptied dust b from miniatur vacuum cleane

Birch bracket fungus

Crampball fungus on ash

Honey fungus

Merulius fungus on oak

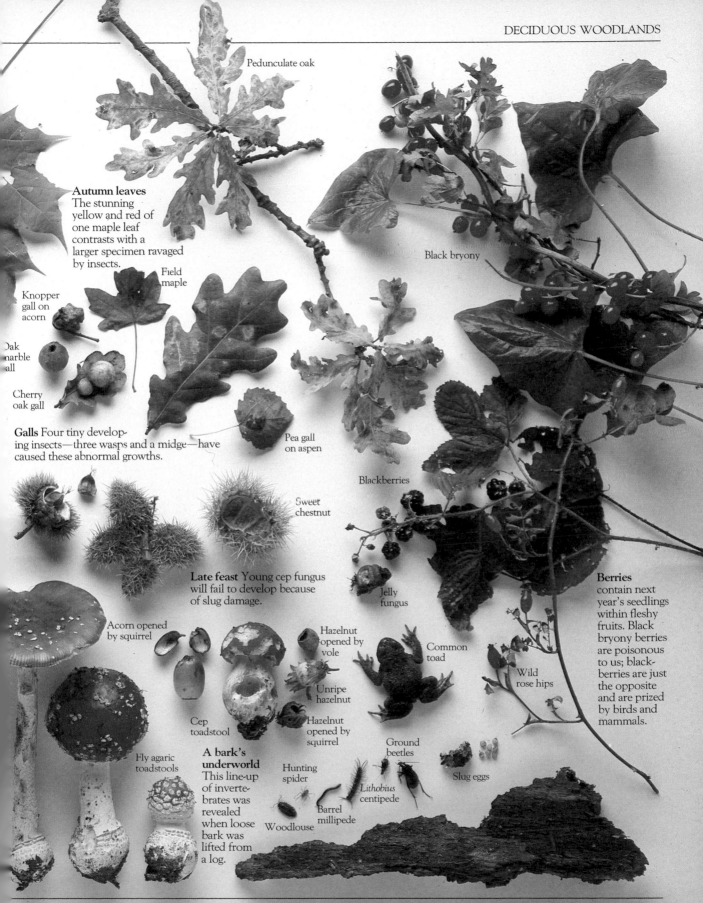

Pedunculate oak

Autumn leaves
The stunning yellow and red of one maple leaf contrasts with a larger specimen ravaged by insects.

Black bryony

Field maple

Knopper gall on acorn

Oak marble gall

Cherry oak gall

Galls Four tiny developing insects—three wasps and a midge—have caused these abnormal growths.

Pea gall on aspen

Blackberries

Sweet chestnut

Late feast Young cep fungus will fail to develop because of slug damage.

Jelly fungus

Berries contain next year's seedlings within fleshy fruits. Black bryony berries are poisonous to us; blackberries are just the opposite and are prized by birds and mammals.

Acorn opened by squirrel

Hazelnut opened by vole

Common toad

Unripe hazelnut

Wild rose hips

Cep toadstool

Hazelnut opened by squirrel

Fly agaric toadstools

Ground beetles

A bark's underworld This line-up of invertebrates was revealed when loose bark was lifted from a log.

Hunting spider

Lithobius centipede

Slug eggs

Barrel millipede

Woodlouse

125

days when the young pigeons have just hatched, both parents produce from the crop a substance called "pigeon milk". The babies plunge their heads into the crop and suck up this fluid. So that the young don't get drenched and bedraggled in the process the head and chin feathers are the last to develop. In the case of the finches the mother and father regurgitate seeds for their new offspring, adding insects to the diet later on.

All this avian house-building, cleaning and baby care does not enter into the curriculum of the European cuckoos, who have evolved a perfect method of getting other birds to do the work for them. The male floats around, making the countryside resound with his beautiful call, and the female of course lays eggs in other birds' nests. Amazingly her eggs resemble in colour and marking the eggs of the bird whose nest she is usurping. When the baby cuckoo hatches it finds itself uncomfortable with the rightful owners of the nest. Its hollowed-out back is extremely sensitive and if young or eggs get into the concavity the cuckoo, to rid itself of the tickling, hoists itself up and jettisons eggs or babies from the nest like a tip-up truck. There are many different species of cuckoo in the world, and not all of them have the same reprehensible habits as the European representative. The black-billed and the yellow-billed cuckoos of North America lead blameless and exemplary family lives just like the majority of other birds.

In view of all the nests brimming over with eager mouths, it is a good thing that deciduous woodlands provide an incredible wealth of food for the birds that live there. There are thousands of different arthropods, snails and many larger creatures. In addition to this living prey, of course, there is the prodigious menu of vegetable foods in the shape of nuts, seeds and juicy berries.

Many of the forest birds are omnivorous, especially the medium-sized and large perching birds like the jays, magpies and the numerous members of the *Turdidae*, the thrush family. There are also birds such as grouse and pheasants who collect a wide variety of foodstuffs in the low growth and the litter of the forest floor. Most woodland birds, however, tend to concentrate on one or two types of food found in one area of the wood, and the variety in their feeding methods is startling.

Magpie

Jay

Nuthatches inch along branches and trunks winkling out insects from the crevices in the bark with their "retroussé" beaks, and they are just as happy insect-hunting head-down as head-up. They will also wedge nuts into crevices in the bark and then drill holes in them to get at the soft centres. They hoard nuts for the winter in this way but, having bad memories, frequently forget where they have stored them. Other great insect-hunters are the woodpeckers, who make the woodland resound with their noisy activities as they hammer away at the trees in search of wood-boring insects. Once they have found a gallery they extract the larvae using their beautifully-adapted tongues. The woodpecker's tongue is a remarkable tool, coated with sticky saliva and armed with barbs. Because of the extraordinary extensor muscle in the head the tongue can be protruded and plunged deep into the tunnels. The patterns of woodpeckers' holes are characteristic of the different species, as are the drumming rhythms. So, without even seeing the bird, you can tell from its feeding area and the sound of its drumming which species of woodpecker inhabits your wood. Like nuthatches, woodpeckers also wedge nuts into crevices in the bark of trees, but unlike the more profligate nuthatch a woodpecker uses the same crevice over and over again, and an untidy pile of empty hulls rapidly accumulates on the patch of ground beneath this kitchen.

The beautifully-feathered woodcocks have extraordinary adaptations of both beaks and eyes for their feeding technique. When the woodcock goes "worming" over the soft moist soil of the forest floor it plunges its beak, sword-like, into the earth. It can detect worms lurking in the soil by the ultra-sensitive tip of the beak. Meanwhile, because of the position of its eyes, the bird can see above and behind for predators while watching in front for surfacing worms disturbed by the probing beak.

The insect-feeders in a forest work every available space in pursuit of food. Robins and nightingales ferret their way through the litter, turning over leaves like busy housewives making a bed. Treecreepers scuttle up the trunks of trees like someone going up a spiral staircase. Fly-catchers choose a special branch on which they sit, immobile, until they sight an insect on the wing; then they launch themselves like acrobats into the air and snap up their prey. The delicate warblers and tits (with the charming American name of chickadees) assiduously glean the leaf surface for insects, and wrens feed in much the same way but at a lower level in the tangle of young trees and dense bushes.

As the seasons change in your wood, note the amount and the kind of food available to the bird population. Insects and soft fruits vanish with the advent of cold weather. Many bird species get round the diminishing food supply by migrating to more lush feeding grounds, but the migratory habits of each species are very complex. The warblers are night migrants, and it is possible that they navigate by the positions of the stars. Thrushes and others fly by day in loose flocks, and it is thought that they steer their course by the position of the sun and familiar landmarks. It is miraculous that the young birds born that same year can find their way over hundreds of kilometres with unerring accuracy. Of course some follow their parents, but others go alone, like the young European cuckoo on its journey to Africa. Some birds don't migrate very far—they simply head for the warmer parts of their regions where they join birds of their own species who reside there. Other birds, such as the seed-eaters, stay at home all year round since their food supply is constant. Finally, there are others who simply change their diet to suit the food supply that is available in their particular habitat.

The observant naturalist will notice that many bird species sometimes leave on their migrations before there is any real change in the food supply. Departures are triggered by different things for different species. It may be the decrease in daylight, or changes in the weather, or even the position of the sun or stars. Keep a close watch on the migratory species that live in your wood and try to work out what factors signal their leave-taking. Note down their times of departure. From your local newspaper or meteorological office you can get the times of sunrise and sunset (to calculate the day length) and the vital information on temperature, precipitation, cloud cover and other atmospheric conditions. Make the same careful notes when your birds return to the wood in the spring, and very soon you will have built up a fascinating picture of the birds' timeclocks. Remember that the job of a naturalist is to observe and record. Some of the complexities of bird migration have only been unravelled in recent years by the ringing projects of ornithologists. They

The amazing tongue of the "yaffle"
A green woodpecker feeds its young on a pulp of regurgitated insect food. Note the way the adult bird supports itself by means of stiff tail feathers and grips the bark with its opposable claws. With a country name of yaffle, it is a bird of open woodland and is often seen feeding on ants on the ground. The sticky tongue can be extended a good 15 centimetres beyond the tip of the beak to slide into beetle larvae holes and ants' nests.

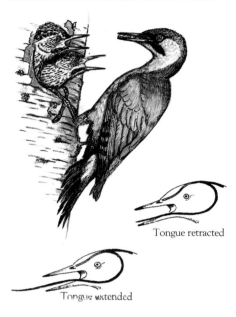

Tongue retracted

Tongue extended

use hair-fine nets called mist nets, which birds cannot see, and put them up in clearings where birds are likely to fly into them. Thousands and thousands of woodland birds have been caught, measured, sexed, weighed, adorned with a leg ring stamped with their own number and the address of the ringer, and then released. From this work we are still finding new facts about the distances birds travel and when and why they go. If you should find a dead bird with a ring on it (remember that it is illegal to trap wild birds unless you have a special licence), send the ring to the address indicated. Add as much information as possible—for example, the date, place, weather and the condition of the bird—and your help may assist in solving another of the great mysteries of bird migration that still puzzle ornithologists.

Large mammals — summit of the forest food web

The naturalist who studies the larger terrestrial mammals of the woodland—badgers, foxes, polecats, wildcats, deer and, in North America, lynx, bobcat and bear—can start to tie up the strands of the food webs that started with the plants and tiny creatures described earlier on. To take a very simple chain: a fox may include quite a large proportion of rabbits in its diet, and these in turn feed on the undergrowth. Of course this is very straightforward compared to the number and complexity of food chains that you could conceive in a woodland, and the ways that different food chains connect together into food webs is almost impossible to imagine.

Look at the handsome badger, for example, with his distinctive black-and-white bandit-masked face. You can see him slouching across the

Out for an evening stroll
A badger emerges cautiously from its sett to forage over the woodland floor. The badger will eat almost anything, plant or animal. In spring a badger will make frequent visits to the ground beneath a rookery to feed on bird casualties, and at this time of year it has been known for a badger to dig young rabbits out of their stops, decapitate them and eat the bodies. It will sniff round a rotten log and then methodically tear it apart with its powerful foreclaws to get at the slugs and succulent insect grubs within. But a recent study revealed that up to half the badger's diet consists of earthworms.

forest floor with his bear-like walk, occasionally by day but more generally at twilight or at night, searching for his food. Now the badger's food consists of anything from birds' eggs to voles, carrion, wasps, honey, slugs, blackberries, frogs and snakes. Just imagine the intricacies of all the different food chains in this wide menu that lead up to the badger! The complex inter-relationships, like a jigsaw puzzle, can start with a microbe and go through many stages to a frog which, when eaten, becomes (so to speak) a badger. Each of these food chains is part of a web as complex as the delicate mesh of a piece of elegant lacework. But of course the story does not stop with the badger. Although they have no real enemies apart from man, badgers—like all mammals—play host to parasites. There are *ectoparasites*, including ticks and fleas that feast on their blood, and various internal *endoparasites* that feed on their tissues. When the badger eventually dies it creates yet more food chains, for its body is fed on by other creatures that form parts of the scavenger food web—crows, beetles and fly maggots. Finally the minute organisms take over and the remains of the badger are turned back into the soil. And here the sequence comes full circle, for in the rich soil provided by the badger's body will grow a moss, a fern, a violet or some other plant that starts the whole cycle once again.

The badgers' life is centred round their home or *sett*, a complex series of tunnels and rooms which they keep scrupulously clean. They are tidy animals, regularly making themselves clean beds of bracken or grass and bluebells. All around the sett holes are dug as lavatories, which also ensures that the nest is kept clean. Near the sett is a tree used as a scratching post for sharpening and cleaning claws. Badgers can live for up to 30 years and setts are handed down from generation to generation, enlarged and improved, so an extensive sett may be very old indeed.

The first badger sett I saw was in the New Forest. The farmer who lived nearby said that it had been there in his great grandfather's day, and the farmer himself was a man of 80, so this would have made the sett well over 150 years old. I took up my station for my first badger watch one evening just as the sun was setting. To my surprise almost at once a badger came wandering through the green twilit wood. He approached the sett in a rather hesitant manner, pausing frequently to sniff the air or to scratch himself. Presently at the sett entrance appeared the black-and-white head of another badger. This one sniffed the air cautiously and slowly emerged. She, for it was a female, went over to the newcomer and they sniffed each other, uttering little grunts and throaty noises. It was then I realized that I was having incredible luck, for the pair were starting their courtship display. First they snuffled at each other's faces and "talked" for a bit, and then the female started running round and round in circles, clockwise, with the male lumbering behind. After she had done about 30 circuits she stopped for a moment, and the male licked her muzzle. Then she set off again in circles, this time anticlockwise, and after 30 more revolutions she stopped and allowed the male to mate with her. He held her by the scruff of the neck while he did this and she gave small grunts and little cries of satisfaction. When they finished they groomed one another and then they had the most charming game, wrestling and pretending to bite each other, rolling over and over in the moonlight for almost an hour. Finally they shook themselves and waddled off together into the shadowy wood in search of their evening meal.

TRACKS AND PRINTS

You may not often see woodland mammals since most of them are shy, secretive and nocturnal, but you can be sure of their presence by the tracks they leave. For good clear prints look especially on the muddy banks of ponds and streams, on woodland rides and in damp hollows.

Making a print plaster cast
You will need to take with you some plaster of Paris in a plastic container, a strip of cardboard and a paper clip. First make a sketch of the print, then clean the area around the print and the print itself with a length of stiff grass. Press the cardboard into the ground around the print and secure it with the clip. Next, mix the plaster and water to a cream and pour it *slowly* into the print. After about half an hour dig up the cast, card-board and surrounding earth and take the lot home in a plastic bag—it is best to let it set really hard. Use an old toothbrush to wash the plaster, then let it dry and finally label it with the identity of the owner and other information.

Rear footprints of some common woodland mammals

Hedgehog
4 cm

Red squirrel
5 cm

Fox
5 cm

Red deer
8 cm

Badger
4 cm

Rabbit

Front 2.5 cm   Rear 12 cm

THE NATURALIST IN
CONIFEROUS WOODLANDS

One of the greatest continuous areas of dense forest in the world is not, as one might expect, in the tropics. It is instead the huge northern belt of coniferous trees called the boreal forest or *Taiga*, which runs across the Old World from Scandinavia right through Russia to Japan, and in the New World from New England across the breadth of Canada to the Pacific. The boreal forest covers an almost continuous area of 2,000 million hectares, merging into tundra to the north and reaching in corridors along southerly-running mountain ridges as it gradually gives way to deciduous woodland in the warmer south.

During the last Ice Age, the land now covered by coniferous woodland was mostly under solid ice. In the region of 10,000 years ago the glaciers retreated from these areas, clawing at them like reluctant fingers. The retreat of the ice uncovered deep cold lakes, dark soggy bogs and gouged-out paths for rivers, and left scars of rocky debris called "moraines". As the ice shrank northwards it revealed a land bridge between Russia and Alaska, at the place where the Bering Straits now separate the two continents. Animals and plants could spread across the land bridge at will, so it is not surprising that Old and New World coniferous forests have many inhabitants in common—the land bridge only disappeared a few thousand years ago.

Coniferous forests are for the most part cold and snowy areas of the world, but there is no permafrost and so trees can grow. The trees best adapted to these conditions are the cone-bearers (conifers)—the spruces, larches, pines and firs. Unlike other trees the conifers (except for larches) do not shed all their leaves at once and face the winter naked. Instead they shed small numbers of leaves continually throughout the year, always keeping a proportion of their greenery so that they can carry out *photosynthesis* (trapping the energy in sunlight) whenever the temperature gets above freezing. Designed for living in such an inclement area, the leaves of these "evergreens" are needle-shaped and waxy in texture in order to reduce moisture loss; this is important because roots cannot readily take up water in cold or freezing weather. The thick carunculated bark protects the main trunk, the backbone of the tree. Pines are shaped like green helter-skelters and spruces and firs are like green spires, shapes which are admirably suited to allowing weighty snow to slide from the downswept branches.

Coniferous trees are superb all-year-round wind deflectors, so the interior of the woodland is often sheltered from even fierce winter blizzards. But within the trees, the perpetually furry canopy allows little light to penetrate to the forest floor, and the dry layer of needles does not encourage seed germination. The result is little thick undercover below the trees—only ferns, mosses and a few herbaceous plants can flourish in such gloomy areas. However, where a watercourse, lake or mossy bog

Grey squirrel

Red squirrel

Grey and red squirrels
The introduction of the grey squirrel to Britain late last century and its subsequent spread are well known, as is the demise of the red squirrel. But whether the larger, more aggressive greys drive out the shy reds is unclear. They have been seen fighting over food, but reds are vanishing from areas where there are no grey squirrels. It seems that, as usual, man is partly to blame. Reds thrive only in natural pine woodlands. We are tearing these woods down, replacing them with ornamental trees and fast-growing foreign conifers for timber. The adaptable greys take to these woods, spreading farther

Tree top predator
The agile pine marten (opposite) hunts birds, squirrels and other small animals through the pine branches. A shy and rare animal, it lives only in remote areas and has taken to treeless rocky areas to avoid persecution.

breaks the forest canopy and creates a wet sunlit opening, other types of plant appear. Here grow the willows, with their long jade-green leaves beloved of the moose, and alders with tangles of wild raspberries that flourish at their feet and provide a larder for fruit-eating birds and mammals. In natural gaps in the green blanket of conifers grow aspens, whose leaves tremble and quiver because they are so loosely connected to the branches—hence the saying he "trembled like an aspen". Often accompanying the aspens are brown- or silvery-barked birches and the paper birch, whose bark peels in great white sheets and flaps in the wind, looking like a sunburnt back. Various wild flowers and bushes—wintergreen, bilberry, heather and juniper among them—thrive in such sunlit groves and provide seeds and berries for the woodland inhabitants. In the most northerly reaches of the Taiga, firs and larches start to peter out but tough cold-resistant birches still grow. This birch species, with its branches twisted and deformed by the torture of the icy cutting winds coming down from the tundra, is called *Betula tortuosa*.

In the dim depths of coniferous woodland many mushrooms thrive on the forest floor, pushing up their bald heads through the carpet of needles. Some are like yellow or orange umbrellas, others like upright fingers. These fungi form a complicated web of threads called *hyphae* which enmesh the top soil and bind together millions of fragments of needles, dead twigs, branches and other dead matter that has fallen to the woodland floor. All this fallen debris forms ideal food for the fungi, which also flourish on the rain of spring pollen from the cones above. The whole creates a complex woven mat that retains and stores chemicals, holding them so that they are not leached away and lost. You

UNDER THE CARPET OF NEEDLES

The floor of the pine forest may not look as interesting as the rich carpet of decaying leaves in an oak or beech wood, but there are fascinating creatures and fungi that thrive on the fallen needles and other debris. Most numerous are the mites, who swarm through the pine litter in their millions. Within the mite world are many specialists—some eat pine needles, some eat fungi, others feed on rotting wood, and there are predatory mites that hunt nematode worms, springtails and other mites. If you dig deeply to expose the soil profile you will find, below the recently fallen needles (1), a thick, partly decomposed peaty layer called acid mor (2). Farther down is a paler greyish band of soil called the leached horizon (3). This has had most of the nutrients washed out of it. It is interesting to draw the soil profile of a pine wood, and compare it with the profile from deciduous woodland (see page 106).

The upper layer
Most litter mites congregate near the surface, where they number many thousands per square metre. They are all tiny creatures a millimetre or less in length and their eight legs show them to be arachnids, related to spiders.

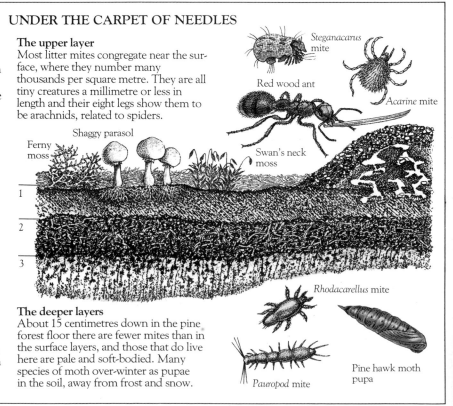

Steganacarus mite

Red wood ant

Acarine mite

Shaggy parasol

Ferny moss

Swan's neck moss

The deeper layers
About 15 centimetres down in the pine forest floor there are fewer mites than in the surface layers, and those that do live here are pale and soft-bodied. Many species of moth over-winter as pupae in the soil, away from frost and snow.

Rhodacarellus mite

Pauropod mite

Pine hawk moth pupa

find few earthworms to help with the breaking down of the woodland litter and, because of the cold, decomposition is slow in the thick layer of needles. But in this coverlet of needles millions of soil mites live, feeding on the web of hyphae produced by the fungi. Collect a bagful of the forest floor and back at home you can extract a fascinating world of tiny arthropods with the home-made Tullgren funnel (see page 264); you will discover that what looks like a dead blanket of needles is, in fact, a teeming mini-metropolis.

Insects of the pine woods

Most of the coniferous woodland insects spend the winter as pupae, protected from the harsh icy winter weather buried either deep in the ground or else inside the trunks of the trees. But when spring comes to the wood there is no great show of tree blossom for insects to feast on. Conifers, like grasses, are wind-pollinated and so they have no need of colourful nectar-laden flowers to attract insects. The conifer "flowers" are tiny and remarkably inconspicuous, so the herbivorous insects of the Taiga turn elsewhere for their food and consume the leaves, the bark skins of the trees or even the wood core itself. The caterpillar of the pine hawk moth, with its yellow and green markings, hangs among the needles and is particularly invisible as it munches its way through them. The strange-looking pine weevil has a curved snout like an elephant's trunk, and through this tube it can suck the sap from the living tree. In this way it does much more damage than its larva, which only feeds on dead rotting trunks.

Coniferous woodland is the home of many species of woodwasp and sawfly, which are primitive relations of the true wasps and bees but without the "wasp waist". The giant woodwasp in particular is an incredible insect. The female has a long yellow abdomen with bluish bands around it and her ovipositor is, in fact, a built-in drill. With this amazing tool she bores a hole about one centimetre deep into a tree. In this she lays her egg. But the astonishing thing is that she has "doctored" the egg with a fungus which, when her larva hatches out, is carried in its gut. Without the fungus the larva would perish, for the fungus enables it to digest its diet of wood. You would think that a larva like this, enshrined in its wooden tomb in the depths of a tree, would be safe from predators and parasites; but this is not the case. Its enemy is an ichneumon wasp, *Rhyssa*. The female of this wasp also has a long drill-like ovipositor. She alights on the trunk of a tree and delicately walks to and fro "listening" with her antennae for the movements of a woodwasp grub feeding deep within. As soon as she finds one she unsheaths her ovipositor like a sword and plunges it deep into the trunk. When its tip reaches the fat grub in its cell she deposits an egg on the grub. Eventually the egg hatches into an ichneumon larva which feasts on the woodwasp grub, using it as a living larder.

The sawflies are also brilliantly equipped for working in wood. The ovipositor of the female is a double-set of little saws which she produces from a sheath, rather like someone opening a Swiss Army knife. With this remarkable tool she slits open pine needles lengthwise, and in the envelope thus constructed she lays up to 20 tiny eggs. When these hatch they are free-living, unlike other larvae of the *Hymenoptera* (the bee and wasp group). In fact they are more like caterpillars, and they are coloured

The persuasive burglar and its victim
The female giant woodwasp is sometimes called a horntail, referring to what looks like her fearsome sting. In fact this is her ovipositor, with which she drills into the trunk of a sickly conifer and lays her eggs inside. When the woodwasp grub hatches, it is fair game for the wonderfully named persuasive burglar. This is a large ichneumon wasp *Rhyssa* that parasitizes the woodwasp grub. The burglar drills through to the grub and, ovipositor sheaths aloft, lays her egg in what will be fresh food for her own larva.

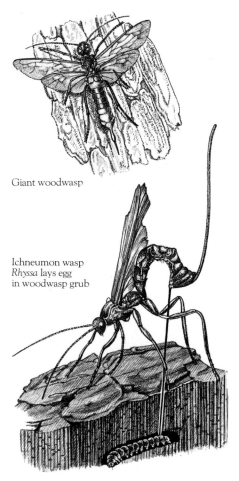

Giant woodwasp

Ichneumon wasp *Rhyssa* lays egg in woodwasp grub

Pine woodland

Many of the pine stands in Britain are maintained commercially for timber, and here the trees are planted very close together and allow little undergrowth. The silence of this sort of woodland reflects the lack of animal life. In a natural pine forest it is different, but such habitats are now confined to small areas of the Scottish Highlands and it is only here that one can get the feel of the vast sombre coniferous forests that once covered much of northern Britain. Our visit was to a small forest just south of Loch Ness on a cool and rainy day in late September. Between the pines, with their strong resinous smell, there are several types of understorey plants rich in animal life.

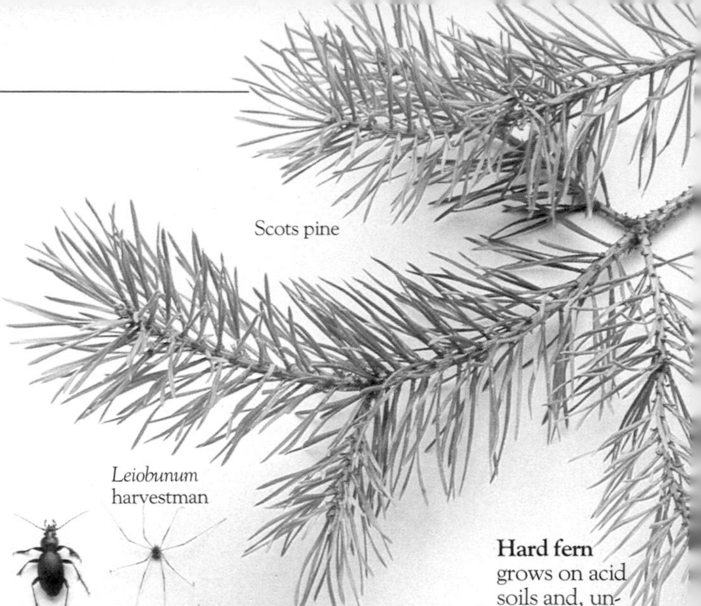

Scots pine

Leiobunum harvestman

Hard fern grows on acid soils and, unusually for such large soft-looking leaves, will stay green and fresh-looking throughout the winter.

Violet ground beetle Despite its name this predatory beetle was found among pine needles in the fork of a tree.

Wolf spider

Hammock web spider

Geometrid moths

Foliose lichen *Hypogymnia*

Spider's nursery tent

Moth caterpillar

Pine sawfly larva

Beard lichen *Usnea*

Black slug

Jelly fungus

Lichen *Cladonia*

Blue russula fungus

Gymnopilus fungus

Lichens Wispy green lichens, hanging like beards from low branches and spreading over dead wood, indicate clean pollution-free air. The damp cool air of the forest interior is ideal for these fungi-algae "plants".

Fungi The floors of pine forests have their own special fungi with many edible types, not only for man but for small mammals, fly larvae, and—as shown here—slugs.

Pixie cup lichen

Campylopus moss

Norway
spruce

Scots pine was
the dominant
tree in our
forest. Older
specimens were
particularly im-
pressive with
huge spreading
flat-topped
crowns.

Cones The
crossbill prises
cone scales
apart with its
unusual
beak; that of
the Scottish
crossbill is
particularly
large, to tackle
pine cones.
Squirrels
simply gnaw
off the scales.

**Pine and
spruce** These
two intruders
into the
natural Scots pine stand
probably grew from seeds
blown in from nearby
plantations.

Scots
pine cones

Cone
attacked
by crossbill

Cone eaten by
squirrel

Understorey plants In a forest
juniper grows as a straggly
shrub and is less prickly than
the erect specimens of open
ground; multicoloured
bilberry leaves signal autumn.
Young shoots of bilberry and
heather are eaten by birds.

Norway
spruce cones

Corsican
pine

Juniper

Common heather

Bilberry

Pine sawfly

The pine sawfly is notorious for the harm it does to pine woods. Canadian forests have suffered because of accidentally-introduced European sawflies. The female lays eggs in pine needles with her saw-like ovipositor, and the larvae eat vast quantities of needles. Unlike the larvae of their relatives the wasps and bees, sawfly larvae look like caterpillars. But the pine sawfly larva has at least seven pairs of "false legs", as well as three pairs of "true legs", while a caterpillar has only two or three pairs of false legs.

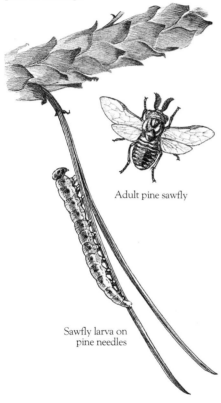

Adult pine sawfly

Sawfly larva on
pine needles

to resemble the pine needles which constitute their food. Pine sawfly larvae live in groups and when disturbed they arch their bodies into an S-shape and wriggle to and fro, weaving and dancing about on the branch in a sort of caterpillar disco.

The larvae of some species of sawfly have a most astonishing method of climbing. When a member of one of these species falls from a branch it does not automatically let out a silken lifeline, as caterpillars do, in order to climb up again. Instead it uses a system rather like that of a telephone repair man. The larva spins a strand of silk around itself, pinning each end of the strand to the branch or twig. Then, supported by this cradle, it wriggles a short distance upwards. Next it makes a fresh band to create a new support and wriggles a bit higher, and so on until it achieves its goal.

Many of these insects do considerable damage to the coniferous forests of the Taiga. One of the worst is the spruce budworm, the small thread-like larva of a rather inconspicuous North American moth that has unfortunately found its way into the coniferous woodlands of Eurasia. The larvae hibernate on twigs of spruce and fir and wake up in May, ravenous and ready to devour needles, buds and flowers. In periodic holocausts they can destroy whole stands of timber.

There is one creature, however, which could almost be described as a "forest ranger" or protector of the woods, and that is the large red wood ant. Colonies of these ants build huge nests in the forest and range widely over the trees in search of caterpillars and other insects to take back to their nests as food. It has been estimated that a good-sized colony of these ants—from 50,000 to 500,000 individuals—can consume up to 100,000 assorted insects and larvae a day. The insect control done by this ant species is so immense that in some countries it is protected by law.

The red wood ants' nests are one and a half metres high and three metres in diameter, giant mounds consisting mainly of pine needles. Underneath the mound is a network of corridors and chambers that goes deep into the earth. The entrance, which is guarded all day, is carefully closed up at night. On warm sunny days in spring the ants from the nests in a region commence their wedding flights. Swarms of winged males and females emerge from a nest like a column of smoke. From another nest another column flutters and twists its way into the air and the two columns eventually meet and intermingle in one great waterspout of ants, generally in some high place like a tree top or a hill. Many of the males and females fall to the ground and die, exhausted by their marriage flight, but sufficient fertilized queens remain to found new colonies. These queens can live for as long as 20 years, and all the thousands upon thousands of worker ants in a nest are the young from one or a very few queens. The various nests can be linked together by ant highways, so a whole community can number millions of individuals.

There are two creatures that share the ants' nests with them, in different and very curious ways. The first is the larva of a beetle who, as an adult, has an underside and head covered with white "fur" and reddish-yellow wing cases with some black spots. The larva builds a curious barrel-shaped dwelling, closed at one end, out of its own excrement. It lives in this peculiar and (one would have thought) rather unhygienic structure, feeding on the red ant pupae. The second creature that has been known to share the ants' nests is the bear. During the winter months the ants retreat deep into the ground and remain there until warmer weather comes. Bears, searching for suitable places to sleep in during the harsh weather, apparently think these huge piles of pine needles very comfortable beds. They dig into them and snooze the winter away, wisely getting up just in time before the ants resurface in the warmth of early spring.

Birds of the northern forests

When spring arrives, the frozen waters of the boreal woods thaw and the thick banks of vegetation at their edges attract many kinds of waterfowl. In addition the forest, like the tundra, produces a myriad of insects in summer, and this rich food source acts like a magnet and draws many different species of insectivorous birds, such as warblers and tits, from the south to swell the numbers of resident birds.

One of the most colourful residents of the Taiga, the large green woodpecker, is so bright in its colouring that it looks more like a bird from the tropics. You may hear them uttering their giggling cries as they search for their favourite food of red wood ants. They burrow into the great nest of pine needles and into the earth below, creating wide breaches in the walls and through into the core of the nest so that their flicking tongues, like sharp spoons, can garner the ants. Another woodpecker species, the three-toed woodpecker (who, unlike most members of the family, has no red on its head but a yellow crown), is adept at chiselling out insect larvae from the depths of the conifer trunks and boughs. It lacks the backward-pointing first toe of other woodpeckers and so, unimpeded by this, it can lean even farther back to deliver its splintering blows to the bark and wood.

Even birds which are normally seed-eaters or berry-eaters feast on the summer glut of pine wood insects. The siskins and pine grosbeaks do this, and the beautiful waxwings skim and dive like flycatchers for various insects. The waxwings are delicately and beautifully coloured in pastel shades of brown with highlights of yellow on the wings and tail. They get their name from the strange wax-like tips of the secondary wing feathers. No one seems to know the reason for these strange appendages, which look like little scarlet candles from a miniature Christmas tree. Waxwings also have an interesting courtship display. The male bird brings the female a "present" in the shape of an ant pupa, or a berry, or even something inedible. With the tip of his beak he presents it to her, she takes it with the tip of her beak and then she gives it back to him. This action is then repeated several times. It appears to be symbolic, in the same way that humans give each other rings when they get engaged or married, since the waxwing's present, even if edible, is scarcely ever eaten.

At the end of autumn the lakes go still under a coating of ice and the insect hordes die or pupate for the winter. The waterfowl and the fragile warblers and tits fly south and the Taiga residents turn to plant material for food. The waxwings return to berries, the finches rely on seed and shoots, and even the woodpeckers now will take seeds. The greater spotted woodpecker is expert at wedging pine cones into a favourite fork or cleft in a tree and then, making the forest ring with the sounds of its task, it hammers off the scales to get at the seeds inside. Under one of these woodpecker "smithys" you can find hundreds of dismembered cones, just as you find broken snail shells round a thrush's "anvil". One

The crossbill, a member of the finch family

The capercaillie's display

The handsome capercaillie (opposite) is giving its mating call. It is the largest and most spectacular of all the grouse family. The huge male frequently utters a very soft song just before dawn, prior to starting the real display. Then he takes up the stance of a gobbling turkey, with wings drooping and tail spread in an upright fan. The song is given 200 times a day, or sometimes as often as 600 times at the peak of the breeding season. It consists of a noise like someone sharpening a scythe, together with cracking and snapping sounds like the breaking of twigs. The bird calls about eight times a minute, and then has a rest before starting over again.

member of the finch family is so well adapted to its life in the pine woods that it does not usually have to worry about food shortage in winter. This is the crossbill, who possesses one of the most curious beaks of any bird in the world. The hooked ends of the beak cross over each other and with this natural form of "staple-remover" the bird can extract the seeds from the depths of cones. Not surprisingly, with such a specialized bill and such an abundant food source, the crossbill lives almost exclusively on a diet of pine seeds.

The ground birds of the pine forests such as the capercaillie and the black grouse (which also inhabit Scottish moorland) and the spruce grouse of North America all have a wide and varied diet. In summer, of course, they take advantage of the numerous insects. Young capercaillie, like woodpeckers, rely on red wood ants for the bulk of their summer diet and this is a curious fact because the ants produce a high concentration of formic acid, which you would think would make them unpalatable. In the winter these grand birds shelter under the downswept "umbrellas" of the conifer branches and feed off pine needles, which sound almost as unpalatable as red ants. All these ground birds have very strong gizzards and long digestive tracts to cope with their tough diet. The capercaillie even grows a hard sheath over its beak in winter to help it in food gathering, shedding the sheath in spring.

Of the purely predatory birds found in coniferous woodland the smallest is the handsome Northern shrike, clad in subtle grey feathering and with the strong hooked beak typical of hunting birds. Its prey is chiefly frogs and grasshoppers in summer and finches and mice in winter. Unlike the other shrikes, who create their "larders" by impaling their victims on thorns, the Northern shrike wedges its victims into the forks of branches. By impaling or wedging their prey, shrikes can feed at leisure and without having to hold the corpse in their claws.

No woodland is complete without its share of owls. The smallest owl to live in pine forests is the pigmy owl of Eurasia, which is only a tiny bit bigger than the diminutive elf owl of the American desert, the world's smallest owl. Pigmy owls "marry", that is to say they form what is called a "permanent pair bond". They generally breed in holes abandoned by greater spotted woodpeckers, and while the female incubates the eggs the male is out hunting small rodents and birds which he brings back to the nest for his wife. She is a very tidy housekeeper and cleans the nest regularly, throwing out pellets, excreta and food remains. These pile up at the bottom of the tree and are a sure indication that these little owls are nesting there. Pigmy owls are very vocal and the small birds they hunt, like coal tits and crested tits, gather round immediately they hear an owl call and do their best to "mob" the owl and drive it away. It has been said that if you go into the wood and imitate the owl's call and you get no response from the small birds, it means there are no owls in the vicinity. It could, however, simply mean your imitation is a bad one!

Another lovely owl who lives among the conifers is Tengmalm's owl. Handsomely marked in brown and ash-grey, it has very pronounced "eyebrows" and a broad white facial disc which shows off the bright yellow eyes. This owl, nesting as it does in holes in trees, has a particularly deadly enemy, the pine marten, which tries to steal its eggs or young. If you find a likely hole in a tree and want to know if the owl lives there, try scratching the bark with your fingernails. If the owl is at home its white

Tengmalm's owl

forests, they mistook them for moose and called them elk (the European name). So now we have the ridiculous situation where a moose in North America is an elk in Europe, while an elk in North America is a red deer in Europe. Careless naturalists who do not observe closely can certainly cause a lot of confusion!

Taller than the tallest Shire horse, the huge and powerful moose, with its slashing hooves, is more than a match for wolves. In fact wolves have to hunt in packs of well over 20 before they can pull down even a very young or sick Alaskan moose. But in spite of its great size and rather fearsome appearance the moose, when hand-reared, is a gentle and steady animal. Stone Age men in Siberia may have used them to pull loads, and recently in Scandinavia and Russia attempts have been made to domesticate them again. However, even with a tame moose you have a problem. They are insatiable mushroom-lovers and will wander into the forest, "kneeling" to get at the tender caps. If they find people gathering mushrooms the moose will frighten them away, and then seize the chance to steal all the mushrooms from their baskets.

The coniferous woods are the home of the grizzly bear, a type of brown bear. Left alone, the grizzly is a placid creature who snuffles about the forest in search of insects, small rodents, ground-nesting birds and their eggs, and nuts and berries. In the spring, however, it will take large mammals like young deer. Grizzlies also eat carrion whenever they can, and as they were seen to feed on cows that had died from some other

A watery vegetarian
The moose loves to browse on soft succulent lake plants, and can hold its head underwater for a considerable time while its fleshy lips gather submerged vegetation. Its long legs and broad flat feet, besides helping it to stay mobile in the deep snows of winter, enable it to wade deep into muddy waters to reach the aquatic plants of summer.

cause, the grizzly received the unfounded reputation of a cattle-killer. In consequence they were mercilessly hunted and poisoned in their southerly range, to such an extent that they have practically disappeared and are now only holding out in the far north.

Among the various races of brown bear (for all brown bears belong to one species—*Ursus arctos)* are the largest terrestrial carnivores, the gigantic Kamchatcan and Kodiak bears. The first time I saw a Kodiak bear was when I paid a visit to Whipsnade Zoo many years after I worked there. The bears were housed in a deep pit which in my day had held tigers. Looking down on the animals from the high side of the pit I was disappointed, as they did not seem very big. Then the keeper took me downstairs to the barred doorway that entered the pit and the bears rose on their hind legs and walked towards us. It was an extraordinary sight. The door was high enough for a man to walk through but the bears towered above it, shutting out the light, and I had to crane my neck to see their heads. On their hind legs I calculated they must be at least three metres high, if not more. I marvelled when I thought that these shaggy giants, although grouped as carnivores, were in fact mainly vegetarians. Their main "meat" source in the wild was the salmon they caught in spring. Nevertheless, they looked so formidable I did not feel that they would take kindly to being offered a berry and might well decide to take one's hand as well.

Most people think of bears when hibernation is mentioned, but bears are not true hibernators. They drowse through the winter, living off their fat reserves. While they sleep their body temperature remains at about 35°C, roughly normal for a mammal, and nowhere near the 5°C or even lower reached by true hibernators such as the dormice and insectivorous bats. With this normal temperature bears can emerge at speed from their caves or twig-covered, moss-lined beds should danger threaten. This fact was only discovered recently, and whoever crawled into a cave to take the rectal temperature of the slumbering bear must have been a very intrepid naturalist.

Another creature that has steadily lost ground in its southerly range is the North American porcupine, a charming creature that at first glance looks more furry than spiky. It is losing its forests and is being persecuted because of the damage it allegedly does to timber with its sharp teeth, but as usual the extent of the damage has been greatly exaggerated. In winter time the porcupine can indeed kill a tree, for the simple reason that, being a rather heavy and clumsy animal, it finds difficulty in moving through thick snow and prefers to stay put. Having ensconced itself in a suitable tree, it proceeds to eat the needles and the bark until eventually only the skeleton of the tree is left. Then, forced into action by a food shortage, the animal reluctantly moves on to another tree. The porcupine is more of a nuisance when it finds a camp, for during the night it has been known to gnaw at sweat-soaked things like canoe paddles, wooden boxes, handles and saddles. When the campers wake in the morning they are justifiably infuriated to discover that half their equipment has been damaged or destroyed.

The female porcupine has one young each year (and a very heavy baby it is—as if a human mother had produced a 16 kilogramme infant, about four times the normal weight). Moreover, the youngster is not sexually mature until its second year. This slow breeding rate combined with

A bear who climbs trees
The black bear of North America is an agile climber. Even large adults are able to shin up trees to raid nests or search for honey. Unlike its larger cousin the brown bear, the black bear breeds only every other year. But like most bears, its feeding habits are best described as opportunist, since it will eat virtually anything that comes to hand.

Members of the mustelid family

Mustelids are adaptable hunters, and have spread into numerous habitats. The fisher is one of the larger mustelids and, like its relatives, it is a swift and agile hunter, relentlessly pursuing its prey of squirrel or porcupine both on the ground and in the branches. The stoat, with its black-tipped tail and white tummy, ranges across Eurasia and North America. It is ferocious, and will pounce on any animal up to the size of a hare.

Fisher

Stoat

deforestation and persecution by humans explains why the porcupine is now found only in the very remote northernmost forests.

This dim but endearing beast is not related to the African porcupines, but it is really a sort of prickly guinea pig. Like other porcupines it has the reputation for being able to shoot out its spines, but this is quite untrue. It defends itself when cornered by slapping at its enemy with its spiny tail. The quills are only loosely planted in the skin and each of them is minutely barbed like an arrow. When they penetrate the enemy they are pulled loose from the porcupine's tail and, because of the barbs, work farther and farther into the enemy's flesh. In spite of its defensive quills the porcupine has one enemy, apart from man, that can kill and eat it. This is the fisher, one of the martens, who considers the porcupine a tasty dish. When it catches one it flips the unfortunate porcupine on to its back and then tears into its spineless and defenceless underbelly.

The fisher looks not unlike a small elongated bear with a tail and has long luxuriant shiny fur. It belongs to that extraordinarily adept group of predators, the *Mustelidae* (the weasel family), which includes the martens, polecats, minks, otters and skunks. The European members of this group are very resourceful and adaptable. The weasels and stoats also inhabit hedgerows and can hunt in open country as well. (In another naming mix-up, a stoat in Europe is a short-tailed weasel in North America—although the original, but now seldom-used name is ermine.) Polecats have long been domesticated—as ferrets—and used for hunting rabbits.

The group of handsome and graceful mustelid predators has earned itself a considerable reputation for bloodthirstiness. Up to a point this is deserved, for their instinct is to kill and then store all the prey they find, regardless of whether they are hungry or not. When one gets into a well-stocked chicken coop it will therefore wreak havoc and will be unlikely to leave a single bird alive. However, mustelids do have a redeeming feature and that is they are wonderful at rodent control. So if you can make your coops or hutches impregnable to attack you will find they will keep down all unwanted rodents. In the Bronze Age it was probable that various mustelids were kept instead of cats, so swift and clever are they in their pursuit of rodents. They are much more energetic and playful than cats, behaving more like little ballet dancers as they romp together. However, unlike cats, their tempers are uncertain and if you try to cuddle one you may get a bite for your pains.

European mustelids range into coniferous forests of Eurasia, and weasels and stoats spread into North America as well. Ferocious hunters, they will take any prey they can tackle. The so-called least weasel, only 17 centimetres long and the tiniest of all carnivores, is so slim it can follow small mice into their burrows, while stoats and polecats can overpower something the size of a hare.

The whole group is amazingly versatile. The pine martens (now, unfortunately, the rarest wild mammal in Britain) and American martens are more skilled than any circus acrobat, leaping and bounding gracefully through the trees in pursuit of birds and squirrels. The fishers and sables are more at home on the ground, chasing such things as mice and red-backed voles; even so, a fisher has been known to move through the trees at the speed of a running man, in pursuit of chipmunks (who are pretty good acrobats themselves). Pine martens are adept at stalking and catching capercaillies—no mean feat. At any time all these adaptable creatures will eat nuts, insects, berries and sometimes even honey to supplement their more usual diet of live prey.

Probably the member of the weasel family with the worst reputation is the wolverine, which ranges right across the Eurasian and American coniferous forests. Its voracity is legendary and has not only earned it the nickname glutton but also made it the sworn enemy of trappers. This is because it patrols the trap lines and steals the dead or dying animals, and even breaks into the trappers' sheds to get at their catches. Although the biggest of the weasel family it is a rather stocky animal less than a metre long and rather dog-like in appearance. For its size it is a ferocious and fearless beast, and has been known to drive bears and mountain lions away from their kills so it could feed on the catch. In the deep winter snow, while most other animals flounder, the wolverine bounds along effortlessly, its broad feet acting like snowshoes. In this way it can successfully catch even moose and caribou. Unlike other weasels it never kills more than it can eat at one or two sittings. In the summertime it noisily snuffles about the forest, feeding mainly on nuts and berries, rodents, birds' eggs and carrion.

In spite of its reputation the wolverine, if hand-reared, makes a delightful pet and treats its master with dog-like devotion. I had a friend near Copenhagen who had two wolverines he had brought up from babies, and they were enchanting creatures. When I was first introduced to them they almost knocked me down, so enthusiastic was their reception. When we took them for a walk in the forest they gambolled around us like dogs, uttering throaty little cries of excitement. In the evening when we got back we sat in front of the fire and both animals tried to climb into my lap. Finding I could not accommodate two fat wolverines, they compromised by each laying their head in my lap and then going to sleep, uttering rich prolonged bubbling snores.

As with so many other creatures the wolverine's last outpost is the coniferous woodlands of northern Eurasia and North America. Fear, hate and even superstition have turned them into outlaws and they are relentlessly persecuted. They are even trapped for their fur because it does not harden in freezing weather and so makes useful face-trims for hoods. Man is the wolverine's only enemy. It is sad and ironic that in Michigan, called the "Wolverine State", no wolverines have been seen for a very long time.

As the wolverine loses ground, the fishers and other martens are also reaching a very low ebb in over half their ranges in Canada. Yet in many of these provinces where there should be most concern over the dwindling populations, these animals are still hunted and their pelts sold. The furs fetch a good price, good enough to attract hordes of what are known as "week-end trappers"—an unpleasant breed of people, one can't help thinking, who are motivated not by the rather feeble excuse of sport or by the necessity of earning a living, but purely by greed. As usual, the animal suffers. One is not being particularly anti-Canadian in adopting this attitude; just anti a certain all-too-common type of human being. At least the Canadians publish the results of their trappings; it is more difficult to get satisfactory figures from Russia, though we know that the Russians have been doing some excellent re-introduction work with sables and beavers. But still, in most parts of the world, fur-bearing animals walk around inside their greatest danger—their skins.

The aptly named glutton
The wolverine's nickname of "glutton" may in part be justified, for it is a voracious hunter and will steal trappers' bait or raid food stores. But man has punished the wolverine far in excess of the damage it has caused, and this heavily-built mustelid is holding its own only in the remote coniferous forests of Canada and Eurasia.

THE NATURALIST IN
TROPICAL FOREST

For a dazzling array of surprises and an inexhaustable fund of enigmas, there is no part of the world to compare with a tropical forest. In South America I have watched what I thought was a tree trunk turn into a tapir and walk away. On a similar occasion in Malaysia, what appeared to me to be the sawn-off stumps of branches on a tree suddenly started moving—they were remarkable insects resembling Chianti bottles with legs. The first time I was taken into the rain forest in West Africa I watched carefully how my guide bent twigs and leaves to mark our trail; the first time I tried to do it myself I was hopelessly lost inside an hour. I have squatted down in a clearing in the Malaysian jungle and within three minutes been covered with a multitude of leeches who had looped their way through the leaves, drawn to me by some mysterious radar. Some surprises are very different. Once, at a campsite in West Africa, I spent a long time carefully tracking down what I was convinced was a new species of mammal giving its mating cry. When I finally located the source of the sound I discovered it was my cook—no musician—singing as he washed the dishes in a nearby stream.

The warm moist parts of the world in which tropical forests grow are the richest, most flamboyant and most exciting areas of our planet from the naturalist's point of view. Stand in a tropical forest and turn slowly in a circle. Within ten metres there is such an incredible profusion of living things that it would take many lifetimes just to work out the natural history of a few of these organisms, let alone the relations between them. Everywhere you look in a tropical forest there is some strange and wonderful adaptation to astound or fascinate you.

The mammoth tropical trees are strengthened by huge buttresses, great flanges of trunk as high as a house and with space enough between them to park a large lorry. The trunks are as straight and slender as wands, some reaching a height of 50 metres before the first branches start to spread out and form the canopy. It is these giants of the forest that provide, on their trunks and arms, living space not only for animal life but for other plants as well. High up in the canopy grow huge staghorn and nest ferns. They grow in basket shapes so that they can catch the leaves, fruit and twigs that fall from above; in time this debris is broken down to form enough humus to nourish the ferns and to provide living space for earthworms or millipedes. Each airborne fern is a little world of its own, a sort of hanging garden. In many parts of the world the life of the forest giants is imperilled by the strangler fig, which winds its muscular arms round a tree and swarms octopus-like up its trunk. Eventually, crushed by the fig, the tree dies, rots and disintegrates to leave the triumphant fig like a monstrous net growing up from the forest floor.

The incredible range of creatures to be found in the tropics is bewildering. Among the reptiles are Komodo dragons, four metres long

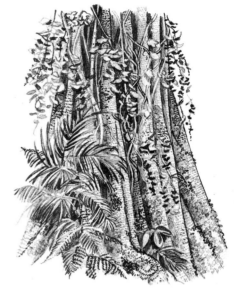

Marvels of engineering
Many of the tall tropical trees have buttresses which rise up like the supporting walls of a cathedral, giving a solid base in the relatively shallow layer of tropical forest soil.

An arboreal great ape
The orang-utan (opposite) is seemingly four-armed as it moves sedately but with tremendous agility through the Sumatran tree tops. This ape, whose name means "old man of the forest", is endangered due to the rapid destruction of its forest habitat.

and capable of devouring a goat, and little island-dwelling geckos who are shorter than their scientific names. There are slender tree snakes that "fly" from tree to tree and winged lizards that do the same. In the mammal world, what a contrast between those mammoth gentle vegetarians of the African forests, the gorillas—who can weigh as much as three grown men and stand almost two metres high—and the diminutive marmosets of the Amazon basin, two of which would fit comfortably into a teacup. There are mound-building birds who use rotting leaves and earth to construct incubators for their eggs, and hornbills who wall up their wives in hollow trees to make sure they sit on their eggs. There are Goliath beetles as big as a man's hand and blue *Morpho* butterflies so huge they look like sizeable birds. I have seen a moving column a kilometre long made of army ants, relentlessly driving vast quantities of insects from the undergrowth to fly or hop in panic ahead of them. Though the fleeing insects may escape the ants they rarely escape the ant-birds, who follow or fly ahead of the ant army and thus garner their food. There is the forest giraffe, the okapi (only discovered in the 1900s, so well did the trees hide it) and the tiny Royal antelope, 30 centimetres high with legs as thin as pencils. There is a great wealth of cats, ranging from the South American jaguar, wonderfully black-spotted on a background as yellow as a chrysanthemum, to the melanic black leopards of Asia with coats like ebony. Nothing you can imagine is too bizarre for the tropics, from frogs in brilliant suitings with deadly poison in their skins to trees that produce flowers without interruption for a hundred years.

Philip Darlington, the famous zoogeographer, once said, "Any young naturalist who thinks he can understand the world and living things and evolution without experiencing the tropics is, I think, deceiving himself, to his own great injury." I agree with him entirely. In fact, if I was a fairy godmother the gift I would bestow on young naturalists would be a month in a tropical forest. It would alter their whole outlook on the incredible world we live in.

The wealth of life in the tropics

It is startling to realize that the number of species in the tropical forests is more than the number from all the other terrestrial ecosystems combined! Take an area of jungle in Malaysia or Amazonia; you will find ten times more tree species there than you would find in the same area of temperate forest. In the tiny tropical corridor of Central America dwell nearly four times the number of bird species that live in the whole of the eastern forests of the United States.

What makes a tropical forest so rich in living things? After all, it has much the same blueprint as any other forest: there is a bed of soil and leaf litter, then the understoreys of low plants and shrubs, and finally the roof or canopy. A tropical forest also works in the same way as other forests: the tall trees form the rib structure and, along with the smaller plants and the animals, are dependent on the soil and climate—which, in turn, the trees help to make. So why the vast difference between a tropical forest and one elsewhere?

The answer lies in the position of tropical forests on the planet. They have fed to them enormous quantities of sunlight and warmth and these, combined with large amounts of rain, allow life processes to continue all year round. It is rather like being in a gigantic greenhouse. Plant growth,

The curious okapi comes from Central Africa

The largest land lizard— the Komodo dragon of Indonesia

South American tree frog *Dendrobates*

on which all other life depends, becomes so riotous it creates hundreds of different microhabitats that provide the conditions in which thousands of different creatures can evolve.

There are seasons in the tropics, but nothing so marked as seasons in the temperate regions to the north and south. The appearance of the wet tropical jungles, so wealthy in animal and plant life, remains fairly stable over time. In the year-round sunlight and warmth the death and rebirth of leaves is a constant process. The seasons, such as they are, are really a matter of more or less rainfall, and the effects are that flowering and fruiting slow down in the drier months. But in some areas of the tropics, where there is the least amount of rainfall, the trees drop their leaves all at once. It is a bizarre experience to walk through one of these forests, bare as any English woodland in January, and feel the sun beating down and hear the cries of hundreds of exotic birds.

As in other forests, a tree forms a vitally important unit for the creatures that live in, on or around it. In the cooler forests of the world a tree is the equivalent of a block of apartments to a human. However, in the tropics each of the giant rain forest trees could be called a whole metropolis, so rich and varied are its inhabitants and visitors.

I remember the first tree of this sort I came across, when I was animal collecting in West Africa. It had a great hole in its base the size of a church door and many cracks and fissures about 20 metres above the ground where the branches started. To all intents and purposes the tree looked uninhabited, but we decided to investigate it anyway for we were pretty certain that it contained at least some surprises. Since we were a party of experienced animal collectors and professional zoologists, the officials from the local forestry department had given us permission to use the "smoking out" method on any suitable hollow tree we came across (smoking out should never be done without permission and unless supervised), so this is what we proceeded to do.

First we had to cover all the cracks and fissures with nets. Then a small fire was carefully lit in the hole at the base, and immediately covered with green leaves. This prevented the tree from catching fire and the smouldering leaves provided a dense smoke. As the first coiling columns of smoke were drawn up into the hollow interior, a veritable zoological garden made its appearance. At the top there were owls, tree snakes and a colony of the curious and rare *Idiurus*, the flying mice that use a flange of skin on each side of the body to glide from tree to tree. Lower in the tree were strange little chocolate-and-yellow bats with faces like bulldogs, and near them lived some of the small green forest squirrels with flame-red tails. Lower still were giant millipedes, like huge shiny red-brown sausages with forests of legs on their undersides, and large whip scorpions that look like flattened spiders, and with their long legs and pincers they can be the diameter of a soup plate. There were also small tree frogs, some grass-green, some striped and spotted with so many bright colours they looked more like exotic sweets than amphibians. Then, at the base of the tree, the smoke drove out several species of forest rat. Some of these rats were wearing what looked like pin-striped suits, the effect being given by lengthwise nose-to-tail stripes made up of tiny cream-coloured spots on a grey background. With these rats was a giant grey pouched rat, as big as a small cat, with huge cheek pouches in which it collects and carries its food. At ground level, too, there were the beautiful and curious brow-leaf

Furry flier
The nights in West Africa are enlivened by the mouse-sized flying rodent *Idiurus*, commonly known as the pigmy scaly-tail. It glides between trunks by means of paired membraneous "sails" that stretch between fore and hind limbs. In flight the long tail fringed with wavy hairs is held outstretched like a rudder. *Idiurus* can launch itself off a trunk and glide to another at least 50 metres away. During the day these small rodents cluster together rather like bats in the hollow crevices of tree trunks. The development of gliding flight among forest animals is not uncommon; flying squirrels, frogs and snakes have all developed flaps of loose skin which enable them to traverse the forest without having to descend to the ground.

toads, the size of saucers, with huge gentle golden eyes and a pattern on their backs that so cleverly resembles a dead leaf they simply merge into the background and disappear. All of these creatures and more had a single hollow tree as their home.

Ground dwellers of tropical forests

The soil is much thinner in tropical rain forests than in coniferous and deciduous woodland. This is because decomposition works so rapidly in the hot humidity of the tropics and trees suck up moisture and nutrients so quickly that the humus has little chance to form. The thin soil is the reason that erosion in the tropics is so swift and devastating, for once the protective sheath of trees is felled the thin poor soil is washed away, leaving only desiccated bedrock. But shallow though they are, the undisturbed soil and litter on a tropical forest floor are teeming with life, many forms of which are not found outside the tropics. In addition to earthworms and many other creatures similar to the ones found in deciduous forests, there are such strange things as blind burrowing snakes and the caecilians, curious earthworm-like amphibians who weave their burrows in the leaf mould around the protective roots of giant trees. Spiders and mites roam the leaf litter, and in pursuit of them in West Africa there is a tiny chameleon no longer than a matchstick. He stalks this territory of crumpled leaves, looking more like a dead leaf than the real dead leaf does. Here you can find snails bigger than apples and shrews smaller than your little finger; cicadas who are ventriloquists and can throw their voices, so that while you watch them their cry seems to come from behind you; and beetles who glow with such a vivid green light it only takes half a dozen for you to read by.

Above the litter layer and ground plants is the understorey or shrub layer. Besides the shrubs and saplings there are *epiphytes* like orchids, climbers like the rope-thick lianas, and the stranglers. All these plants seem to be bursting upwards like rockets, jostling each other for position so that they can reach the sunny heights of the forest canopy. The trees themselves act as ladders for the climbing vines and carry gardens of epiphytes on their broad limbs. Where there is a clearing, or on the banks of a river, these various understorey plants embrace the trees so thickly that they make the forest look impenetrable, hanging down from the branches in lush green curtains.

The world's greatest roof garden

The green upper reaches are the most exciting and wonderful area of the tropical forest. The vast sunlit garden that forms the canopy is a world of its own, and here live creatures that never descend to the ground. Their whole world is high in the sky, where the tops of trees interlace and form highways and bowers and proffer a rich bounty of bark, leaves, flowers and fruit for the forest's inhabitants.

While the canopy is one of the most richly-inhabited regions of the forest it is also the one that causes the naturalist the greatest frustration. There he is, down in the gloom among the giant tree trunks, hearing the noises of animal life high above him and having half-eaten fruit, flowers or seeds rained on him by legions of animals high in their sunlit domain—all of which he cannot see. Under these circumstances the naturalist develops a very bad temper and a permanent crick in the neck.

Tropical hanging basket
Orchids flourish high up in the canopy where there is abundant light. These attractive epiphytes have strong clinging roots and long dangling ones with air spaces at their tips which literally "suck in" moisture from the atmosphere. As a safeguard against drying out, orchids store water in swollen bulbs. Other epiphytes, like bromeliads, have a rosette of leaves which traps water and plant debris and forms a humid compost heap at the centre of the plant. In the rainwater pool that collects in the rainy season mosquitoes, tree frogs and salamanders can complete their entire life cycles far above ground.

However, there was one occasion when I managed to transport myself into the forest canopy and it was a magical experience. It happened in West Africa when I was camped on the thickly-forested lower slopes of a mountain called N'da Ali. Walking through the forest one day I found I was walking along the edge of a great step cut out of the mountain. The cliff face, covered with creepers, dropped away for about 50 metres, so that although I was walking through forest, just next to me and slightly below was the canopy of the forest growing up from the base of the cliff. This cliff was over a kilometre in length and provided me with a natural balcony from which I could observe the tree top life simply by lying on the cliff edge, concealed in the low undergrowth. Over a period of about a week I spent hours up there and a whole pageant of wildlife passed by. The numbers of birds were incredible, ranging from minute glittering sunbirds in rainbow colouring, zooming like helicopters from blossom to blossom as they fed on the nectar, to the flocks of huge black hornbills with their monstrous yellow beaks who flew in such an ungainly manner and made such a noise over their choice of forest fruits. From early morning to evening, when it grew too dark to see, I watched this parade of creatures. Troops of monkeys swept past, followed by attendant flocks of birds who fed eagerly on the insects that the monkeys disturbed during their noisy crashing through the trees. Squirrels chased each other, or hotly pursued lizards, or simply lay spread-eagled on branches high up in the trees, enjoying the sun.

Very early one morning I got to a spot where a big tree had just fruited. The whole of its canopy was full of squirrels feeding on the fruit and small birds feeding on the insects attracted by the fruit. The squirrels, like all canopy feeders, were very wasteful, taking a bite from a fruit and then dropping it and moving on. The fruit dropped to the forest floor far below, where I could hear a sounder of Red River hogs grunting and

A striking-looking balancing act
An Indian pied hornbill perches on a favourite lookout point. The bird, an immature male, has its bony casque only partially developed. The great bulbous casques on the foreheads of mature males are thought to aid amplification of the bird's resonating call. In some species of hornbill the male barricades the female into the nest hole using a mixture of mud and droppings. Only a small entrance hole is left through which the male feeds her on fruit, insects and small reptiles.

squealing over the squirrels' left-overs. Presently a crashing sound, like surf on a rocky shore, told me a troop of monkeys was on the way, and the squirrels (who obviously also knew what the sound meant) all disappeared. Preceding the monkeys came a troop of hornbills, their wings flapping and whooshing in ungainly flight. They landed untidily in the fruit tree and started to wheeze and honk as they selected their food. Almost immediately the monkeys arrived and to my delight they were a group of my favourites, the green-grey putty-nose monkeys, each with a big white heart-shaped spot of fur on its nose that made it look as if someone had hit it with a snowball. They are agile delicate monkeys with an odd, rather petulant cry like a bird. It was a troop of about 30 and there were a number of enchanting babies among them. While the adults settled down quietly to feed, the babies—like all baby animals—quarrelled, explored and annoyed their parents. I watched one baby making her way along a branch to where a great bunch of fruit was growing. Just before she reached it a hornbill arrived in a whoosh of wings and did a crash-landing among the fruit, right in front of the baby monkey's white nose. The sudden appearance of this great black bird, five times her size and with a huge and deadly-looking beak, completely demoralized the baby. She turned to run, missed her footing and, with a wail of fright, fell from the branch. Fortunately, she landed safely in a tangle of creepers about five metres below.

It was fascinating to watch this procession of creatures and to find that certain ones fed together, while others avoided each other. I could also see that through the tangle of branches there were well-defined highways, as familiar to the forest creatures as the streets of your town or village are to you. Finding the cliff that lifted me into the tropical forest canopy was a great privilege and taught me a lot about tree top life.

Black-and-white colobus monkey of tropical Africa

In temperate forests the predominant canopy creatures are the birds, sharing the branches with relatively few other animals. But in the tropics they have to share their elevated world with a host of mammals, reptiles and amphibians, as I saw from the cliff face of N'da Ali. In Borneo the tree canopy gives living space to proboscis monkeys, with their extraordinary noses hanging down to their chins, and the flying frog, which has greatly enlarged webs on its feet that it uses like a sort of parachute to glide from tree to tree. In the central African forests live giant squirrels and troops of black-and-white colobus monkeys vying with chameleons and tree snakes. In the forests of Amazonia the birds share the giant trees with numerous canopy-dwellers: sloths who spend all their lives upside-down and who grow algae on their fur as a disguise; troops of tiny marmosets and squirrel monkeys; spider monkeys who hang by their tails; and, of course, those great opera singers of the forest, the howler monkeys.

The first time I heard the howler monkey chorus was when I was in Guyana, camped on the banks of a river. I didn't know it but I had slung my hammock between two trees right under a group of red howlers. At five o'clock in the morning, just as it was getting light, the whole troop of some 25 animals started to tell the whole forest that this was their territory. They were just above me in the trees, and the power and vibration of their calls almost blew me out of my hammock. It was a wonderful sight to see. Among the vivid green foliage and pink flowers in the tree above me these magnificent animals, whose fur shone like a cross

Male proboscis monkey from Borneo

between spun gold and copper, swelled their throats to the size of grapefruits as they sang their protective chorus.

The wealth of bird life in the tropics is extraordinary. Of course the finches, warblers and similar birds that dwell in other forests all have their tropical counterparts. In addition, there are also groups of birds that are mainly confined to the tropics—the great multicoloured concourse of fruit- and seed-eating parrots and parakeets, for example, and the glittering array of nectar-feeding honey creepers and hummingbirds.

Some birds forsake the tropical tree tops and migrate in the dry season, not in typical north-south movements but in practically any direction that will take them to a better food supply. In Madagascar, for example, there are five different species of bird that fly due west and spend their "winter" in Africa. And a species of oriental roller flies the incredible distance of 3,200 kilometres from "summer" in Thailand to "winter" in Borneo; the entire migration, therefore, occurs within the tropics.

The folly of destruction

The tropical forests of the world are untouched treasure houses of different forms of plant and animal life which could be beneficial to man. Already from these vast forests have come things vital to our medicine, industry and agriculture. Various extracts from the serpent-root plants of South-East Asia, for example, are used in treating cardiovascular disorders, intestinal diseases, hypertension and even schizophrenia. A South American armadillo is being used in the study of, and hopefully in finding a cure for, leprosy. A small weed from Madagascar helps in the treatment of leukaemia. Natural rubber was first discovered in the tropical forests. Many varieties of our food crops—coffee, nuts, pineapple and so on—were developed from their wild cousins. Even the idea that everyone should have "a chicken in his pot" would not have been possible without the domestication of the Asian jungle fowl. As well as producing all this, of course, a tropical forest is the weather-creator of its region. The huge array of trees absorbs moisture like green blotting paper and then releases it gradually. Eventually this moisture falls as rain, is reabsorbed, released again, and so the process goes on.

You would think that, with all the beneficial things our tropical forests have given us and the incredible potential for things they *could* give us in the future, they would be one part of the planet we would care for. Unfortunately, this is not the case. As usual, man is acting in a profligate and unthinking way in destroying these forests. The incalculable loss is ours. As the forest is felled, as-yet undiscovered medicines, foods and aids to industry vanish. The loss of the forests, the weather-controllers, creates floods as well as water shortages in the tropics and causes death and disease and misery to millions.

The current rate of felling, burning and clearing of tropical forests—for firewood, timber, cultivation, cattle-raising and human settlement—is horrifying. An area half the size of Great Britain, which is about equal to the full size of Lee's home state of Tennessee, is cleared *every year*. At this rate the tropical forests of the world, with all their treasures and potential benefits to the human race, will vanish in 85 years' time. This means that your children's children will not have the privilege of knowing what a tropical forest was like, and they will always wonder what treasures we, their forebears, deprived them of by our greed and stupidity.

Exploiting the riches of the forest
The beaks of tropical birds are almost as varied as their plumage. Those of parrots and macaws are very strong and horny, and are unique among birds in being hinged. They work like efficient pliers in cracking hard thick-shelled nuts. The super-efficient tongue and beak of the hummingbird extracts nectar from long-tubed flowers. This energy-rich food is ideal fuel for birds whose wings beat at an average rate of 60 times a second.

Scarlet macaw from South America

Purple-crowned fairy hummingbird of South America

THE NATURALIST ON THE
MOUNTAIN

Many different terrestrial habitats, ranging from rich forest to bleak tundra, can be visited and experienced within a short distance of each other if you are lucky enough to live near a mountain. All the variations in temperature and rainfall which are responsible for the different ecosystems can be found from bottom to top of a mountain. As you gain height the atmosphere becomes less dense and the air cools down and deposits its moisture as rain or mist on the lower slopes. So as you travel up the mountain you can expect to pass from layers of tropical or deciduous forest into a belt of evergreen conifers, which flourish in the cold dry air, then into a tundra-like belt which finally merges into the snow and ice of the highest slopes.

The whole framework of nature is immensely complex, of course, and it is impossible to describe part of the natural world as though it obeyed a simple series of rules. A mountain cannot be thought of as a sort of wedding cake, with perfectly demarcated and decorated tiers. The different types of vegetation run into each other and the demarcation lines become blurred, and things can be further complicated by the way the mountain lies in the world. Lee and I saw an example of this in the rain forests of Costa Rica. We climbed a small mountain that was positioned right in the path of the fierce winds sweeping through a gap in the taller mountains. On one side of this small mountain were lush forests and on the other side, on which the wind pressed, the plants were completely different, smaller and clinging closely to the mountainside. Walking along the ridge of this mountain (taking care not to be blown away) we could see the extraordinary demarcation line caused simply and solely by the force of the wind. As another example, the windward side of a tall mountain range may be covered in lush forests, encouraged by the rain the rising air has let go, whereas on its other face, the lee side, there is a "rain shadow" area of desert and scrub. The temperature differences are important, too; in temperate zones the coniferous forest starts lower down on the cool shady north-facing slope compared to the sunny south-facing slope.

Plant life above the treeline
Once you have climbed past the protective belt of trees on your mountain, you come out into the alpine tundra. ("Alpine" tundra, of course, is tundra encouraged by the cold of altitude; the word "alpine" simply means on a mountain—and any mountain, not just the European Alps.) This thin bleak land is subjected to the intense beating light of the sun, precarious sliding soil and enormous swings of temperature every day. In consequence, to protect themselves in their blistering homeland, many of the plants grow in plump cushions or thick mats. Each clump is so tightly woven that it is like a miniature trap, catching whatever soil

The delicate-looking edelweiss grows at heights of more than 3,500 metres in the Alps

The beautiful apollo
An apollo butterfly (opposite) feeds on hardhead, a member of the thistle family. Each mountain or mountain range has its own race of apollo, and over 600 races have been described by lepidopterists. Apollos fly slowly, fluttering and gliding down the mountain slopes like wonderful giant snowflakes.

The layers of the mountain

Every mountain is different, but a general rule applies to the plant cover at different altitudes. The diagram below shows the belts of vegetation you would encounter travelling up the south side of one of the taller Alpine mountains (altitudes for the colder north-facing slope would each be about 300 metres lower). Many plants and animals are common to both alpine and arctic regions, those on the mountain tops having been stranded since the time of the last Ice Age.

Snow field to summit
4,200 metres

Alpine tundra grading into rocky scree above

Alpine meadow
2,600 metres

Coniferous forest

Deciduous woodland of foothills

is blown by the wind. The plants crouch on the ground, holding off the sun and wind, and become like little incubators for insects and other small creatures. The inside of a cushion plant can be 20°C warmer than the air outside, which may mean all the difference to the survival of a beetle or butterfly.

The alpine plants have to husband their resources and so generally they are slow growers. Their point in life is not to waste the brief summers showing off flamboyant luxuriant greenery, but to spend the time steadily building up food stores in their roots in order to survive the biting cold seasons. They bed themselves cosily in the eiderdown of snow (which is warm compared to the air temperature outside), and they do not become frozen solid because the liquid moving sluggishly in their cells is so thick it acts like antifreeze put into a motor car. The leaves themselves have a sort of haze or mist on them, which is in fact a fine hairy fuzz that acts as a heat trap and deflects the wind.

It may take years for mountain plants to store up enough energy to burst briefly into flower, but when they do, the alpine meadow becomes so intricately and beautifully coloured that it is transformed beyond belief. It burns with spots of blazing colour—pink and blue and yellow on a background of dark-green leaves. The first time I saw this marvellous flowering of alpine plants I was driving through the Alps. I had never studied the flowers at close quarters so I stopped the car and walked off the side of the road. I looked down and found myself ankle-deep in an incredibly colourful pageant of pigmy flowers whose names I did not know. Some looked like tiny roses designed for the smallest pixie you could imagine, and others were pansies with minute blue faces no bigger than my little fingernail. The temptation to pick them was overwhelming; you wanted to carry these handfuls of intense colour home and enliven your rooms with them. But when you know that one of their seeds may take many years to grow into a plant and produce the next generation, you must resist the temptation. Sit quietly in a meadow like this and exult in the colour and the miniature beauty around you that can exist in such a harsh environment.

The plants in the alpine tundra have numerous problems connected with their physical environment, and they also have some trouble with the animals who live there with them. A perfect example is the story of the Northern pocket gophers of the Rocky Mountains. These charming-looking little rodents have one fault, which is that as gardeners they are, if anything, too energetic. In the Rockies are lovely meadows of sedges and cushion plants (the latter, as their name implies, look somewhat like an old-fashioned pincushion). These two plants operate very happily together because the sedge has a wide and spreading, but not very deep, network of roots whereas the cushion plant has one very deep anchoring taproot. When gophers move into an alpine meadow such as this, they busily excavate their tunnels and come across the roots of both plants, which they consider to be a great delicacy. This feasting on the roots does the plants no good, and in addition the soil that the gophers throw up to the surface smothers all the low growth. The small soil particles are swept away by the wind, leaving only gravel—which does not hold water very well, and so the sedge cannot re-establish itself because water has filtered down too deep for its roots. With the demise of the sedges and cushions the meadow is invaded by taller stronger plants like the sky pilot, yarrow and harebell, whose roots are not eaten by gophers. Their food supply dwindling, the little rodents move on, and after a time the cushion plants reappear. They can grow now that their roots are no longer used for lunch by rodents, and their long taproots penetrate the gravel and are thus able to get water. Dust and plant material slowly collect in the dense wig of cushion plants, and a layer of humus is gradually formed. As the soil builds up the sedge is able to return and as it does so it gradually edges out all the other plants except the cushion plants—and the cycle is complete, and the meadow is once more back where it started. This process, which can take as long as a hundred years, shows that nature is flexible enough to cope with and repair damage that is done naturally. It is the sudden savage assault by man that it cannot cope with.

Small creatures of the heights

At alpine heights the invertebrate fauna is rather sparse, but you can still find springtails and short-winged beetles and grasshoppers out on sunny days. There are some species of small butterflies such as the mountain ringlet whose flight paths, you will notice, are very low to the ground. They hug the plants and try to move, as far as possible, below the strong sweep of the wind. Most of these insects are diurnal, and the morning after a freezing night they are, for the most part, in a trance, relying on the warmth of the sun to melt them into vigour. You can see them in the sun's first rays, making tentative movements and as reluctant to get out of bed as you or I on a frosty morning. Mountain insects, like the plants they live among, have a very long development time. Many of the butterflies take two years to complete their cycle, and some of the grasshoppers take three.

One of the mountain insects I would dearly love to meet is the Himalayan grasshopper. High up in the thin air of one of the world's greatest mountain ranges, this flamboyant creature makes its home. Painted like a clown with scarlet cheeks, blue-black head, brown wings spattered with yellow drops and huge blue leaping legs, this magnificent grasshopper zithers his way through life and leaps with gay abandon among the alpine flowers at an altitude that makes us humans gasp and wheeze and pant because of the thin atmosphere.

Reptiles and amphibians, being cold-blooded like the invertebrates, are scarce in high cold regions, though there are some species which manage to exist in such inhospitable areas. In Europe, in the Alps, live the alpine newt and the glossy black alpine salamander. In the Himalayas there is a lizard, the toad-headed agama, that can proudly boast being the highest-living reptile species—it has been found at heights of 5,000 metres (over 16,000 feet).

Mountain birds

On the whole, it is the warm-blooded creatures—the birds and mammals—that fare best in the cold of altitude. In the high meadows of the European mountains dwells the alpine accentor (a dunnock-like bird), the snow finch and the beautiful wall creeper which, when it settles on a rock face and shows its dazzling magenta wings, looks like some huge butterfly. In the spring and summer these birds hawk insects in upper reaches of the mountains, while in winter they drop down to the lower parts of the range. Some of the larger birds spend most of the year

High-altitude amphibians
The alpine newt (below) lives in cold plantless ponds and slow-flowing streams. Its swimming larvae often fail to "grow up" in body shape although they become sexually mature and can breed while still "tadpoles"—a curious reproductive habit, widespread among tailed amphibians, known as *neoteny*. The dark coloration of the alpine salamander (bottom) is thought to be an adaptation for absorbing heat. This amphibian gives birth to live young in damp crevices; it may be found at altitudes of 3,000 metres in Europe.

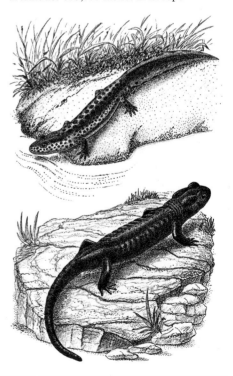

Mountain

The great French naturalist Fabre noted that walking up a mountainside seemed to him to be the equivalent of travelling from the sub-tropics to the Arctic circle. The mountain he was referring to, and the one we visited on a sunny September day, was Mt Ventoux, near Avignon in southern France. The typical "zonation" of plant and animal life is sometimes difficult to appreciate as you travel upwards, mainly because of the irregular terrain and the difference between the shady north and sunny south slopes. Towards the summit, at about 1,800 m (6,000 ft), the plants become pigmies—compact low-growing forms with strong roots and small tough leaves. This design is ideal for withstanding the rigorous high-altitude climate of cold strong winds, intense ultraviolet radiation and lack of water on the rocky scree-strewn slopes.

Scree lichens are the first step in colonization of the rocky summit. They are slow-growing, need little water and obtain nutrients from loose rock particles and wind-borne dust.

Cones, probably eaten by red squirrel. Squirrels living in montane forests are often dark-coloured, even black. The pine seeds and scales have been consumed.

Rowan Mountain ash is its other name. Conspicuous red berries attract hungry birds, especially blackbirds and thrushes, which eat the flesh and spread the seeds over a wide area.

Male cones These larch cones are ma female cones grow in pairs and all point in the same direction.

Mossy saxifrage

Juniper berries take two years to ripen; these purple ones are nearing maturity (in their first year they are bright green). The juniper on the mountain formed small compact shrubs against the rigours of the cold and wind.

Stemless carline thistle

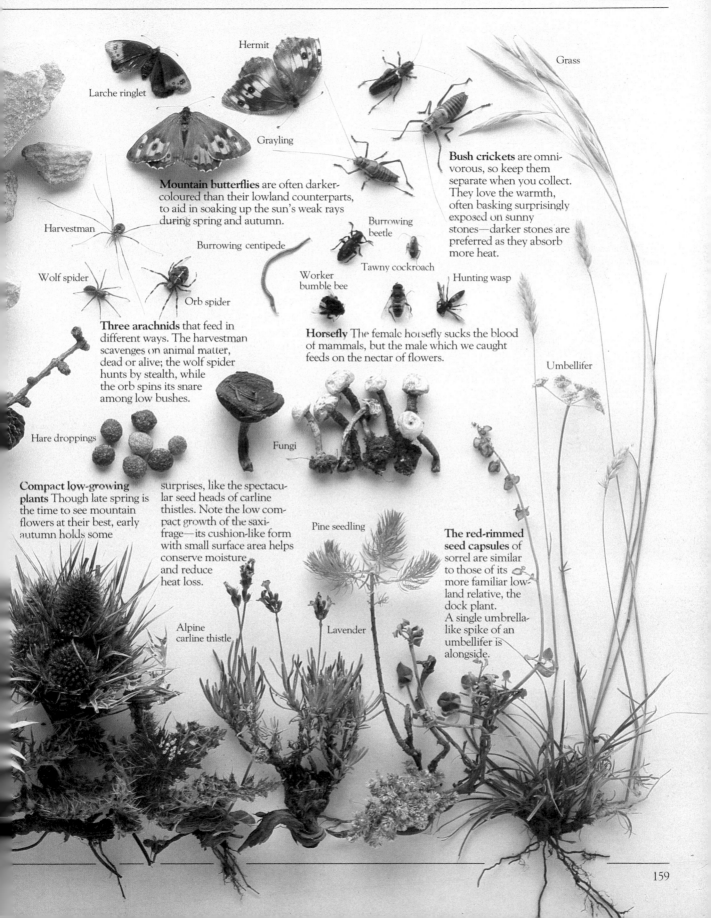

Hermit

Larche ringlet

Grass

Grayling

Mountain butterflies are often darker-coloured than their lowland counterparts, to aid in soaking up the sun's weak rays during spring and autumn.

Bush crickets are omnivorous, so keep them separate when you collect. They love the warmth, often basking surprisingly exposed on sunny stones—darker stones are preferred as they absorb more heat.

Harvestman

Burrowing centipede

Burrowing beetle

Wolf spider

Orb spider

Worker bumble bee

Tawny cockroach

Hunting wasp

Three arachnids that feed in different ways. The harvestman scavenges on animal matter, dead or alive; the wolf spider hunts by stealth, while the orb spins its snare among low bushes.

Horsefly The female horsefly sucks the blood of mammals, but the male which we caught feeds on the nectar of flowers.

Umbellifer

Hare droppings

Fungi

Compact low-growing plants Though late spring is the time to see mountain flowers at their best, early autumn holds some surprises, like the spectacular seed heads of carline thistles. Note the low compact growth of the saxifrage—its cushion-like form with small surface area helps conserve moisture and reduce heat loss.

Pine seedling

The red-rimmed seed capsules of sorrel are similar to those of its more familiar lowland relative, the dock plant. A single umbrella-like spike of an umbellifer is alongside.

Alpine carline thistle

Lavender

Golden eagle

Lammergeyer

above the treeline, among them the alpine chough (a sort of crow), various partridges and, on the Himalayas, the lovely highly vocal and swift-flying snowcocks. In the Alps of New Zealand there is even—of all unexpected things—a species of parrot which lives and breeds above the treeline. This is the large and very beautiful kea.

Some years ago, when I was in New Zealand making part of a series of television films on conservation, we found that the kea was a very controversial bird. It was universally disliked by all the sheep farmers, who accused it of not only killing lambs but also attacking full-grown sheep. Apparently the birds were supposed to tear open the backs of the sheep to get at the fat around the kidneys which, the farmers said, the keas considered a delicacy. The sheep, naturally, die after this treatment. Ornithologists, on the other hand, said that this was rubbish and that keas feast only on dead sheep or, at the most, on very sick sheep unable to get away. With this controversy going on (and still going on) it was essential that we take some film of wild keas. Now, everyone told us this would be easy. All you had to do was go up into the mountains, set up your cameras and shout "Kea, kea, kea", and the birds, being insatiably curious, would fly down to investigate. Day after day we went into the mountains, set up the cameras and shouted "Kea, kea, kea", and night after night we got back to the hotel with no film of the reluctant birds. Very depressed, we were discussing our problem one morning when we were overheard by one of the hotel's maids. Did we want to film keas, she enquired, because if so a flock of them came every morning to the back of the hotel, to be fed on kitchen scraps! So, after shouting ourselves hoarse in the mountains, we found the keas were on our doorstep. After this revelation, we finally took our film of these lovely large parrots with their sage-green plumage, brilliant scarlet-and-yellow underwings and heavy hooked beaks.

The bigger mountain birds soar and glide over the slopes, using as support the spiralling currents of air that sweep up through the valleys. Lifted and pushed to and fro by these endless wind tides the birds can, with their tremendously keen eyesight, seek out their prey or any carrion there might be. Eagles sometimes nest at the treeline but hunt above it, thus making their task of feeding their babies easier, since the food is carried downstairs, as it were, instead of up. In the Eurasian mountains the griffon vultures slide through the air of the valleys, and in the South American Andes you get the gigantic condor, which has the biggest wing "sail area" of any bird in the world. In these desolate high mountains, when the air is very still and there is no sound, you can hear the condors gliding past, their wings cutting the air with a sound that is reminiscent of the whistling of sword blades.

In Europe, the biggest and (to my mind) most handsome bird of prey is the bearded vulture or lammergeyer. Soaring on its two and a half metre wing span it is an impressive sight. Its hooked beak, fierce red eyes and darkly-bearded face give it something of the look of a 16th century pirate. One of the lammergeyer's favourite delicacies is bone marrow, and it is very adept at obtaining this. It carries a bone up to a great height and then drops it on to the rocks below, to be split open by the impact. The bird then recovers the smashed bone and scoops out the soft marrow with its tongue. This lovely vulture nests in the winter, piling a huge tangle of twigs and branches against a mountainside, and the outside of the nest is padded with great shawls of sheep's wool to protect the nestlings against the bitter weather.

Mammals of the higher slopes

Mountain mammals come in all shapes and sizes but the small ones are generally shaped like furry balls, since they lose less heat with this design. In the cold season some of these small mammals hibernate (the marmots and the ground squirrels do this) while others stay active in their warm underground dwellings (like the gophers and voles). Also in this latter group are the pikas, relatives of the rabbits but looking more like fat round soft-furred guinea pigs. These Asian and North American animals are very energetic farmers, which enables them to be active the year round. The Russian name for pikas is "hay-stackers", and this is exactly what they do. Towards the end of summer they become terribly busy at their hay-making, cutting down the grasses with their sharp teeth and arranging them carefully in small bundles to dry on the sun-warmed rocks. If it starts to rain, it is said that the pikas feverishly move their precious bundles to a protected spot, and return them to the sun when the storm is over. When their hay has been cooked well and is nice and crisp, the pikas store it carefully in a dry place as food for the winter.

A small mammal which is a native of the foothills of the Andes is the sprightly mountain chinchilla, who rather confounds the "furry ball" theory with its long ears and tail. I remember driving through Argentina until we came to the great towering wall of the Andes. Here, although it was summer, the high mountains all wore splintered caps of snow, and the air was crisp and cold even though the sun felt hot. I was here to try to film and catch the mountain chinchilla. (Its relative, the chinchilla, from which the coats are made, is a small dove-grey animal which is also

Autumn harvest
A hoary marmot prepares for winter by gathering great bundles of dried grass which it takes down into its burrow to make a substantial nest and winter food reserve. Marmots live at high altitudes and escape the worst weather by hibernating for up to eight months of the year. During the warm days of summer they spend most of the time feeding on roots and grasses, building up large reserves of fat to tide them over their long winter sleep.

Bighorn sheep from North America

European chamois

found in the Andes, but is now very rare owing to persecution for its fur.) The mountain chinchilla is an odd-looking animal with a body about the size of a rabbit, a rather rabbit-like face and powerful hind legs like those of a kangaroo. They live in colonies in the high rocky valleys, feeding on the sparse vegetation and lichens found there and bounding about through the boulders exactly like miniature kangaroos.

After a couple of days' search we found a small valley, with a tiny stream running through it, which was the home of about 20 mountain chinchillas. We made camp a little way off from the valley itself, and then set about the task of trying to film and catch our quarry. There followed the most unrewarding two weeks I have ever spent. We tried every method of filming them, and in desperation we even built a hide of rocks. After two weeks all we had was a few seconds of film showing the backside of a mountain chinchilla disappearing into a tumble of boulders. The annoying thing was that when we went into the valley without our cameras the chinchillas were quite tame. They allowed us to approach reasonably close as they bounced through the rocks, pausing to feed and sit up and clean their whiskers. Attempts to trap them were about as successful as our attempts to film them. No kind of trap and no kind of bait would tempt them, and in that terrain of great boulders, splintered rock and patches of dense undergrowth it was impossible to chivvy them into nets. Our two weeks up, we were forced to leave. We were delighted to have seen them, of course, but disappointed that we had not achieved our goal. A naturalist's life is not always a success story.

The larger mountain-dwelling mammals do not hibernate. Things like wapiti (American elk) and bighorn sheep (with their lovely curling horns that look like Roman helmets) in North America, and the various beautiful wild sheep and goats of the Eurasian and North African mountains, move up and down the slopes according to the season and the availability of food. Because their movements are dictated by the growth of plants, so the movements of predators are dictated by the drift of the ungulate prey. As the herds move up and down the mountains they are followed by wolves, bears and various cats. On the long mountain backbone that stretches from North America right down to the tip of the South American continent there hunts the feline which has the distinction of having more common names than any other cat. This is *Felis concolor*—the puma, also called in different parts of its range the cougar and the mountain lion.

One of the ungulates that doesn't deign to go through the boring procedure of trekking up and down mountains following its food supply is the so-called Rocky Mountain goat, which is not a goat at all but related to the European chamois. This stocky snow-white animal has a long lugubrious face that always makes me feel it should be wearing a gold pince-nez, for it resembles a rather ponderous Victorian vicar or man of letters. This feeling gets even stronger when you watch these animals moving about—they are incredibly sure-footed, but slow and dignified, making their solemn way through their world above the treeline as they feed on lichens and twigs. In the winter their hooves and sharp stiletto-like horns turn the glistening blue-black of a gun barrel. Against the dazzling white of a sunlit snowfield all you can see of the animal are its hooves, horns, black lips and dark soulful eyes. As usual with any animal that has a worthwhile skin, the Rocky Mountain goat was so persecuted in the 19th century for its wonderful milk-white wool, softer than cashmere, that it was nearly exterminated.

Creatures of the mountain forests

Wherever your mountain is situated in the world, the living things on top of it bear a striking similarity to each other. The lichens and cushion plants, the large gliding birds using the winds as roads to soar through the valleys, the small round furry mammals and the large agile ones are found on the tops of nearly all mountains. But as you start to descend from the higher slopes you get differences, for then the mountains take on characteristics unique to the country in which they have been born. There are still broad similarities, of course; the first belt of plants you are likely to meet at the treeline are the twisted and stunted evergreens, but exactly which trees they are depends on where you are in the world. In North America and the Alps they will be dark-green aromatic conifers, and in the Himalayas the bright-blossomed rhododendron forests. In the Mountains of the Moon that straddle the equator in Africa, the first tall trees you find may not be trees at all, but giant heathers; when walking among them it is almost as though you had suddenly become a beetle making its way through a Scottish moor.

Farther down the mountains of temperate regions you get belts of deciduous forest, but again they are quite different from one another according to the part of the world. For example, in the Chinese province of Szechwan this deciduous belt includes firs and rhododendrons, and lower down still these give place to bamboo forests. This fantastic area in the west of China is full of curious and flamboyant beasts. There is that great comic character, the black-and-white giant panda, which will perform complicated gymnastics and stand on its head with great solemnity, blissfully unaware of the fact that zoologists argue ferociously as to whether it should be classified as a bear, a racoon, or something all on its own. Oblivious to this important controversy, the giant panda plods its way through the mountains, munching up its diet of bamboos. The lesser panda also lives here; much smaller than its giant relative, it lacks the charming Winnie-the-Pooh character and looks like a cross between a racoon and a fox, with chestnut-coloured fur and ringed tail.

On the mountain ranges in the tropics, one descends from the tundra-like conditions on the top, through the evergreen belt, to what is known as the cloud forest. Here the air is continuously saturated with moisture and the trees live in a perpetual swirling mist of clouds. It is a tremendously rich area for the naturalist—the trees are overcoated with moss as thick as a carpet, festooned with bizarre *epiphytes* and decorated with tumbling waterfalls of multicoloured orchids. Giant tree ferns also live here, sprouting up like huge shimmering fountains. The plant life in general is riotous and exuberant, and so too are the birds that feast off this rich vegetation. The New World hummingbirds flash from flower to flower like handfuls of scattered opals, and in the depths of the forest the extraordinary cock-of-the-rock, as orange as a winter sun, does his remarkable display. The cock bird comes down to a special "stage" and performs his courtship dance, earnestly observed by a group of both males and females. When one male is exhausted with dancing, his place is taken by another, so that there is a continual "cabaret" of these beautiful tangerine-coloured birds.

Fabulous birds

The cloud forests that clothe the mountainsides of Central America are home to the marvellous quetzal. A pair of birds are illustrated, with the male resplendent in his elegant plumage and long tail feathers. These feathers were prized by Aztec rulers and used for decoration, the birds being kept in large aviaries and their long plumes plucked annually. Today, the quetzal is still held in great esteem; it is the national bird of Guatemala and the *quetzal* is the unit of currency in that country.

THE NATURALIST BY
PONDS AND STREAMS

The ponds and lakes of our planet are the great reservoirs of fresh water, and the rivers and streams that criss-cross the land are their veins and arteries. It is a strange fact that, of all the water in the world (the oceans, the water trapped in glaciers and the vast snow and ice ranges at each end of the earth), only a minute fraction, equivalent to one drop in a bottle of wine, is found in these inland waters—and yet they are so vital to us. Every type of terrestrial ecosystem has its essential water supply. There are the icy pools of the tundra; the deep cold lakes of the northern forests; the rushing tumbling mountain streams that, on reaching the plain, turn into wide placidly moving expanses of water; the great brown rivers that feed the rain forests; and the surprising oasis springs in the vast areas of barren desert. All these inland waters are important to us but so, in its own context, is the tiny pond in the wood near your home.

Water is one of the most important commodities in the world. Land plants search and grope for it with their roots, terrestrial and arboreal animals come to it to drink or feed, or simply to bathe or play. Some organisms cannot reproduce without it. Water exerts a tremendous influence over the whole web of life on land. In American deserts, for example, the seeds of the ironwood and smoke trees can only germinate after they have been tumbled and abraded by the flash floods disgorging down the dry river beds, the *arroyos*. On top of all this, fresh water contains its own teeming world of plants and creatures—from the beautiful floating flowers of water lilies to the giant shiny water beetles.

I think the first time I really realized what a complex world exists in water was when, in Corfu, my friend and teacher Theodore showed me a single drop taken from an old water-butt that stood outside the kitchen of our villa. As soon as I peered through the microscope, I could see an extraordinary jungle revealed. There were strange *Cyclops*, looking like misshapen shrimps and carrying great bags of scarlet eggs slung on each side of them like sacks of onions on a donkey. Crowds of *Paramecium* whizzed past, the rows of short hairs (called cilia) on their one-celled bodies working like minute oars. Things like *Daphnia*—the "water fleas"—revealed the most marvellous colours inside their transparent bodies, and the gorgeous diatoms were of such an extraordinary complexity of shapes that they looked as though a master architect had designed them. All these creatures and more were living in a single drop of water in a rain-barrel outside my house, and I had been completely unaware of it until I looked through the microscope.

A good naturalist is always prepared, so whenever you are out exploring the countryside—even if you are not planning to visit a pond or stream—be sure to take a couple of stoppered test-tubes in your pocket. The water in whatever ditch or puddle you come across, from holes in hollow trees to the leaf-choked gutters of your roof, will give

Springtime courtship
Lured at last to his watery nest, a female stickleback succumbs to the attentions of her mate. The stickleback's courtship is a complicated affair; the male attracts the female with his gaudy nuptial dress, enticing her to the entrance of the nest by a curious zig-zag dance much studied by ethologists.

A jewel of the river bank
One of the most exciting experiences for a naturalist is to come across a kingfisher (opposite) perched above some quiet stretch of river. Its plumage is a spectacular mix of shimmering hues, and as it dives for prey all the colours blend into one illuminating flash. The spear-like beak is most efficient at taking small fish like minnows and sticklebacks from just below the water surface.

AQUATIC COLLECTING

To collect aquatic plants and creatures you need some extra equipment in addition to the items on page 20. Besides the plankton net on page 190 and the nets and drag shown below, a tall narrow glass or plastic trough is also useful for examining mud or weed samples.

The drag
This is a three-pronged grapnel affair like an anchor attached to a long cord. You swing it round your head and cast it into the centre of weedy ponds or lakes, then draw the weeds towards you so that you can examine them for specimens. Always replace the weed when you have finished. A drag is easily made from three bent pieces of strong wire (from a coathanger, say) bound with softer wire.

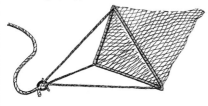

The dredge net
The strongly built dredge is pulled along the bottom to scoop up weeds, mud and the creatures that live there. The net itself is usually of tough nylon, and the lower frame is covered with a piece of hose-pipe split lengthwise for protection. Do not use the dredge in rough or rocky areas since it is bound to get jammed.

Dip net

Scoop net

Dip and scoop nets
The dip net, as its name implies, is used to capture small fish and other mid-water swimmers or floaters. The scoop net is a little longer and stronger, and can scoop up samples of mud or fine gravel from the shallows or pick up lengths of floating weed.

hours of enjoyment with the aid of a comparatively simple microscope. Even if you are no artist, the shapes and colours of these minute water-dwellers will make you want to become one.

It was in Corfu that Theodore taught me the art of investigating freshwater habitats. We would plan to leave as early as possible, and the night before I would make sure that all my equipment was in order in my special aquatic collecting box. This was copied from a design evolved by Theodore. It consisted of a metal box divided into specially-shaped compartments: some for test-tubes, some for jars, some for metal boxes with perforated lids to house such things as water beetles, snails or tangles of interesting pondweed. Then there would be the all-important dip net and the plankton net—a small net with a little bottle hanging on the end—and also a grapnel and line, the drag, for hooking and pulling in weed close to the shore. The collecting box itself had no lid, but fitted snugly into a stout canvas container with a flap that covered the top. This sheltered the container from the sun's glare but at the same time allowed air to circulate. For this sort of expedition (provided the weather is warm) you need the simplest of clothes because, almost inevitably, you end up getting wet. I used to wear a short-sleeved shirt, shorts and sandals. It is best to choose clothing that, if it does get wet, dries easily and does not hamper you with its cold and clammy weight.

Your first approach to any body of water, whether it be a quiet lake or a fast-flowing stream, should be done with caution since you never know what animal life might be visible at the water's edge. There could be a fox or a stoat or weasel drinking, or perhaps a heron fishing. You might be lucky enough to see a herd of deer or, if you live in North America, a beaver pulling a tree trunk through the water to repair its dam or its lodge. You may see a muskrat or a coypu having a light snack of water plants before going to bed. One of the inhabitants of mountain streams that I would dearly love to see, but never have, is the Pyrenean desman. This curious little mammal looks like an aquatic mole with an extraordinary long and whiffly nose, almost like an elephant's trunk. In France, where it is found, it is called the "trumpet mouse" because of its strange nose, which it uses like a snorkel when swimming.

As you get nearer the shore, look for tracks of birds and mammals on the muddy shore, and there may be burrows of sand martins or bee-eaters or kingfishers in the river bank. There was an area in Corfu where the channels of water ran like a gigantic chessboard. These were old Venetian salt pans but in my day the squares of land were filled with crops like maize and water melons. The channels themselves, full of fresh water, were an absolute paradise for a naturalist, since they contained water snails, terrapins and a host of other creatures. I always thought that it was a test of how good you were as a naturalist if you could approach the banks of these channels without disturbing the terrapins, snakes and frogs basking in the sun.

Go very slowly as you near the water itself, so that the sudden sight of you or your reflection and the vibrations from your footsteps do not frighten away the tadpoles, newts, water beetles and small fish that like to haunt the shallows. Morning and early afternoon are probably the best times for dabbling in ponds and quiet streams, since with the warmth of the sun the small creatures are much more active and there is plenty of light for you to study them. Don't forget, as you work your way along the bank, to check the stems of the vegetation for the cast skins of aquatic insects. Sometimes you might be lucky enough to find that strange creature the dragonfly larva, who looks so like a Martian monster, climbing the stem of a bulrush. If you do, be patient and watch. You will see an amazing transformation from this drab brown ugly-looking creature to a glittering blue or red dragonfly. It is difficult to believe that one could have come from the other.

At the end of our day's trip, we found it very fruitful to sit quietly on the banks of the pond or stream in the dusk and simply watch the wildlife. Various waterfowl would fly in after feeding in the fields all day, and small bats skimmed the surface for a quick drink. Gradually we saw the emergence of the night animals—otters on their way to find a succulent trout, or a badger going to bathe his masked face in the stream. On a spring evening the frogs or toads would flock to the ponds and lakes to mate and sing their magnificent twilight chorus.

A watery world

Imagine what it would be like to live in water. You would be like a bird, for as a bird is supported by the air so the water gives buoyancy to aquatic organisms. Most aquatic plants, for example, do not need stiff woody stems to keep them upright—they use the water as support, simply letting their leaves float. Currents and eddies, the "winds" of the underwater world, have their uses, too. Minute organisms need not exert themselves to get from one place to another; many just float or drift, and the delicate weak-looking cilia of the protists and the tiny bristles of little crustaceans are quite sufficient for their Lilliputian journeyings. Water also protects its inhabitants from wide swings in temperature, because it heats up and cools down much more slowly than the surrounding air.

Despite these advantages, there are of course problems with living in water. The plants and animals who like to live in one place must work out various means of anchoring themselves, so that they don't drift away. Larger animals like water beetles, fish and even otters also have a problem: moving rapidly is much more difficult in water than in air. These creatures have evolved streamlined shapes in order to help them push through the water to escape enemies and chase prey.

If you are an aquatic plant, probably the most important thing to you is the transparency of the water. Any plant must have enough light to carry out photosynthesis, and the deeper you go into water, of course, the less light you get—especially if the water is turbid. If you are an aquatic animal, on the other hand, then making sure of your oxygen supply is the most important thing. Icy water contains only one-twentieth the oxygen that air has, and warmer water contains even less. Some creatures get their oxygen by breathing air through "snorkels"—mosquito larvae, dronefly larvae and water scorpions do this. Others such as diving beetles and water boatmen have invented a rather ingenious aqualung-type method. They carry down bubbles of air pressed against their spiracles (insect breathing holes, like our nostrils), and return to the surface for more bubbles when those are exhausted. Still other creatures extract oxygen directly from the water by diffusion through thin membranes into the body. Some devices for doing this are very elaborately constructed, such as the gills of a fish and the feathery gills protruding from the tails of damselfly larvae. But this system poses

Two common amphibians

Both the common frog and the crested newt come to water to breed. Frog courtship relies on vocal croaks to reinforce the "gathering of the clans". Once a male finds his female he holds her tight and, as she lays her massive load of eggs, he fertilizes them. Compared to frogs, newts are relatively mute. The male attracts the female with his handsome livery, used together with a highly-charged vibrating tail display. Newt eggs are laid singly on long-leaved water plants neatly folded over by the female newt.

Common frog

Crested newt

Pond

The pond is now becoming a vanishing habitat, choked by human refuse and filled in by land developers. Creatures such as the once "common" frog are becoming rarer. Yet ponds are fascinating habitats, and our pond in Shropshire, visited on a warm and sunny day in July, was no exception. Look in the shallow water and you will see various creatures moving across the bottom or swimming past you: newts with their wonderful tail movements, whirligig beetles glinting like diamonds on the surface, and great jelly clouds of frog spawn or else innumerable wriggling tadpoles. Being a small volume of water, a pond warms up quickly during the summer and gives an ideal home to a variety of life. Permanent inhabitants such as water snails and beetles are joined in season by numerous larvae, such things as dragonflies and amphibians using the pond as a sort of nursery.

Food and housing The rasp-like radula (tongues) of pond snails have left etche brown trails across the upper leaf, while caddis probably ma the neat circular ho

On the sunny side of the pond you get the large yellow flowers of flag iris, whereas in the shade this plant grows only impressive sword-shaped leaves and no flowers.

Bottom dwellers The large dragonfly and beetle larvae are voracious predators and can easily tackle a stickleback. The caddis is altogether more re- tiring, feeding on pond debris and hiding in a tiny case.

Yellow flag iris

Marsh bedstraw

Lesser water parsnip

Dragonfly larva

Caddisfly larva

Water milfoil

Diving beetle larva

Horse leech

Golden-ringed dragonfly

Soft rush

Water horsetail

Branched bur-reed

Aerial hunters
Dragonflies are very swift and can outfly almost any other insect prey. They and their daintier cousins the damselflies use their huge eyes to locate quarry.

Large red damselfly

Damselfly larva

Empty damselfly larva case on reed

Pond snails
The great pond snail needs a well-established pond with plenty of aquatic vegetation, but the ramshorn can tolerate stagnant as well as running water.

Ramshorn snail

Great pond snail

Stickleback

Metamorphosis
Within a couple of weeks the just-legged tadpole will be more like the froglet, ready to make its debut on dry land.

Rushes and reeds
Branched bur-reed has broadsword leaves, small yellow-brown male flowers and large female flowers that are just swelling into spiked burred fruits.

Common frog tadpole

Water milfoil

Common frog froglet

Young smooth newt

Common rush

Eggs of great pond snail

a problem, in that the water flowing over the membrane is absorbed too, and dilutes the body fluids. So, extraordinarily enough, each of these freshwater creatures must have some method of getting rid of the excess water. Protists such as *Amoeba* and *Paramecium* use a very simple method, and one that is great fun to watch under a microscope. Water collects in a bag called a vacuole, which you can see getting bigger and bigger until suddenly the creature contracts it and the water squirts out of its body. Kidneys do the job in the case of the aquatic vertebrates like the fish, for example, and the amphibians who "breathe" water through their skins when submerged (as well as using their lungs in air).

The surface of the water presents a whole different world. Here, if you are light enough, you can actually walk on the "skin" of the water. This skin is formed by surface tension, which means that the molecules of water are more attracted to each other than to the air above them. This, in effect, forms a ceiling to the pond. The "magic trick" of floating a needle on water has been performed by pond skaters and water crickets long before magicians invented it. Surface tension also aids creatures such as water boatmen, who zoom up from the depths of the pond and hang from this ceiling, their tails protruding through it to gather air.

As you can see, a watery existence is just as complex as living on land—you must be concerned with moving about or anchoring yourself, getting enough light, and procuring oxygen and food. All these facts of life are influenced by where you live—in still water or flowing water, in a warm tropical lake or a cold pool of northern latitudes, in the shallows or out in the depths.

The rich life of the shallows

When I was a young naturalist in Corfu, I spent a lot of my time wading about in ponds or dabbling in the shallow coves of small lakes. The shallow water of a pond or the rim of a lake (called the littoral zone) is by far the richest freshwater habitat in terms of numbers of species. Rooted here are aquatic plants with completely submerged leaves, reaching out in long filaments like green hair or with branched green filigree work. This bushy construction means that the leaves are bathed in a great volume of water, which aids in the exchange of oxygen and carbon dioxide. Some move gracefully like green ballet dancers in the currents of water, while others like the stonewort are stiff with a lime crust. The plants rooted a little closer to the shore have their leaves floating on top of the water, spread out in discs to catch the sun, as in some of the water lilies and pondweeds.

Some aquatic plants are really rather astonishing, since they have three different kinds of leaves. The arrowhead, for example, has very grassy-looking underwater leaves, flat oval floating leaves and also arrow-shaped aerial leaves pointing skywards. Of course, a plant like this can be a useful home to a variety of animals, for it offers so many different apartments, as it were, in which to live. Aerial leaves are convenient resting places for flying adult insects who have spent their early life as aquatic larvae—such insects as caddisflies, china mark moths and dragonflies. The dragonfly, a voracious winged hunter, is probably one of the most alert insects you can find near water, as it sits on its hunting platform of reed or leaf, watching for prey. Its fast and powerful flight and extraordinary eyesight allow it to hunt down almost any other winged insect, and also to

A three-tiered aquatic plant
Arrowhead likes rooting in deep rich mud in ponds and by the sides of canals and slow-flowing rivers. The submerged strap-like leaves are the first to appear in the spring, followed by the floating oval ones and finally by characteristic arrow-shaped aerial leaves. The plant bears both male and female flowers, the female ones lacking petals.

recognize intruders (for each dragonfly jealously guards its own territory) and tell them apart from a suitable mate. You will probably be lucky enough to see two dragonflies flying "in tandem". This is a mating chain, part of their ritual of reproduction.

Baby frogs who have just got their legs and have absorbed their tadpole tails into their bodies find aerial or floating leaves comfortable resting places. On the stems just above the water you might find the strange mud cocoons of the whirligig beetles and the silken cocoons of the brown china mark moths, which are covered with pieces of leaf.

The larvae of the china mark moth are, of course, caterpillars. Now one might think that under the water is a strange place to find a caterpillar, but this curious creature is very well adapted to live there. Underneath a floating leaf the larva dwells inside a sort of sandwich that it has made for itself, by cutting out a piece of leaf a couple of centimetres long and attaching this to the underside of the leaf with silk. Inside its envelope, the larva is perfectly dry and is thought to breathe the oxygen given off by its leaf case during the process of photosynthesis. China mark caterpillars feed on leaves, as other caterpillars do, but they have to get their meal by poking their heads out of their little water-tight cases. Their bodies are so hairy that, when they do this, the hairs prevent the oxygen from escaping and water from entering the case—a really remarkable adaptation.

Many other creatures inhabit the upside-down world under a floating leaf. You may well come across minute bryozoans, the "moss animals" as they were once called. Although colonies of some bryozoans bear a close resemblance to patches of moss, they are in fact true animals who can move. They travel as a colony, however, and it is unlikely that they would achieve any Olympic records since they only move about two centimetres in an hour. They feed by waving the tiny cilia on their tentacles to waft floating organisms into their mouths. If you want to retrieve a hat or something similar that has fallen into the water, you will know that by scooping water towards you with your hand you can create currents to bring the hat within reach. This is exactly the principle these simple animals use to get their food.

Another small and simple creature living on submerged leaves is the translucent thready *Hydra*, looking like an elongated sea anemone. A hydra is a fascinating thing to collect and keep in your aquarium or even in a large jam-jar. In common with its relatives, the sea anemone and the jellyfish, it has in its tentacles many stinging cells called nematocysts, which paralyze its prey. Hydra are rather difficult to see when you first lift a leaf out of the water, since they collapse into little blobs of jelly, but put the leaf into a jam-jar and if it contains any then you will see, after a few hours, thread-like creatures a centimetre or so long, each with a wild hairstyle of tentacles. Their bases attached to the glass, the hydra will be waving their wigs in the hope of catching water fleas or tiny worms.

From the underside of leaves in spring and summer you can collect the gelatinous spawn-like masses of pond snail eggs, looking like miniature sago puddings, and the very handsome ramshorn snails. These are particularly interesting because they are really descended from land snails and so they still breathe by lungs, not by gills like the truly aquatic snails. Sometimes you can see a ramshorn float to the surface of a pond "foot upwards" in order to take a breath of air. But if it becomes aware of your

Diving belle
You may be lucky enough to catch a water spider in a dip net or with a drag. Try to capture it and place it in an aquarium (with a tight glass cover) to watch how it constructs a silken home filled with air bubbles carried down from the surface. When hunting, the spider carries its own oxygen supply in the form of tiny air bubbles trapped by its covering of body hairs.

presence, it will not stay near the surface for long. It will suddenly expel the air and plummet like a stone to the bottom of the pond.

Under some floating leaves, you are bound to find little dark lumps that are the predatory planarians. These deadly flatworms slide along, devouring snail spawn and any other animal matter they come across. They are active mainly at night, gliding about feeding with a tubular pharynx which protrudes from underneath the body. This apparatus is so powerful that the animal can even suck out little bits from a living prey—so you can see that, if you were a snail or something similar, a planarian is the last creature you would want to meet on a dark night. If you want to collect these curious and interesting creatures, hang a little piece of meat on a thread in the shallow water. By the next morning you should find quite a little collection of flatworms feasting there.

Lily pads, of course, are favourite haunts for many aquatic creatures, and birds like moorhens and coots can frequently be seen turning them over in order to glean the harvest of snails and other small animals clinging beneath. But most of the animals on the undersides of floating leaves also live on submerged leaves and stems. This is where your hook-and-line drag comes in handy, though it can be a frustrating business when it gets itself jammed on rocks or roots. Dragging in a tangle of water plants can be just as rewarding as lifting leaves; in a quiet clear lake, for example, you may pull in long fragile filaments of green algae or some colonies of the so-called "river sponge" encrusting a plant stem. This sponge, misnamed by an unobservant naturalist since it actually lives in lakes, may in turn be embedded with the spiky larvae of sponge flies. You may also drag in various leeches who have been waiting in the weed for rides from fish, frogs or snails, and you might get lively liquorice-coloured water beetles and bright-red water mites. Remember

ON AND IN THE MUD

Rotting vegetation and decomposing animals provide a rich supply of food for pond animals. Caddis larvae trundle along the bottom sorting out settled organic debris, and freshwater cockles filter tiny food particles from the water siphoned into and out of their bodies. Food poses no problem, but breathing can be difficult in the oxygen-poor water. *Tubifex* bloodworms cope by having their thin-walled bodies packed with the pigment haemoglobin, which picks up oxygen. The worms project from the mud and wave their bodies in sinuous writhing; in large numbers they turn the bottom red. The larva of the phantom cranefly goes one better—it lives below the surface of the mud but its elongated tail sticks up like a submarine periscope above the water's surface, where it sucks in the air it needs to breathe.

Freshwater cockle's fleshy foot anchors it in the mud

Tubifex bloodworms protrude from their sunken tubes

Phryganea caddisfly larva scours the muddy bottom for food

Phantom cranefly larva lies buried just below the surface

that some leeches—among them the horse leeches—do not suck the blood of their victims, but swallow their prey whole.

Creatures at the bottom of the pond

Among the stones, mud and decaying vegetation at the bottom of the pond live various kinds of flatworms and leeches, but this is really a special area where you find the immature forms of various insects. If you sit at the edge of a pond or lake and study a small area in front of you, once your eyes have become accustomed to looking through the water you will see various creatures crawling about the bottom. Caddisfly larvae creep about in their silken cases, decorated (as disguise or protection) with sticks, stones, sections of leaves or grains of sand. Also on the bottom are the nymphs (larvae) of dragonflies and damselflies—probably the fiercest predators in shallow waters. In fact, one naturalist said that if a dragonfly nymph were the size of an average red setter, it would be one of the most ferocious and dangerous animals on earth. Watch one waiting patiently on the bottom of the pond, and you can see what he meant. A mayfly larva or a tadpole, or even a small fish, might pass by and immediately a monstrous hinged grappling iron (like those huge grabs you see on building sites) is extended from underneath the nymph. It shoots out, the big claws on the end close like fingers round the prey and the nymph drags it back, to be sucked dry.

You can use a net to collect the bottom-dwellers, but if you use your dredge or scoop you will also catch a range of microscopic creatures that glide over the surface of the mud. In this tiny world are things like *Amoeba* and its relative *Arcella*, which makes a shell so that it looks rather like a miniature moving bowler hat. You will also dredge up creatures which actually live in the muddy ooze, like orb- and pea-shell cockles, larvae of various flies and midges, and worms that are related to the terrestrial earthworms. Tip your catch into a flat white tray with some water in it, leave it in the shade and after a while you will be able to see how the creatures of the ooze live at the interface of water and mud.

The mud layer is the winter resting place for a number of plants and animals. You may find various insect larvae and pupae, hibernating snails, diving beetles and whirligig beetles, and the "winter buds" of water plants. Using your scoop during the winter months can be as productive as during the spring and summer.

Out in the open water

The numbers of drifting organisms in a fairly large pond or a lake are enormous. There are plants such as duckweed and frogbit that coat the surface like a lawn, and then there are the submerged plants like the extraordinary carnivorous bladderwort and the hornwort, whose entire life cycle, including flowering and pollination, occurs under the water. But the majority of the drifters are minute. They form the plankton— a strange cocktail-party collection of miniscule plants, animals and other organisms such as protists.

Diatoms, desmids and green algae—the phytoplankton—live (as their name implies) by photosynthesis. Anywhere that sunlight penetrates the water, they can exist. In the middle of a big lake they form the basis of the food web, because few larger plants grow out in open waters. On the animal side *Paramecium*, rotifers and minute crustaceans like *Daphnia*

Predatory diving beetle
Both the adult (above) and the larva (top) of the lesser diving beetle share similar habits. They are voracious pond predators, tackling animals which are often much larger than themselves. The adult is shown starting on a meal of young tadpole, while the larva is just finishing off a small fish. Both creatures breathe at the surface—the spiracles (breathing holes) of the adult are situated at the tip of the abdomen, which is thrust out of the water as the beetle reaches the surface, while the larva hangs with its two tail filaments (which contain spiracles) adhering to the surface film.

Fish of still or sluggish waters
The hunched shoulders and deep flat-sided body of the bream are unmistakable. Its narrow shape enables it to swim through closely-spaced water stems. When feeding the mouth is extended into a downward-pointing tube which sucks up mud for molluscs and worms, or alternatively blows water to stir up the soft sediment. The pouty-lipped carp mumbles at the bottom for mud-dwelling prey. It is of a much heavier build than the bream, attaining great size and age. Carp like warmish water with plenty of weeds and have adapted to the river outlets of power stations. The pike is a powerful predator, built like an arrow for sudden bursts of speed. The eyes possess good binocular vision and the teeth are backward-pointing to secure prey.

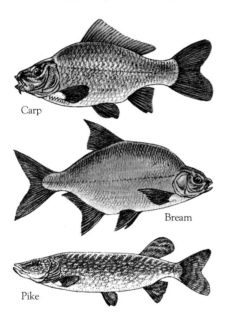

Carp

Bream

Pike

and *Cyclops* make up the zooplankton. These feed on the phytoplankton (and, when they get the chance, on each other) to take up the next strands in the web. Being so tiny, most zooplankton are at the mercy of water currents, but some do swim very weakly and where the water is still they can move slowly up and down to whatever layer happens to suit them.

In ponds and lakes the creatures that are strong swimmers are, for the most part, predators. There are many voracious and swift-moving bugs and beetles—backswimmers and water boatmen, giant water bugs with their mantis-like clasping arms and their hypodermic-like sucking mouth-parts, whirligigs and big shiny diving beetles. Beware when you catch any of these insects because they can give you quite a nip. They use their powerful jaws to feed on practically any underwater animal they can overpower. A giant water bug, for example, can catch and eat a fish twice its own size. The beetles and bugs, although they are rather scrunchy to eat, are of course preyed upon in their turn by large fish and the herons and ducks who feed in the water near the banks.

The amazing thing about fish is that, like bats and bees and many other animals, they have senses which are not shared by us humans, and so they live in a world that is difficult for us to imagine. Fish have the extraordinary ability to feel the vibrations in the water, and this is tremendously useful to them. Vibrations are made by prey and predators, by the swimming movements of their partners if they are fish that live in schools, and even as water moves around a stone or plant or as a clumsy naturalist treads too heavily on the bank. The extraordinary sense organ is the lateral line, seen as a thin dark line along each side of the fish. It is really a sort of groove with little pores, each pore holding tiny jelly-coated knobs of hairs which bend to and fro with the vibrations in the water. How marvellous it must be not only to have eyes that can show you what is going on around you, but to be able to feel what is going on around you at the same time, like the sensation of the wind on your body or fingers touching you.

Fish rely on their eyes as well as their lateral lines, and many males put on resplendent colourings in the breeding season to bedazzle the females. From March to July the male stickleback, for instance, swaps his normally subdued colouring for a brilliant reddy-orange underside, vivid green back and electric-blue eyes. Certain fish can see clearly above water as well as below, like the archer fish from Asia. This odd fish gets its food by going up to the surface of the water, spotting a fly or a grasshopper on a leaf above and then shooting it down with a machine-gun fire of water droplets. I once had one of these fish spit into my eye, and I can vouch for the strength and accuracy of their archery.

Coping with the currents
The most important thing for life in streams or rivers is, of course, that the water is flowing. Moving water has a low but even temperature and holds much more oxygen than still water. In fact, the great majority of stream organisms could not survive in the quiet warm backwaters of a pond. The problem with a fast-moving stream is that drifting organisms would be smashed against the rocks or dragged along the stony bed. The current washes the bed of a stream clean, leaving only stones or pebbles in which aquatic plants, even if they could cope with the current, would find it difficult to take root. The only plant life is diatoms, green algae,

and perhaps some water mosses and liverworts clinging to the rocks and making them crusty or slimy. The result is that the main base of the stream's food web is not drifting phytoplankton or big aquatic plants, as in lakes and ponds. The basic food supply is imported from upstream—dead leaves and branches from trees, remains of dead animals and rain run-off from rich soils of the forest.

As the leaves tumble down with the current they are attacked by bacteria and fungi, shredded by caddisfly and stonefly larvae and collected and eaten by the net-spinning caddis larvae and the larvae of blackflies and mayflies. A few other small creatures live in the turbulent waters. These include the "water pennies" (larvae of the riffle beetle) who eke out their existence by scraping the algae from rocks, much as a cow browses in a meadow.

Most of the creatures that live in fast-running waters are cleverly adapted for their environment. Some of them are flattened, which allows them to crawl into sheltering crevices. Stonefly and mayfly nymphs and their predators, the flatworms, all have sturdy flattened profiles. Some caddisflies glue their cases to stones, or weight them with outriggers of pebbles. One of them—the net-spinning caddis—has no case but holds on with tiny hooks, waiting for foodstuff to get washed into its silken net. Blackfly larvae have not only hooks but a safety rope of silken thread as well. These strange insects have a fascinating method of emerging from the stream when the time comes to change into an adult. The pupa takes in oxygen from the water, but enough not only to allow it to breathe but also to form a bubble between its body and its case. When the adult insect hatches it rides like a little silver satellite up to the surface of the water, where the bubble bursts and there is the insect, wings dry and ready for take-off.

The main predators in fast-flowing streams are the fish. The muscular streamlined brook and brown trout feed on insect larvae in the spring and switch their diet in summer to insects that fall into the water. A smaller hunter is the little miller's thumb, who lives under stones and feeds on insect larvae and trout spawn and fry. But probably the most curious underwater predator is the dipper, a thrush-like bird. Dippers can close their nostrils and "fly" under the water, and even walk along the bottom in pursuit of insects and tiny fish.

From stream to river

As a tumbling watercourse finds its way to level ground, the current slows and allows some plants to root between the rocks. Water crowfoot, with its long leaves, and watercress can often be seen trailing their shoots downstream. Farther along, the current slows even more and begins to deposit the silt carried down from the hills. Many aquatic plants can grow in this new mud, and they provide a variety of creatures with homes. Plankton starts to appear and parts of the stream community now take on the characteristics of a pond or lake. Even so, many river organisms require well-aerated water and you find creatures who could not survive in absolutely still water. Some of the operculate snails (so named because they have a permanent door, the operculum, hinged to their shells) breathe by gills, unlike the pond snails with their lungs. Fish such as grayling, barbel and bitterling, and crustaceans such as freshwater shrimps and crayfish, also require relatively oxygenated water. Now, you

Creatures of clear fast-flowing waters
The torpedo-shaped dace is a swift shoaling fish of streams, preferring clean gravelly shallows. Dace dart actively after any small insect or worm taken with the current. The brown trout is a magnificent fish with beautiful speckled markings which make it difficult to see as it hovers over the river bed in dappled sunlight. The crayfish is unusual in being our only large freshwater crustacean. It is a nocturnal scavenger that prefers streams which run over chalk and limestone.

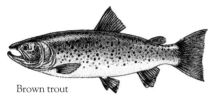
Brown trout

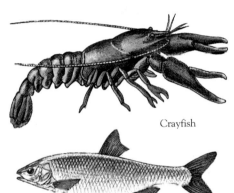
Crayfish

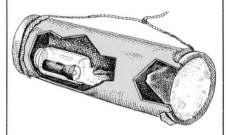
Dace

AQUATIC LIGHT TRAP

Some aquatic creatures are, like moths, drawn to light. You can capitalize on this by making a light trap and placing it in a stream until late evening, then hauling it up to examine the catch.

Making the trap
Turn on a small torch, seal it in a water-proof jar and place it in a length of old earthen pipe. Fashion a cover for one end and a funnel for the other out of fine-mesh wire net. Lower the pipe by means of a strong cord, and position with the funnel pointing upstream.

Fast stream

Our collecting visit to a south Devon brook on a
showery April morning did not seem promising.
The swift-running stream looked deceptively empty,
for when we began to search carefully we discovered
that the stream and its banks were teeming with life.
Due to the swiftness and turbulence of the current, the
stream offers highly-oxygenated water and a constant—
if low—temperature. Reeds, ferns, mosses and other
damp-loving plants grow in profusion on the banks,
trapping silt and debris swept downstream and offering
sanctuary from the current to the animal life.

Understone inhabitants It takes a
sharp eye to spot four caddisfly
larval cases constructed from tiny
gravel grains, and a solitary
freshwater limpet.

Hemlock water dropwort

Isoperla stonefly larva

Golden-ringed
dragonfly larva

Net spun by
caddis larva

Freshwater
shrimps

Cranefly

Caddisfly larvae
in their cases

Opposite-leaved golden saxifrage

Colony on a rotting stump A
streamside rotting log cloaked
with golden saxifrage, moss and
liverwort. Though closely related
to mosses, liverworts are much
more flattened and the body or
"thallus" has broad green lobed
leaves. Both mosses and liverworts are primitive plants and do
not produce flowers but spores,
which will only germinate if they
land in a damp humid place.

Moss

Liverwort

Under the surface The dark mass strung
between hemlock water dropwort leaves is a
food net, woven by a caddisfly larva to trap
current-borne debris.

Mosses need damp
conditions; this cypress-
leaved moss is clinging
tenuously to a streamside
rock, receiving its moisture
from the occasional spray
or splash.

Opposite-leaved golden saxifrage

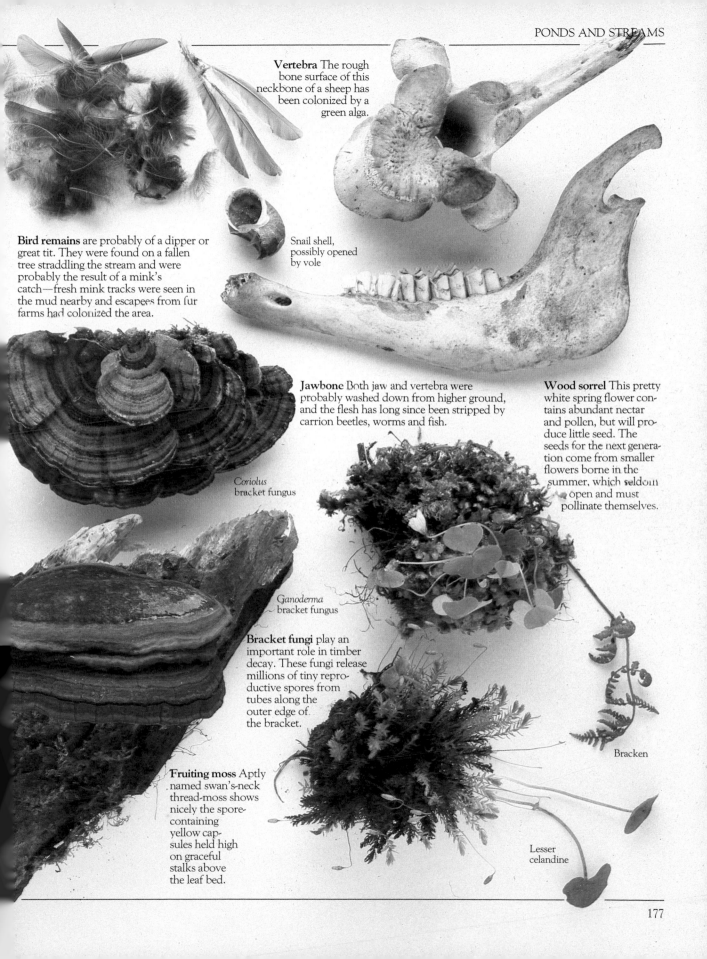

Vertebra The rough bone surface of this neckbone of a sheep has been colonized by a green alga.

Snail shell, possibly opened by vole

Bird remains are probably of a dipper or great tit. They were found on a fallen tree straddling the stream and were probably the result of a mink's catch—fresh mink tracks were seen in the mud nearby and escapees from fur farms had colonized the area.

Jawbone Both jaw and vertebra were probably washed down from higher ground, and the flesh has long since been stripped by carrion beetles, worms and fish.

Wood sorrel This pretty white spring flower contains abundant nectar and pollen, but will produce little seed. The seeds for the next generation come from smaller flowers borne in the summer, which seldom open and must pollinate themselves.

Coriolus bracket fungus

Ganoderma bracket fungus

Bracket fungi play an important role in timber decay. These fungi release millions of tiny reproductive spores from tubes along the outer edge of the bracket.

Bracken

Fruiting moss Aptly named swan's-neck thread-moss shows nicely the spore-containing yellow capsules held high on graceful stalks above the leaf bed.

Lesser celandine

Streamlined for swimming
An otter slithers down a river bank into the water. As soon as it submerges the long guard-hairs of its coat become wetted and lie flat over the soft underfur. When swimming the whole body moves in one fluid graceful movement, the short powerful limbs sculling the water with their webbed feet while the flattened tail serves as a rudder. Otters hunt from dusk to dawn. Eels and trout are favourite fare, and crayfish are taken in large numbers. Though you are unlikely to come across an otter in the wild, as they tend to be secretive, look out for their cigar-shaped droppings ("spraints") deposited at regular spots on the river bank and serving as terri-torial markers.

have to be really fast in order to catch a crayfish. These crustaceans see extraordinarily well with their stalked eyes that can swivel to watch what is approaching from behind as well as in front. When danger threatens, their evasive tactics are excellent. The fan-shaped tail snaps closed on the abdomen and, jet-propelled, they shoot through the water backwards. It is best to search for crayfish at night, when they are out prowling along the stream bed, hunting for small animals and carrion and stopping off to browse on water plants. A piece of rather smelly meat anchored in shallow water will be sure to attract them, but remember that if you do catch and keep them in an aquarium, these animals will need plenty of aeration.

Creatures who require plenty of oxygen will not be found in polluted streams, because any sort of pollution—whether organic (from sewage dumping, for example) or thermal (from the cooling towers of power stations)—decreases the amount of oxygen in the water. I know of one stream that runs through the garden of some American friends of mine. Ten years ago, this brook was full of crayfish, terrapins and a host of other life. With the spread of industry higher up the valley, it is now a completely dead piece of water. This, unfortunately, is the profligate way we are treating one of the most important commodities in the world.

The interdependence of land and water

One of the most exciting rivers I have ever been on is the Essequibo, in northern Guyana. This wide deep-brown river meanders slowly between a large area of grassland on one side and thick tropical forests on the other, giving the best of both worlds. I was staying with a man called Tiny McTurk, who had been born in the area and lived there all his life. He knew the river and its banks intimately, and I could not have wished for a better guide.

The first day of our stay, Tiny took us down the river. We travelled between giant trees festooned in creepers and with huge bunches of multicoloured orchids hanging from them. We disturbed storks and herons and the black-and-white scissor-bill, whose scarlet beak has the bottom mandible longer than the top; as it skims the water (looking as though it were cutting cloth) the lower mandible scoops up small fish and water insects for food.

After a time, we rounded a bend in the river and came upon a mammoth tree decked out in a profusion of pink and yellow blooms. McTurk cut the outboard engine and our big canoe drifted towards the bank. As the echo of the engine died away, McTurk grinned and told us to listen to the tree singing. Sure enough, we heard a great vibrant humming like a giant dynamo. At first I thought it was bees, but peering up into the tree with my binoculars I saw that the noise came from hundreds of hummingbirds. They hovered there, feeding on the flowers, and the noise was the sound of their wings. It was a breathtaking sight to see this myriad of tiny birds gleaming like gems in the sun as they flipped from flower to flower. Now and then they engaged in fights that shook the flower petals free, and the petals drifted down to carpet the brown waters. Some were eaten by fish, others sank to the bottom to form the river humus downstream. It was a wonderful example of how a river works: the waters give sustenance to a tree which, in its turn, provides flowers and fruit and shelter to goodness knows how many animals; the tree also drops petals and fruit into the river, to be carried downstream where they fertilize other parts of the forest; and the river spreads the seeds of the tree, to ensure dispersal of future generations.

Presently, we turned off the main river into a wide placid tributary more like a canal. Here the surface of the water was covered with the leaves and flowers of the giant Victoria water lily, each pink flower as big as a wedding bouquet and each green leaf the size of a car wheel. McTurk warned us not to put our hands into the water because of the infamous piranhas. These particular piranhas were large egg-shaped silvery-white fish, each with a fearsome bulldog mouth full of razor-sharp teeth. They hunted in schools, and when one caught a prey, the unfortunate victim was literally torn to pieces by the rest of the school before it could escape. It has been said that a gang of these ferocious fish can reduce the body of a capybara (a guinea-pig-like animal as big as a St Bernard) to a skeleton in a couple of minutes. Needless to say, we kept our hands in the canoe.

Piranhas, of course, have got a very bad reputation, but in the main this is unjust. Like most animals that people love to hate, their ferocity is not the whole story. A wonderful ecological study was done recently in the Amazon River basin, and it has shown that some piranhas are gentle vegetarians. Besides exonerating these fish, the study revealed some fascinating results. The Amazon and its tributaries contain probably one-quarter of the world's fresh water, and the main river pours the amazing amount of 200 million litres of water into the Atlantic *every second*. The study shows clearly the interdependence between the water and the land through which it flows. Great areas of this basin are flooded for most of the year and the fish can swim out of the river beds and through the forest, gathering under trees to feast on the fruit and seeds as they fall into the water. Delicious rubber tree seeds explode noisily from their pods like little rifle shots and then fall into the water, to be gobbled up immediately by the fish. The naturalist who studied the fish said that you can actually hear, in this strange water forest, the constant sound of "pop, plop, gulp . . . pop, plop, gulp".

Many significant points emerged from the Amazon study. Firstly, the fish who graze in the forest help to bring extra fertilizer to the river in the form of their rich faeces—much as African hippos graze away from the water and fertilize it on their return—and the big caiman (South America's answer to the crocodile) do the same when they hunt land mammals. Next, as the fish swim across the flooded plain and up and down smaller streams, they may act as gardeners, passing seeds through their bodies and thus "planting" them over a wide area. The study also showed that people who live in the Amazon basin rely to a very large extent on fish for protein, and three out of every four fish they eat have themselves been nourished on the seeds and fruits of the forest. By allowing the wanton destruction of the Amazon forests, the governments of South America may indirectly cause their people to starve.

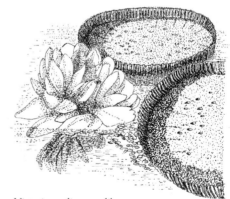

Victoria regalia water lily

The familiar phrase "Don't miss seeing the wood for the trees" is a good dictum for the naturalist, who must understand how an ecosystem works as an integrated whole. A friend of mine, a fine naturalist, heard the story of the interdependence of the land and water ecosystems in the Amazon basin and remarked, "This warns us not to miss seeing the fish for the trees".

Black skimmer

THE NATURALIST IN
MARSHLANDS

The marshes of the world and their related habitats—swamps, bogs and fens—are very curious halfway houses, extraordinary amalgams of land and water. One could call them fresh wetlands, to distinguish them from the coastal wetlands like estuaries and salt marshes where the influx of sea water has a major influence on the types of wildlife that live there.

Most fresh wetlands harbour an extraordinary variety of life, because after all they offer the best of both worlds to plants and animals— plenty of water and plenty of sunshine. Of course, there are some exceptions, like the thickly-wooded swamps of North America. These can be very dark, with the huge cypress trees, tupelos and swamp black gum trees standing with boughs interlocked to form a shady canopy.

I remember when I visited South Carolina, my friends there discovered I had never been in a cypress swamp. Straight away we climbed into canoes and set off, literally boating through the forest. There were many beautiful birds but most were difficult to distinguish in the high furry tops of the cypress, and I was not sufficiently versed in their cries to be able to identify them by sound. What were also beautiful, but at the same time deadly, were the cottonmouth moccasins, one of the most venomous snakes of North America. These reptiles lay coiled up on the dead trunks of the cypress trees scattered higgledy-piggledy throughout the swamp. They would stay quite still, watching us unblinkingly, and then as we got too close they would uncoil, slide into the brown waters and disappear. I was fascinated by their beauty and grace, but I began to wonder if they were quite such comfortable things to have around when we suddenly found our way blocked by several great fallen cypress bodies. There was nothing for it but to get out of the canoe and manhandle it over the trunks. It was a creepy feeling standing waist-deep in dark water into which had just plopped several huge and irritable-looking cottonmouths. Fortunately, nothing happened and we continued our wonderful trip through the cypress swamp.

Swamps like those in North America and the tropics probably hold any number of secrets for the naturalist, since very few of them have been properly studied. Quite recently, for example, a major discovery was made in the Great Dismal Swamp in Virginia. Its evocative name sums up the terrain, and it was in this swamp that runaway slaves used to hide at the time of the American Civil War. Just recently, when some scientists were doing a mammal survey of the swamp, they rediscovered the Southern bog lemming. This charming little rodent was found to be alive and well, though thought to be extinct since the turn of the century.

Swamp to forest—the process of succession

Fresh wetlands are usually associated with the edges of bodies of water, such as the shallow rims of lakes or ponds, or where wide rivers

White for danger
A threatened water moccasin stands its ground and opens its mouth to expose venomous fangs and the white lining that gives it the name "cottonmouth". This snake spends many sluggish hours lounging about in the North American swamps, but when hunting it is galvanized into action and can chase, catch and swallow fish under water.

A change of life style
A southern aeshna dragonfly (opposite) makes its dramatic entry on to land after spending two years under water. When fully grown the aquatic larva (called a nymph) crawls up a stem and, after a short breather, struggles out of its skin. The small feeble wings gradually expand as blood is pumped to them, and within an hour the insect zooms off in a flash of blue-green iridescence.

occasionally flood their banks. Industrious beavers can create wetlands by damming up streams, causing the water to halt in its course and spill back over the forest. But most wetlands are stages in the slow progression of nature which is called *ecological succession*. This means that one community prepares the ground for another, different sort of community, which gradually takes over. By the process of succession, a lake can eventually turn into a forest. It works like this: the lake slowly becomes clogged with plant and animal debris and turns into a swamp; this gradually turns into a marsh; the marsh in turn progresses to a fen or bog or wet meadow; and eventually this is succeeded by woodland. The last stage of this process is described on page 41, where an abandoned hay meadow, for example, slowly turns back to original forest. Of course, a forest is not necessarily the end result. The final habitat depends on, among other things, temperature and rainfall; be it woodland, desert or scrub, though, the final stage is called the *climax* community. As a slight variation on this straight-line or *linear* succession, the story of the Northern pocket gopher, the gardener of alpine meadows (see page 156), is an example of *cyclic* succession. Whatever kind of ecological succession it is, the speed of the changes can be so slow as to be almost imperceptible and each stage may take hundreds of years to merge into the next.

The swamp life

When they die, the aquatic plants that grow in a lake add their bodies to the bottom, enriching the water and gradually decreasing the depth. As the lake becomes shallower, emergent vegetation begins to grow very well. The stiff but flexible bodies of the reeds, sedges and other plants rising above the water are able to bend with the wind, while their roots or rhizomes (underground stems) spread out like hands and anchor them in the mud. The rhizomes of many reeds and bulrushes, being stems, produce buds and so there is a tangled web of rhizomes supporting a dense crowd of plants. Here, in what is now called a swamp, aquatic worms busily plough the soft slushy mud and water snails creep above the surface to graze on the stems and leaves of the plants. The swamp is a nursery for baby fish and tadpoles and there is an abundance of mayfly and dragonfly nymphs, although the individual species you find here will differ from the ones found in lake waters. The ducks, heads down and tails up, strain mud through the finely serrated edges of their bills in order to glean food, and the reeds are felled for both food and bedding by things like coypu and muskrats. A rich swamp is a most important breeding area for waterfowl such as swans and other water-loving birds like the herons and rails, who can hide their nest platforms in among the expanse of rustling reeds.

One of the strangest bird babies I have ever encountered is that of the hoatzin, a crow-sized bird which lives in South American swamps. The nestling has remarkably well-curved claws at the front of its wings, which are in fact the "remnants" of its thumb and first finger. This set-up is very like the one possessed by the prehistoric *Archaeopteryx*, the first bird so far discovered in the fossil record, which evolved from a reptile. Indeed the baby hoatzin, when it hatches, behaves more like a reptile than a bird. As soon as it is dry it can scramble about the branches near the nest, using its hooked wings to climb with. Not only that, but within a few

Natural dispersal
As a water rail pads about between clumps of reeds, it unwittingly acts as a means of transport for mites, snails and the seeds of water plants which stick to its feet. In this way both plants and animals are transported from one patch of swampy ground to another. The distances are magnified during the bird's migration, when a water mite may suddenly find itself transported to a new swamp many hundreds of kilometres away.

minutes of hatching, should danger threaten, the baby will drop out of the tree into the water below and submerge itself until it is safe to reappear. This ability to climb like a reptile and swim like a fish is a very extraordinary adaptation, and when I watched these strange birds in the swamps of Guyana I felt as though I were watching the prehistoric ancestor of all the birds suddenly come to life.

The world of the marsh

The death and gentle decay of the reeds, bulrushes and other swampy plants serve gradually to build up the ground. It is of course still waterlogged, but not necessarily covered with water all the year round, and so is now called a marsh. Rushes, sedges and grasses grow, making the marsh look like a wet grassland dotted with shallow pools. Like an old meadow, the marsh is filled with perennial wild flowers—the marsh iris (called, appropriately enough, the yellow flag), golden kingcups, meadowsweet and loosestrife. Unlike the wild flowers of a forest, marsh plants do not flower particularly early. There would be no advantage in doing so, since they get plenty of sunlight throughout the growing season, while forest plants must garner light before the tree canopy grows.

A marsh can support an enormous quantity of animals, because the lush vegetation attracts a lot of insect life, especially flies and craneflies. The smaller animals are similar to those found in a wet meadow—frogs, grass snakes, shrews, mice, and their hunters such as the weasels and stoats. It requires a fair expanse of marsh to provide enough prey to support the larger predators like the short-eared owls and the marsh harriers, and also to allow room for creatures like ducks who, although they flock together gregariously at other times of the year, are not colonial breeders and need large areas within which to hide their nests.

There is nothing more rewarding than rigging up a hide in a swamp or marsh, in order to watch the private lives of the birds that live there. In a marsh in Corfu I once had as many as six hides built at strategic points so that, according to the time of day, I could view the activities of a wide variety of nesting birds. The ducks, geese, swans and rails have young ones that are called *precocial* (rather like precocious children). This means that, unlike most other birds, their babies hatch, dry their feathers and then, within hours of being born, are wide-eyed and strong enough to swim after their mother and feed for themselves. Of course, their mother still protects them and gives them shelter and warmth under her feathers when they sleep. Tired baby swans (cygnets) ride on their mother's back, using her as a stately white galleon. There is a strange rail-like bird called the finfoot, found in the swamps of Mexico, which has two little pouches of skin under its wings, and in these it carries its young.

Bird behaviour and communication are fascinating subjects to study. When a duckling gets lost, for example, it peeps loudly so that its mother will know where to find it. Unfortunately this also tells predators such as the marsh harrier where the duckling is. These hawks, in fact, listen out for the calls of lost ducklings as they quarter the marsh in search of food. But, to counter this, ducklings don't get lost very often because as soon as they hatch they instinctively follow the first thing seen moving. In the wild, of course, this is inevitably the mother. Such behaviour is called *imprinting*, and was perfectly demonstrated by the famous naturalist Konrad Lorenz in the 1930s. He accidentally imprinted 10 newborn geese

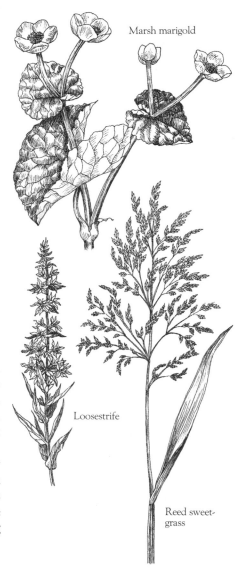

Marsh marigold

Loosestrife

Reed sweet-grass

Marshland

Large areas of freshwater marsh such as our collecting area of the Camargue, in southern France, represent some of the last truly unspoilt habitats in Europe. In summer, marshes are uncomfortable places for human visitors, principally because of the countless biting insects but also because the swampy ground is difficult to traverse. The day of our visit it was late summer, typically warm and sultry, and just right for the mosquitoes! For the prepared naturalist, however, marshes are exciting places indeed. As a river (in our case the Rhône) approaches the sea it slows down and deposits thick silt in which grow the predominant marsh plants—reeds and sedges. Such plants can tolerate continual submersion of their roots and flourish in the organically rich silt and mud.

Catfish

Medicinal leech

Aeshna dragonfly larva

Gambusia mosquito fish

Feathers The high turnover of visiting moulting birds, who have few predators in marshy areas, means abundant feathers. If feathers had been the result of a hunter's attack they would be broken off or bent somewhere along the central shaft.

Mallard flight feather

Grey heron back feather

In the water The whiskered catfish and mosquito fish (which devours mosquito larvae) are both introductions from North America, the latter for "biological control" of the innumerable mosquitoes.

Water bug

Pond skater

Great diving beetle

This large fearsome-looking spider spins a web of white silk between the reeds. We saw its characteristic behaviour when threatened—the spider sits at the web centre and swings to and fro, vibrating the web as though on thin elastic bands.

Hemianax dragonfly

Seeds of fennel-leaved pondweed

Wild boar droppings

Stranded carp The skeleton, ribs and scales of this fish are a common sight in areas where the marsh dries out. Its flesh is picked off by gulls, crows, kites and harriers.

Seeds form plentiful food for many birds and grazing mammals. The seed-containing droppings were found on soft mud, with deep tracks of heavy wild boar criss-crossing the area.

Hunter-killer One of the largest of the myriad of dragon and damselflies which abound in the marsh—an ideal habitat for their larvae and prey.

Scales of carp

Young stripeless tree frogs

Water net alga

Reedmace

Reeds or rushes? Not the familiar bulrush on the right—the reedmace is its correct name. The mistake is said to come from Victorian painter Alma-Tadema's work "Moses in the bulrushes", when he erroneously painted reedmace, not bulrushes; the latter are much less impressive plants with thin leaves and clusters of spiky flower heads. The wrong name has stuck ever since. The reedmace and common reed are cosmopolitan species and form the basis of flora in marshes throughout the world.

This year's youngsters, these tree frogs are just starting to make their way from a watery early home up into bushes and small trees at the marsh edge. Bright green colour gives excellent camouflage as they sit in the branches waiting patiently for winged and crawling prey to come within reach. In the Camargue region a sky-blue variety of the common tree frog is sometimes found.

Common reed

Duckweed Leaves of greater and lesser duckweeds float and, as their name suggests, are eaten in large quantities by ducks.

Feeding signs Mud-dwelling invertebrates form food for vast numbers of wading birds. These beak marks were probably made by a ruff as it probed for worms and crustaceans in the thick glutinous marsh mud.

Marsh grass This quick-spreading floating grass is much consumed by the famous Camargue horses and cattle during summer months.

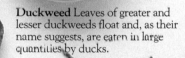
Dog's-tooth grass

on himself—that is to say, he inadvertently let himself be the first moving creature they saw upon hatching. Naturally, the baby geese assumed he was their mother. He had to spend an exhausting summer playing "mother goose" to his offspring, looking after them, swimming with them and talking to them in their "language". It was certainly a full-time job, but from Lorenz's studies of his foster brood came the basis for his work on instinctive behaviour, which set the foundations of *ethology*, the study of animal behaviour. So you see that even a humble goose can be of great importance in its own right.

Life on the Camargue

Near our house in France lies the great marshland known as the Camargue, a low-lying wedge of sand and mud between the Petit Rhône and Rhône rivers, just before they empty into the Mediterranean. Here in the calm shallow waters dotted with patches of reed beds lives a great conglomeration of wildlife, from frogs to flamingoes. This area is also famous for its little fierce black fighting bulls and the beautiful white Camargue horses. The bulls are used for fighting, but the object of the fight is for the "razateurs" (the local village youngbloods) to remove one or more little coloured tassels, the "cockades", from the bulls' horns. This they do with a thing like a curry comb. It is a very dangerous sport and many razateurs have been badly injured, but the bull is never injured and after fighting for about 20 minutes he is ushered out of the ring and released back in the marshes of the Camargue. Some of these bulls become very famous for their skill, and the people in Provence follow their fights much as boxers or wrestlers are followed in other parts of the world. I thought this subject would make a splendid film, so I wrote a script about a bull called Marius and the story was of his whole life, from birth to death. Naturally, in order to film the sequences of his early life, we had to go deep into the marshes where the bulls and cows live and breed in the great reed beds. It was a wonderful and exciting experience, because we had to ride in on the white horses—the bulls would attack if they saw a man on foot. Of course, we were eaten to death by millions of mosquitoes that hung like a vibrating veil round us and our horses, but we filmed the bulls and in addition we saw a lot of the fascinating wildlife. There were bee-eaters hawking insects through the sky, gleaming jade green and blue; thousands of cattle egrets like snowflakes; purple herons and common ones; hosts of small fry, such as warblers; and in the patches of open water between the reeds, a great concourse of flamingoes looking like drifts of pink and scarlet rose petals. Beneath our horses' hooves the frogs and pond tortoises scuttled away and on the reeds the green, and sometimes blue, tree frogs clung to the foliage and sat there gulping. We saw several coypu, and once a wild boar crashed out of our way, snorting indignantly. It was a perfect example of how important the marshes are, for this great area was obviously supporting millions of different forms of life—some of them, like the insect-eating birds, of great benefit to man. Fortunately, the bulk of the beautiful Camargue has been declared a wildlife sanctuary.

Bogs and fens

As the slow process of ecological succession continues, the marshy ground builds up, layer by layer. Drainage becomes better and the

Birds of the Camargue

The marshy landscape of the Camargue is a wonderful place to watch birds, especially as one does not have to wade knee-deep in mud to see many interesting species. Both birds illustrated can be seen from a car or along tracks that cross the reeds. The marsh harrier is most often observed flying slowly a few feet above the reeds, scanning the ground for water voles and the young of marsh birds. Bee-eaters are bright delightful migrants from Africa. They often fly in small groups, calling with excited liquid trills as they hunt insects. Once captured, large prey like dragonflies are battered against a branch before being swallowed.

Marsh harrier

Bee-eater

decomposition of plants becomes more rapid, further enriching the soil. Gradually a proper meadow develops (and eventually, of course, a forest). But in rainy or low-lying areas the soil can never quite dry out, and this means decomposition is slowed right down. Partly decomposed plant and animal matter gets packed down, crushed under the weight of plants and animals that continue to grow and die on the surface. In these waterlogged acidic conditions, *Sphagnum* mosses flourish and hold rain water like sponges. Their method of getting nutrients actually makes the water more acidic. Eventually a bog develops and the squashed and compacted semi-decomposed material turns into peat. The coal, oil and natural gas deposits which we now depend on, buried deep under ancient forests or grasslands, were themselves once peaty mires in which lived the fantastic creatures of millions of years ago.

Life in a bog is a harsh one. Apparently, *Sphagnum* moss is not very tasty, and the other plants support only grouse and hares. The main herbivores are insects, and life here is bizarre as well—one can see craneflies devouring liverworts; and in return there are plants such as sundews that devour insects. But some of the peatlands are not as lonely and desolate as the bogs can be. In the East Anglian fens, for example, the soil is so rich in minerals that it can support not just *Sphagnum* but a host of different plants. These, in turn, form the base of a large food pyramid which supports a rich animal life. The fen pools are teeming with life, and their borders are rich in reeds and sedges which provide nesting sites for, among others, the various marsh warblers who sling their delicate cup nurseries among the reeds. Coots, moorhens and grebes are also present, and they weave their floating nests at water level. On the higher peaty ground around the fen you get an enormous assortment of marsh plants, and trees such as alders and willows will thrive in this sort of soil.

The first attempts to drain the fens were in Roman times, but they had little overall effect. However, in the early 1600s the 4th Earl of Bedford and a group of speculators decided that the rich fenland, if drained, would be (agriculturally speaking) of enormous benefit and profit. The Dutch, being well versed in the construction of dykes and the drainage of wetlands, were called in and one Cornelius Vermuyden was employed to do the job. He was responsible for cutting the Old Bedford River between Earith and Denver, and then he constructed the New Bedford River (which is 30 metres wide) parallel to his first cut. (It has always seemed astonishing to me that history books never give you this sort of interesting information, but merely a dull catalogue of battles and the posturings of kings, queens and politicians.) Of course, the draining of vast areas of fenlands like these can end in trouble. Drain the land and it dries, shrinks and sinks. In Holme Fen, someone had driven an iron post into the peat until its tip was level with the ground. This was in the mid-19th century; today, the tip of the post is nearly three metres above the level of the peat. Even though large parts of the fens are now nature reserves, the habitat is still reeling from man's impact. Large-scale drainage of the surrounding fields has lowered the level of the water table, and although the marshes and bogs themselves have not been touched, they are gradually drying out and the water-loving plants and specialized marshland insects are disappearing. Unless urgent action is taken, this marshland habitat is in danger of being lost for ever.

Flying marshland insects
The ubiquitous mosquito is found wherever there are still stretches of water. Marshes harbour many millions. On humid evenings in the Camargue one can be attacked by thick black swarms of these insects desperate for a diet of blood. Only the females are blood-suckers; the rich protein diet is needed to boost the production of eggs which are laid as tiny rafts on the water surface and develop into wriggling larvae. In contrast, the attractive marsh carpet moth tends to be restricted by the range of meadow-rue, the plant on which its larvae feed. In Britain it is found only in the fens of eastern England.

Marsh carpet moth

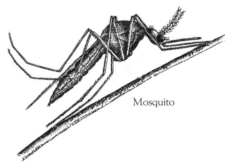

Mosquito

THE NATURALIST ON
COASTAL WETLANDS

Estuaries and their attendant salt marshes—which are together often called coastal wetlands—are an odd sort of love affair, almost a marriage of the land and the sea. This intermingling results in one of the most productive ecosystems in the world, and one that is of enormous interest to the naturalist. Unfortunately, it appears such habitats are of little concern to governments, even though the wetlands in general are of such great environmental importance.

The whole construction of coastal wetlands and the way they "work" are fascinating. As regularly as a pendulum, the tide swings the sea and pulls it in, carrying its cargo of tiny organisms and the rich debris of decay. The sea pushes up the mouth of the river and spills over the mud flats and on to the salt marshes flanking the higher estuary. As the tide swings back again, it cleans the marsh as it carries away accumulated debris. Now it is the river's turn. It pushes its way into the sea, and in due season it floods the banks and carries its own cargo of silt collected and tumbled down from far inland.

The estuary and mud flats and salt marshes, where river water meets sea, form a confusing world. It is sometimes dry, sometimes wet, and when it is wet it can be either fresh water, salt water or a mixture of the two. In this half-and-half world the organisms from the river, the land and the sea look for both food and shelter. Of course, some are more dependent on the land and the fresh waters of the river system, while others are more dependent on the salinity of the sea. Nevertheless, these two apparently opposite habitats interlock like pieces in the most intricate jigsaw puzzle.

There are many different types of coastal wetland. Each one is designed by the nature of the rivers and the tides that feed it, as well as the climate. An estuary in the tropics is very different from one in, say, northern Europe. In North America, on the west coast, estuaries and coastal marshes are narrow because the rivers tumble head-long down the steep slope of the continental shelf. On the other hand, the fresh water being pushed out by the mighty Amazon spreads out in a great brown thumbprint across the blue sea, and you can tell you are approaching the river mouth from as much as 100 kilometres away. In front of the mouths of some rivers, sand bars form and protect the water behind them. In these sheltered estuaries, less influenced by the tides, a brackish bay can develop. The Dutch Waddenzee is a good example of this in north-west Europe, as is the lovely Pleasant Bay off Cape Cod in North America. In the tropics these brackish bays become bordered with mangroves, those marvellous trees whose basket-like roots hold on to the mud and so help build up the land.

In rivers that meander slowly across the coastal plains, the tide can enter the river mouth in what is called a "bore". These are tidal waves,

Nilsson's pipefish, a
northern European species, in eel grass

Taken by the tide
The jellyfish opposite has been swept inshore by the warm waters of the Gulf Stream and pulled by the tide into the mouth of a sheltered Irish estuary. This individual has caught a sea gooseberry (a close relative of the jellyfish) in its stinging tentacles.

varying in size and vigour, that surge up the river and sometimes over its banks. You can see this extraordinary action of the water on the Severn, where the bore can be almost two metres high. The bore on the Yangtze river in China can be as much as five metres high, and it is very dangerous if you are out in a boat.

The ever-changing estuary

Everything in nature moves and grows and changes, and an estuarine habitat is no more static, no more uniform than a forest or lake. In some cases the tide comes rushing in as a great mass—as it does with the push of the bores in rivers like the Severn, or with the sea flooding into a vast estuary like the Thames. The tide pushes back the river and thoroughly mixes salt water with fresh at all depths along the river bed. But the Tees in the north of England and the great Mississippi are both estuaries where the tide surreptitiously creeps in as a "salt wedge" along the bottom. At the surface, fresh water still pours out into the sea, so you have something resembling a layer cake where the mixing of salt and fresh water is not complete.

Estuary and salt marsh plants, as they die and decay, create a debris which is brought seaward by the river, the smaller creeks and the retreating tides. Some of it is fed on by the small creatures that inhabit the area but a lot of it is swept out by the tide to enrich the coastal waters. The cocktail of salt and fresh water creates a sort of nutrient trap, making the estuary up to 30 times more fertile than the open sea. This extra-ordinarily high level of foods in the water supports a dense population of each species of animal, such as countless spire snails and diatoms, and millions of baby fish. But the curious thing about estuaries, mud flats and salt marshes is that, in spite of the rich water and the specialized

NETTING AND SAMPLING

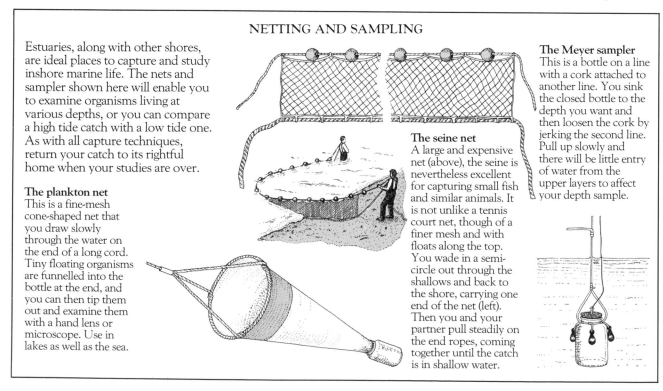

Estuaries, along with other shores, are ideal places to capture and study inshore marine life. The nets and sampler shown here will enable you to examine organisms living at various depths, or you can compare a high tide catch with a low tide one. As with all capture techniques, return your catch to its rightful home when your studies are over.

The plankton net
This is a fine-mesh cone-shaped net that you draw slowly through the water on the end of a long cord. Tiny floating organisms are funnelled into the bottle at the end, and you can then tip them out and examine them with a hand lens or microscope. Use in lakes as well as the sea.

The seine net
A large and expensive net (above), the seine is nevertheless excellent for capturing small fish and similar animals. It is not unlike a tennis court net, though of a finer mesh and with floats along the top. You wade in a semi-circle out through the shallows and back to the shore, carrying one end of the net (left). Then you and your partner pull steadily on the end ropes, coming together until the catch is in shallow water.

The Meyer sampler
This is a bottle on a line with a cork attached to another line. You sink the closed bottle to the depth you want and then loosen the cork by jerking the second line. Pull up slowly and there will be little entry of water from the upper layers to affect your depth sample.

adaptations of the organisms that live there, the actual number of different resident species is low compared to the numbers you would find in a forest or lake. In one sense, life in the coastal wetlands is as difficult as it is, say, on the tundra; instead of worrying about light and dark and water and ice, in the estuary you have to adjust yourself to mud and water, salt water and fresh water, and the tremendous storms and floods that are constantly reshaping the coastline. In addition, in the winter and spring the rivers are fed and swollen from the inland rainfall. This increases their bulk of water, and with it their cargo of silt, thus making the estuaries less salty but more turbid than they are in summer and autumn. All these changes affect life in an estuary, so when you are exploring one you must bear in mind the type of estuary, the time of year and the rhythm of the tides.

Two ways of life — floating and fixed

Estuaries are wonderful places for catching and examining drifting and swimming organisms that move in and out on the two pulsating flows of water. Here you should use a plankton net for small organisms and a seine net for larger ones. If you are lucky enough to have or be able to borrow a boat, make sure that it is securely anchored. In Corfu I had a lovely, almost circular flat-bottomed boat called *The Bootle Bumtrinket*. She was a splendid vessel, made for me by my brother, but she was so flat-bottomed that even when I anchored her the slightest wind or smallest wave would twirl her round like a top and make observations very difficult. With a boat, it is also very important to understand and memorize how the tides work in your area. You might become so absorbed in the work you are doing that you can be swept out to sea if you are not careful.

One of the safer and less laborious ways to study the drifting and swimming organisms is to find an estuary with a pier. You can make the piles that hold up the pier into a sort of thermometer for the to-ing and fro-ing of the tides, marking them at the levels of the various highs and lows of the monthly cycle. When the tide is out, your collecting apparatus may have to be fitted with long handles or ropes, but at least you are on a secure foundation while you are doing your researches and you don't have to worry about being swept out to sea.

A useful gadget for collecting water and plankton samples at different depths (particularly important in a "salt wedge" estuary) is the Meyer sampler. A marvellous time for using this sampler, especially if somebody owns a microscope, is when the waters of the estuary are said to come into "bloom". What happens is that the phytoplankton, which consists mainly of tiny plant-like diatoms and dinoflagellates, multiply their numbers rapidly when the light level increases in the spring. This is followed by an increase in zooplankton, which are mostly minute crustaceans that float about or drift up from the mud at night to graze on the phytoplankton. Under the microscope, you will be able to see how water samples from different depths contain different proportions of these strange microscopic monsters.

In summer, you might find jellyfish browsing on the plankton, or you might be lucky enough to catch a really voracious plankton predator called the sea gooseberry, which is one of the comb jellyfish. Drop it into a jar of water and a gooseberry-shaped body with eight iridescent plates

The plankton hordes

The top few centimetres of sea water swarms with millions of microscopic plants and animals which make up the plankton. Dinoflagellates like *Noctiluca* and *Polykrikos* have whip-like flagellae to aid movement. The long streamers of tiny *Elphidium* are used to trap prey even smaller than itself.

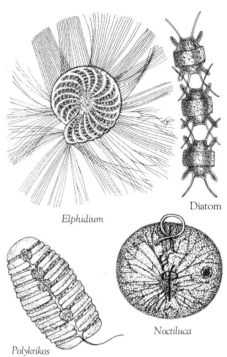

Diatom

Elphidium

Noctiluca

Polykrikos

Estuary and salt marsh

An estuary is a sort of junction, and our sunny June visit to an estuary site near Southampton demonstrated this point admirably. The plants and animals are alternately bathed in salt water from the sea and then, with the ebbing tide, in fresh river water coming from inland. Few plants and animals are adapted to this great change in salinity, but those that are have this curious mixed habitat to themselves and live there in enormous numbers. On the landward side of the saltings and mud flats, salt marshes form up. These are made of plants with varying degrees of tolerance to salt water. Beds of grasses and reeds are the principal vegetation, with mud-dwelling invertebrates and of course wading birds as the most obvious fauna.

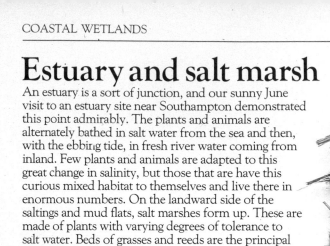

The dark cockle is discoloured due to hydrogen sulphide, a natural gas from mud, reacting with polluting iron salts.

Small shells Tiny Jenkins spire shells are typical of sheltered estuaries—up to 40,000 may live in one square metre of mud. The larger rough periwinkles are more characteristic of upper rocky shores.

Natural wastage A small sample of matted salt marsh debris collected by the tide. Dead and dying seaweeds are studded with empty shells of snails, cast crab skins and other natural refuse.

Chitons on flint

Barnacles and keel worm tubes

Acorn barnacles

Freshly moulted crab carapace Once shed, its owner, soft and vulnerable, retreats to some safe spot while his new carapace toughens.

Encrustations Any solid object finding its way on to the shifting estuarine muds soon becomes coated with barnacles and tube worms.

Flat wrack

Seaweeds live in the lower tidal part of the estuary; if the sea carries them up the shore they dehydrate and die.

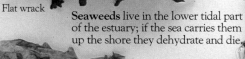

Sea lettuce

Enteromorpha seaweed

Oyster

Whelk

Cockle

Sea
club-rush

Common
salt marsh
grass

Common grasses of
the higher estuary and
salt marsh grow in
thick swards and pro-
vide excellent cover
for nesting birds.

Shore debris The three molluscs were
probably attacked originally by birds;
the sea has done the rest. Among the
large numbers of feathers to be picked
up many are from the ubiquitous
gulls, but the ones here
may be from a resident
redshank or even a whim-
brel on its way north to
breed in the Shetlands or
Orkneys.

Cord grass is a vigorous early
colonizer of bare estuary mud
Its long anchoring roots help
stabilize the mud and it spreads
rapidly by sending out underground stems called "tillers".
These attributes, combined
with its erect form which slows
down tidal water and allows
silt to build up, give a firm
base for other plants and
animals to exploit.

Sea aster Here are
the short basal
leaves; the plant
has tall purple and
yellow flowers.

Glasswort

Expandable leaves Sea purslane,
like aster and glasswort, has thick fleshy
leaves which allow it to lose or absorb
water as the salinity changes.

193

High water
Green seaweeds
Barnacles
Mussels

Intertidal
Sponges
Hydroids
Snails
Red seaweeds
Worms

Low water
Plumose anemones
Dead man's
fingers
Hydroids of
jellyfish

High-rise communities
The underneaths of piers and wharves may
seem inhospitably cold and dripping but they
are well worth investigating to see creatures
normally well hidden low down on the shore.
Owing to their steep sides few plants can
settle and so encrusting animals predominate,
showing a vertical zonation of species from
high to low water marks. Look out for
beautiful plumose anemones and the fuzzy
forms of soft corals known as dead man's
fingers. Wooden piers are often pock-marked
with the burrows of boring molluscs and
crustaceans.

and a pair of long trailing tentacles will become visible. In the water they
are so transparent, like little round bags of cellophane, that the only way
you can tell they are there is because of the lovely colourful play of light
on their "combs" which they beat through the water as they swim along.
They are indeed so delicate—much more so than thin glass—that many
are destroyed during storms, and it can be difficult to catch one intact.
On its trailing tentacles, the sea gooseberry does not possess stinging cells
like true jellyfish. Instead it uses adhesive cells on the tentacles to catch
and immobilize its prey.

The added benefit of working from a pier, besides safety, is that when
the tide goes out you can study the various creatures that encrust the
piles. They, incidentally, have attached themselves for the very same
reason that you are using the pier—to prevent being swept out to sea.
The two-shelled mussels and oysters are nearly always present, and you
will find a wonderful and complex community built up around these
bivalves. There are sponges, bryozoans, bristle worms and crustaceans
living on and around them, and you might even find an oyster crab (or
pea crab as it is sometimes called, because of its shape) which was first
described by Aristotle. This funny little creature steals food from the
oyster's or mussel's gills and uses the shell as a house. Once the female
oyster crab has chosen her bivalve, she remains faithful to it and only
leaves briefly during her mating season.

For many species of fish the estuary is a sort of cross-roads, through
which they pass either on their way upriver to spawn or else on their
way out to sea to spawn. Salmon and sea trout, for example, spawn in
the rivers and yet mature at sea; eels do it the other way round since they
spawn at sea and mature in the rivers. For many species of fish who come
into the area the estuary itself is a vital spawning ground and nursery for
their fry, which prosper on the rich plankton. Apart from these busy
commuter-type fish, there are others who make their home in the estuary
throughout the year—the flounders, for example, and the little sticklebacks and various gobies who feed on small worms and crustaceans, and
the weak-mouthed grey mullet who will eat almost any tiny thing. The
fish who live in the intermixed world between salt and fresh water can
only exist by having specialized kidneys and gills. These organs are able
to switch between excreting more water and less water, and between
absorbing more salt and less salt, as conditions dictate. Such adaptations
are vital in order to survive in their forever-changing environment.

Inhabitants of mud flats
With the moving battle in the estuary between fresh and salt water, life
is difficult for mobile creatures since they are constantly pushed back
and forth. To avoid this problem, many estuarine organisms are found
anchored in the mud. Cord grass and eel grass take root in the shallow
parts, and also perhaps widgeon grass, whose little triangular seeds are
one of the favourite foods of wild duck. In the calmer areas grows sea
lettuce, an alga, and this and eel grass are eaten by ducks and geese. As in
a lake, various tiny algae grow on the submerged estuarine plants and in
their turn provide a rich pasture for such small browsing herbivores as
snails and grass shrimps.

Most bottom-dwelling estuary creatures retreat beneath the mud or
under stones when the tide is out. As the water comes in, however, a

remarkable selection of tubes, legs, heads and other appendages appear and make the muddy bottom look like some strange underwater animal forest. The bivalves extend their extraordinary siphons like elephant trunks, while the ragworms and different crustaceans such as crabs and prawns move over the mud, feeding on algae and the rich debris. The tiny spire snail constructs for itself a sort of Kon Tiki raft of mucus, but in this case it has the added advantage of acting not only as a vessel on which the snail progresses but also as a sticky trap to catch food.

Since it is difficult to view this underwater world unless you set up your own marine aquarium, perhaps one of the best ways to study it is to wait until the tide is out and then walk over the mud flats, looking for signs of creatures that are now beneath the surface. You have to be rather cautious about this, however, because some mud flats look deceptively easy to cross but are in fact almost like quicksands. I remember when I lived in Bournemouth, on the shores of Poole harbour, that when the tide went out it left what looked to be like a perfectly easy walking surface. On one occasion I decided to go out to one of the many little uninhabited islands (uninhabited by humans, that is) which dot the surface of the harbour. After ten minutes, when I was almost up to my waist in mud, struggling to get my legs out, I discovered that walking over the flats was not nearly as easy as it looked. I did make the island eventually, and because it was uninhabited I saw some interesting birds' nests and other creatures on it. Sensibly, before my trip I had checked on the tides; if I had not, and had been caught by the incoming tide wallowing about in the mud like a hippopotamus, I could easily have been drowned.

Once the tide has revealed the mud flats to you, you have another opportunity to become a nature detective and find out where creatures are living. For example, one of the bivalves, the peppery furrow shell, shows you where it lives by making a star-shaped pattern on the wet mud. It does this with its long siphon, which protrudes from the burrow and gropes around in search of food. Where the mud surface looks grainy there might be thousands of minute spire snails still browsing before they burrow down into the mud, and where the surface has a lumpy appearance there may be hundreds of *Corophium*. These tiny crustaceans create little walls of mud around their burrow entrances as they emerge and search for food when the tide is in. You can sometimes get a miraculous result by scooping up a great lump of mud and putting it into a clean plastic canister full of sea water, and watching what emerges once the sediment has settled.

It is not just small creatures that you find on the mud flats. There being so many mud-dwelling organisms, the estuary can be rich in the larger animals which prey on them. You will see great strings of birds such as the waders—knot and godwits—and the bigger gulls and herons that range over the flats for any smaller creatures they can find. Surprisingly enough, even stoats and weasels and, in winter, otters have been seen hunting for crustaceans. In North America, you might see something like a diamond-backed terrapin or even a racoon in search of crabs to add to its usual diet.

In warm North American and tropical estuaries, the fiddler crab plays an important role as an item of diet for innumerable creatures. Here on the mud flats these curious little crustaceans scuttle about, so that as you

Feeding signs in the mud
Estuarine mud is a wonderful place to discover the daily dramas of shore life. Apart from the criss-crossing tracks of waders, other more irregular scrapings appear. The star-shaped pattern is the work of *Scrobicularia* (the peppery furrow shell), a bivalve which extends its sucking siphon over the mud from a central position. Curious conical dimples are the feeding trail of a flounder as it searches for molluscs to crunch up; a redshank has also passed this way. The sweeping pattern is from a shelduck's beak as this bird dabbles for small snails with a motion not unlike that of a car's windscreen wiper.

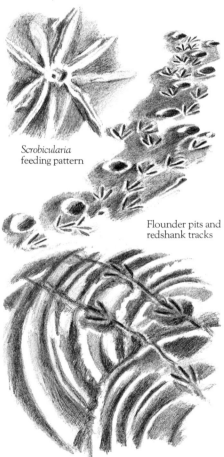

Scrobicularia
feeding pattern

Flounder pits and
redshank tracks

Shelduck
feeding marks

Britain's biggest wader
The curlew (opposite) is a common sight on estuaries during the colder months, probing in the mud and sand for small invertebrate food. When nesting this bird used to be confined to moorland but has recently spread to meadows, marshland, estuary reed beds, sand dunes and even river shingle.

walk and they run ahead of you it is like being in a helicopter pursuing the great animal herds of East Africa. When disturbed, they run for their burrows; these can be identified by the marble-sized pellets of mud which they pile up at the entrance—they have to keep excavating their burrows to keep them ready for use. In this respect fiddler crabs on the mud flats act in the same way as worms on land, for as they burrow they constantly turn over the soil like gardeners.

Sitting quietly and watching fiddler crabs, they are really rather entrancing beasts. Each male has one small claw, which is in fact his "spoon", and one large claw which is a sort of "flag" that he waves. Moving slowly over the mud flats, the crab's small claw is constantly busy putting mud into the complicated mouth-of-many-parts, which sifts it carefully to find the edible contents. (From our point of view, it would be rather like trying to find a mouthful of peanuts while eating a bucket of gravel!) The larger claw he waves about for two reasons; first, to tell

PHOTOGRAPHING WILD ANIMALS

Taking photographs of the creatures you observe in the field is an excellent way of recording and "preserving" without disturbing nature. With care you can produce accurate pictures of animals that appear only fleetingly, and do so without disturbing them. For good results, find out all you can about the species you want to photograph. It is essential to know the creature's habits, and to find out where it is likely to stay, so that you can set up your hide within view, or prefocus the camera lens on the right spot. Whatever you are photographing, remember the natural history photographer's golden rule: Never cause more than the slightest disturbance to your subject. The welfare of a living creature is more important than a tiny piece of celluloid.

Birds and mammals
Larger creatures are often very elusive. You must learn as much as you can about the species you intend to photograph. Find out their daily habits, where they feed, and what their movements are, and watch them quietly and inconspicuously (see page 140). When stalking nocturnal mammals, use a torch covered with red cellophane—the beam will be all but invisible to them. But do not take a photograph unless you are sure it is the picture you want, since the flash will disturb most species. You can set up the camera on a tripod and "press" the shutter from some distance away using a remote control cable release or an even longer air release.

35 mm camera body

200 mm telephoto lens

400 mm telephoto lens

Tripod

Remote cable release

Torch with red filter

Choosing equipment
A 35 mm SLR is the best camera to use. You can see the actual image produced by the lens, and obtain interchangeable lenses. Most models come supplied with a 50 mm standard lens, which is useful for some subjects, but for distant subjects a longer focal length (such as 200 mm or 400 mm) is best. If you have lenses of good quality, you can also use them with a tele-converter, to magnify even more distant subjects. A sturdy tripod is virtually essential.

Hides
If you cannot conceal yourself completely, it may be necessary to build a hide near a regular feeding or nesting site (see page 140). When you do this, you must not disturb the subject. Assemble the hide gradually, over a period of days, or put it together some distance away and move it gradually into place. Take plenty of equipment into the hide. If you can get hold of two cameras it is worth taking them both, in case one camera jams, or unexpected action occurs at the end of a film, before you can reload.

Curlew portrait
To produce the photograph of the curlew (opposite), the photographer constructed a hide under the overhanging roots of a nearby tree. He made a curved wall with boulders, turf and logs, adding a large ivy-covered branch that acted as a curtain through which the lens could protrude, so that the curlews would not suspect anything. Of course, a 400 mm telephoto lens was used to produce a large image of the birds.

Sea holly

Sea lavender

Sea blite

SWEEP NETTING

The sweep net consists of a sturdy wide-mouthed frame and a strong linen or nylon bag. As its name implies, you sweep the net through low herbage—rough reeds or grass, or light shrub—to collect insects and other small creatures feeding or resting on the plants. To avoid damage to the vegetation, do not sweep an area more than is absolutely necessary.

Operate the sweep net in an arc, using both hands

other male crabs that this is his territory and so they should beware, and second, to try to attract female crabs. Besides waving, he also taps the larger claw on the ground, sending out a sort of crab Morse code.

It is unfortunate that the coasts of Britain and Europe do not have such entertaining creatures as the fiddler. Here the crab fauna is, on the whole, fairly poor, there being only the little shore crab that is likely to be seen out on the mud flats.

Between estuary and land — the salt marsh

As you walk up from the mud flats into the salt marsh area on either side of the estuary, you will notice that up to the mean high tide level only a few species of plant grow. In Britain and Europe, for example, you first encounter the glasswort (also called the marsh samphire) and the cord grass. Cord grass can secrete salt, and if you see it in the slanting sunlight of early morning or late afternoon it glistens with salt crystals. Their stems and roots have hollow tubes which carry oxygen down from the leaves, a system which prevents the thick mud from suffocating the lower parts of the plant.

Glasswort is so named because the plump green stems, built like strings of tiny sausages, were at one time used in glass making. The plumpness of the stems is an adaptation for water storage, because sea water tends to drain the plant of its body fluid. I well remember how tasty this succulent plant is; when walking over the salt marsh in Corfu I would break off and chew a handful. You can in fact collect them and eat them in a salad, or even pickle them.

Farther into the salt marshes, towards the land, grow the sea poa and sea blite, and then even farther in are the sea lavender, sea purslane and thrift. All around are the snake-like channels of the tidal creeks which drain the sea as the tide retreats, and here and there are tidal pools. Well beyond the mean high tide you will find beds of reeds and rushes, particularly where the fresh water from the river's creeks begins to predominate. The rushes are sometimes called black grass, because in autumn they turn such a deep dark green that from a distance they almost look as though they had been burnt. Towards the back of the salt marsh, where dry ground begins to rise, is a fringe of shrubs and trees. These provide cover and nesting sites for the birds and mammals who use the lower marsh as a feeding ground.

A salt marsh, like a prairie, changes colour according to the season. In the spring it is mostly green; in summer it is patterned with the purple of sea lavender, the yellow of purslane and the pink cushions of thrift; by autumn it has changed again, and on the landward side you get the border of dark-green "black" rushes with, on the seaward side, a border of bright pinky-red glasswort.

I remember, when we lived in Corfu, one of my brothers was a very keen hunter. Although I disapproved of this, I had to admit that it did provide me with both material for my home museum and also a very good record of what kinds of birds and mammals were about. At the beginning of winter, he and several friends would hire a *benzina* (a Greek fishing boat) and go over to the great Butrintit marshes, which were a paradise for sportsmen and naturalists alike. These marshes, stretched between the Greek and Albanian borders, were a wonderful example of the two worlds of river and sea intermingling. The mud flats and estuaries were alive with shore birds such as dunlin, greenshank and plover, and the marshes behind were teeming with mallard, pintail, widgeon, shoveler, heron, bittern, snipe and woodcock. At that time of the year it was too cold for reptiles and amphibians, but there were plenty of mammals in the shape of wild boar, wild cats, martens, otters, water voles and various mice.

The widgeon rests on estuaries by day and flocks inland at dusk to feed

Once, when I was in my brother's good books, he allowed me to accompany him. It was a thrilling experience. The voyage took several hours and was rough, wet and cold, with the highest Albanian mountains white with snow in the background. When we arrived the wild-looking Albanian beaters were there, wearing red tarbooshes and long thick cloaks of sheep's wool, with their enormous and very fierce dogs ready to help them. While my brother and his friends went off into the cane thickets to hunt wild boar (I was not allowed to go, to my annoyance, as it was too dangerous), I investigated the mud flats. I found marvellous treasures ranging from shells to strange half-frozen insects and, best of all, half a dolphin's skeleton. When the hunters returned they had with them the corpses of two huge bristly fierce-looking wild boar. I was allowed to gut them (so I could find out what they had been eating) and I was also permitted to keep the tusks, hooves and a skull for my museum.

The avocet's upturned beak sweeps the shallows for small marine life

My brother and the other hunters then went duck and snipe shooting. I was allowed to accompany them, and we pushed our way deep into the reed beds in our flat-bottomed punts. Very soon the punts were full of snipe, woodcock and a great variety of wildfowl, and I really had my time cut out as I tried to sketch and record all the different species and their colourings. We were going to eat some of the birds for dinner that night, and I asked my brother if he would let me skin them properly so that I could add the skins to my home museum. He agreed, and that was how my museum came to acquire five species of duck, two species of snipe and two beautiful woodcock skins that looked like piles of autumn leaves. After our visit to the marshes, my brother would always bring back specimens from his trips, even if I did not accompany him.

Large flocks of dunlin over-winter on British estuaries

The richness and profusion of the wildlife in the Butrintit marshes and estuaries in those days had to be seen to be believed, but it may no longer be the same. Nowadays, with complete thoughtlessness, man is practising wholesale destruction of the coastal wetlands by dumping wastes into the rivers, by barraging the estuaries and by causing erosion inland which silts up the mouths of the rivers. Also, the wetlands are being subjected to various forms of so-called "reclamation", and the inevitable offshore oil slicks add to the toll of losses in plant and animal life. In the great salt marshlands of Waddenzee the numbers of tern and duck have been severely reduced because of poisoning by heavy metals and other toxic compounds dumped into the Rhine by Dutch, German, French and Swiss industries. Millions of birds, fish and other marshland and mud flat creatures, if they escape direct poisoning, are faced with food and housing shortages as their habitat is either polluted or "reclaimed" by man. Yet it is a fact that, of the fish and crustaceans eaten by human beings, two-thirds depend upon the coastal wetlands for their survival. When governments talk glibly about reclaiming what they really mean is decimating; like fresh wetlands, the coastal wetlands have evolved in their own special way and are particularly vulnerable. Bear in mind, too, that you cannot "reclaim" something that was not yours in the first place.

THE NATURALIST ON
CLIFFS AND DUNES

Corfu is shaped like an ornate curved knife. The northern half is the handle and consists of mountains and sea cliffs, while the blade curves south and east and is formed of magnificent sand dunes. This meant, as a boy, I was lucky enough to be able to visit both cliffs and dunes—the last outposts of land before you come down to the sea.

On the high rocky coasts the wind and the waves have been sculpting the cliff face for thousands of years. Plants that cling to the cliffs have a very harsh existence, with salt spray near the bottom and exposure to the sun and the wind at the top. In addition, different kinds of rock present different problems. Granite weathers very slowly and produces little soil; only mosses, lichens and algae can cling to its steep bare surfaces. Chalk and limestone, of course, weather and change very much faster, but even on cliffs made of these rocks the soil can only accumulate on ledges and in crevices, which in turn can often be broken away by rock slides. But you do find plants hugging such cliffs, and some of them—sea campion, sea beet and thrift (which forms lovely great cushions of pink flowers)—tolerate the salt and so are found at all levels on the face. The so-called rock rose (not really a rose) and wild thyme are found near the top.

Despite their hazardous world, the colours of the cliff flowers can be very beautiful. Between the groups of pink thrift you see golden flashes of gorse and rock rose and blue stars of squill. If there is plenty of good plant cover on a broken cliff, a lot of small animal life will be living there in the shape of insects, spiders, centipedes and millipedes. But cliffs are relatively inaccessible and the only large animal life you find will be the birds, who simply rest there while on migration or else nest there, fairly safe from all terrestrial predators.

Cliff-nesting birds

Many species of birds nest in huge colonies on top of the cliffs, or else on the ledges and crevices of the cliff face. Each nesting pair has its territory around its nest. This vigorously-defended area is quite small, and is measured by the distance the owners can stab with their beaks.

Various species of gulls nest on rocky cliff tops, building rather scruffy nests. Their young, who are beautifully camouflaged, leave the nest not long after hatching. The gannets, with their long lance-shaped beaks and brilliant white feathering, also nest in colonies at cliff top level. These birds, who are related to the pelicans, build much more robust nests than gulls since their young stay in the nest for several months. The puffins, their faces and beaks painted like clowns, dig complicated nest burrows and thus have to choose cliffs with enough soil on top.

Nesting on the cliff face is a much more hazardous occupation. Some of the smaller colonial birds such as many of the auks—guillemots, razorbills and the like—do not bother to make nests but make sure that

The tender tree mallow
The tree mallow is a beautiful coastal plant with soft velvet-like leaves. It is common in the Mediterranean, and in Britain has become naturalized in the south west where the climate is mild.

Clamouring cormorants
Cormorants (opposite) nest in small noisy colonies at the base of a cliff, where the nests are usually flattish mounds of dried seaweed. These birds are strong underwater swimmers, able to catch several fish at a time before flying low over the waves back to the nestlings.

Seabirds of the high cliffs

Kittiwakes are unusual for cliff-nesters in constructing elaborate cup-shaped nests. One to three eggs are laid and the young take several weeks to develop before leaving the security of their cliff-edge home. Guillemots lay a single egg; the pear shape prevents the egg from rolling off the edge, aided by sticky droppings (guano) which accumulate on the densely packed ledges. Young guillemots develop fast, flying down to the sea with their parents just over two weeks after hatching.

Kittiwake

Guillemot

their eggs are tucked back safely on ledges or in crannies, as do the rock doves with their untidy nests. Rock doves are relatives of our city pigeons; you can often see the latter nesting in small nooks on buildings or on ledges, in effect using the buildings as a cliff. Kittiwakes, who are also cliff face breeders, make deep nests by plastering water plants into tiny crevices with mud and then trampling on them, packing them in tight and forming a hollow in which to lay their eggs. Their aggressive territorial displays do not involve much moving around but instead they have prolonged duels with their beaks. When one kittiwake feels that he is the loser, he gives up by hiding his "sword" under his wing.

For sheer beauty, the guillemot (or "murre" as it is sometimes known) lays the most spectacularly coloured eggs of all. Although a few might be plain white, some are cinnamon-coloured, some are streaked with rust-red stripes, and some have green and black squiggles all over them like mad writing. Others are heavily blotched with chocolate on a cream background, others again might just have blotches at the thick end of the egg and a few squiggles towards the pointed end. On a cliff with thousands of guillemots nesting, you will never find two identical eggs. It is by this coloration that a guillemot knows exactly which egg, out of the thousands on the cliff, belongs to it. Moreover, the guillemot egg is so designed—elongated and very pointed at one end—that if it is blown by the wind or touched by a bird (remember guillemots do not make nests) it rolls round in a circle instead of falling off the cliff. The famous jeweller Fabergé used to make fantastic Easter eggs out of precious stones and gold, and I once saw an exhibition of his work in London. The Easter eggs were spectacular, but not one of them came anywhere near to the beauty of a guillemot's egg.

Cliff sides with big rocky outcrops are excellent nesting places for sea eagles. They pile up huge great stacks of sticks and branches and then carefully line the inside with soft plants. In fact, the bald eagle of North America builds the biggest nest of any bird—the largest one found measured three metres across and six metres deep. Although the New World vultures like the condors do not build nests, but simply choose a suitable ledge on which to lay their eggs, some of the Old World vultures nest on cliff sides, in common with the eagles. I well remember an incident involving a cliff-nesting vulture that taught me a lesson on the dangers of studying cliff wildlife.

How not to watch cliff birds

One day in Corfu I had gone for a long walk in the north of the island with my dogs. At noon, hot and tired, I paused on the edge of a high golden-brown rocky cliff that plunged straight down into the blue transparent sea below. The dogs and I lay in the shade of the myrtle bushes and the one small olive tree, and had our lunch. Then the dogs slept and I dozed, but suddenly I sensed something and was wide awake. A shadow had flashed over me with a great "woosh" of wings, and as I opened my eyes I was just in time to see a huge bird swoop down below the cliff. Greatly excited, for I thought it might be a golden eagle, I crawled carefully to the edge of the cliff and looked down. It was not a golden eagle but something even more spectacular, partly because it was so unexpected. It was an enormous griffon vulture, and it was perched on a ledge some 12 metres below me, busy feeding two fat downy babies. I was so excited I almost fell over the cliff. My first thought was to get the babies as pets, but then I remembered my family had unanimously introduced a new law: no more pets. Our villa was already overflowing with everything from sea horses and toads to hoopoes and eagle owls, and the pressure on my family was considerable. They would never allow me to have two griffon vultures, even if I tried to persuade my mother how useful they would be in cleaning up kitchen scraps. Still, I thought, if I could get down to the nest I could meet and photograph the babies.

As soon as the mother vulture had flown away, I crawled along the cliff top and tried to find a way down to the nest. There was no way that I could see, but I noticed that the ledge on which the nest was built was just wide enough to clamber along, and that it continued along the cliff face for some distance, gradually sloping upwards until it came within about six metres of the top of the cliff. Now, I thought, if I had a rope I could swing down and then edge my way along until I reached the babies. But I knew I could not accomplish this alone, and it would be foolhardy and hazardous to try. I immediately planned another trip to the cliff and enlisted the aid of a young fisherman friend of mine called Taki, who was very strong. When I explained to Taki what I wanted him to do, he said I was mad and that if I fell off the cliff my mother would never forgive him, and would probably have him hanged for murder. But at length, after much argument, he agreed to help.

Looking back, the second cliff trip was actually a very silly and dangerous enterprise. Neither of us had any experience of climbing; our ropes were just ordinary ones, which were used for tethering donkeys; and above all I had absolutely no head for heights. Taki had insisted that we employ two ropes, one fastened around my waist and one which I would climb down. I realize now, of course, that if I had fallen, the rope around my waist—far from saving me—would probably have cut me in half. Anyway, we tied the ropes to the small but sturdy olive tree and, taking a deep breath and closing my eyes, I swung out over the cliff edge.

To my surprise, climbing down the rope was much easier than I thought it would be. Within a minute I was standing on the ledge. Keeping my eyes averted from the awful drop on my right, I edged along until I reached the baby vultures, who greeted my intrusion with interest rather than alarm. I knew I would have to be quick, because I did not want to be caught on the ledge by the parents. I took several photographs of the babies, and swiftly collected some interesting things from the nest site. There was a dog's skull, various other bones, a splintered tortoise shell and the partial skeleton of a large fish. Stuffing these into my collecting bag, I made my way back to the rope and laboriously climbed up it, landing on the cliff top hot but triumphant. But the next day my blood ran cold when Taki told me what had happened on his return. He had tethered his donkey with the rope I had climbed, and as soon as the donkey had put some strain on it, the rope had snapped like a piece of cotton. The shock and realization taught me never to do stupid things like that in future, even for a photograph of one of the most exciting cliff-dwelling birds that I have ever seen.

The griffon vulture, a scavenger of southern Europe.

The process of dune building

On a broad smooth beach, the wind constantly works on the loose sand. It piles up into drifts and dunes and eventually blows it away in

Sea cliff

A walk along a cliff path serves as a real tonic, even for the experienced naturalist. The spectacular scenery, the wind-blown plant life, and the constant activities and cries of seabirds all combine to make engrossing viewing. But the cliff face itself holds many fascinating secrets, as we discovered when we visited St David's Head in Wales and found a not-too-steep slope to clamber along in safety. The plants and lichens we found were mostly short and squat, growing in mats or "rosettes" to help reduce wind resistance and conserve moisture. Shells of marine animals and similar debris, dropped by seabirds, littered the slopes and made "cliff-combing" an exciting alternative to beachcombing.

Shaped by the wind, this sprig of heather is flat-topped and one-sided. Large areas of heath often form the dominant vegetation immediately behind cliffs.

Gull skull picked clean by crows and carrion beetles.

This young adder was found sunning itself on a path close to the cliff edge.

Fox fur Foxes patrol the irregular cliff terrain, hunting rabbits and scavenging on dead seabirds.

Empty mussel shell, the remains of an oystercatcher's meal.

Old droppings already breaking up will eventually decompose and enrich the cliff top soil.

Rabbit fur Warrens riddle cliff top paths, the light soil between rocks being easy to work.

Cuckoo spit on grass

Seed heads of carline thistle, a plant of maritime grassland but also found inland on dry limestone soils.

Undigested remains, probably gull pellets, show hair from carrion and shell fragments.

Crabs provide succulent meals for seabirds. The main body of this shore crab has been dug out; a leg is all that remains of a larger edible crab.

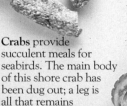

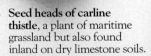

Odds and ends, mainly of crab. The limb muscles have been eaten, leaving hard pincer tips and shell fragments.

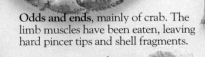

Moss

Yellow scales

Black shields

Cladonia lichen

Ramalina lichen

Pixie cups

Verrucaria lichen

Lecanora lichen

Cladonia lichen

An abundance of lichens At least 18 species of lichens are common to rocky pollution-free coasts. They thrive on rock faces and ledges exposed to maximum sunlight and moisture from rain and sea spray.

Wall pennywort is an original cliff-dweller that has moved on to a new habitat of dry stone walls, hence its name.

Buck's-horn plantain, identified by its jagged leaves, grows on the short turf of cliff tops.

English stonecrop

Thrift grows in pretty rosettes on ledges and may also be found on shingle and salt marsh.

Large isolated plants of sea beet grow near the base of spray-washed cliffs. The young leaves are edible.

The sequence of sand dune development
Like waves, dunes are constantly on the move; as one builds up, matures and flattens out, young ones replace it from up front. This natural succession is shown below. Dunes develop as wind-blown sand is halted by tough pioneer grasses whose anchoring roots bind the sand together (1). Plants like marram grass extend their root systems as the sand builds up around them (2). Stable dunes (3) favour the growth of mosses and lichens, whose remains provide a rich source of organic material for many flowering plants—these dunes are termed "grey" due to the discoloration of sand by the added humus. As dunes merge with the surrounding hinterland, areas of thin turf are interspersed with dense thickets of bushes (4) like sea buckthorn, providing safe nesting sites for many birds.

enormous amounts, but if any plants can gain a foothold in even the smallest corner of the shifting sand they can start a sand dune community. Such a community invariably begins with the humble and lowly sand couch grass—a very tough plant indeed. Its leaves can withstand the stinging salt spray, and it braves the wind which piles sand on the lee side of its tough short blades. Its spreading rhizomes can cope with dry sand that may be blistering hot during the day and icy cold at night. As they grow, the rhizomes form a creeping net just under the surface which binds together the loose grains of the embryo dune. Helping the sand couch there are other plants, like the prickly saltwort and the sea rocket which straggle along in thin lines, holding the sand in place.

The pioneer dune plants slowly form a buffer zone, making up the foredune which helps to protect the area behind it from wind and spray. In the shelter of the protective barrier you will find that marram grass is able to grow, with its leaves rolled into tubes which trap valuable moisture. Marram grass has the unusual ability to put out roots from its stem, which means it can climb "step-by-step" up the growing dune, binding the sand together as it goes and encouraging the dune to enlarge. From the stems and debris of marram grass and the other plants, a sparse layer of humus forms and collects on the lee side of the dune. Here is a slightly kinder environment in which grow such things as sea purselane, sand sedge, fescues, ragwort and a number of other species. These plants still need features that protect them from drying out in the wind and sun, or from being uprooted out of the loose bed of sand by the wind. They tend to have hairy, waxy or fat leaves to reduce moisture loss and long spreading roots for anchorage and collecting water. Also, they usually grow close to the ground; in the case of sea holly, it sends out new shoots the moment it is covered by sand.

As the dunes grow, between the larger ones appear sheltered damp hollows called "slacks". In the slacks you may find orchids and other marsh flowers growing. But near the sea, the combination of wind, sun and spray continues to influence the plant life. There are great sandy patches, and the growth of trees is prevented. The dunes have now established themselves, and with the grassy plants comes an interesting selection of animal life.

Dune dwellers
For an animal, the dunes must be a strange world that is neither one thing nor the other. Yet in this apparently unfriendly terrain many creatures live happily. There is a wolf spider who hangs across the entrance to her nest burrow a curtain of silk, which she draws tight, like any suburban housewife, when she becomes alarmed. On the south coast of England, along the edges of the sandy dunes, lives the largest earwig to be found in Britain—the tawny earwig. Here, too, is the home of the sand lizard, which needs the warmth of the open sand in which to lay its eggs. This reptile is now so rare in Britain that it is protected by law. Rabbits find the dunes a useful habitat, for the sand is easy to burrow in and there are plenty of surface plants to eat. The brilliant black, white and fox-red shelducks, who rummage for molluscs and worms on the shore at low tide, use old rabbit holes as nesting sites. Skylarks also nest on the dunes, as do certain species of gulls, terns and plovers, many of whom simply make a nest by scraping a hollow in the sand and then go off feeding in

the shallows along the shore. And, of course, you will inevitably find on the dunes your resident hedgehogs and foxes, who again use the ubiquitous rabbit holes for shelter, while for food they raid the nests of the ground-nesting birds.

On the dunes in Corfu, I found many species of animal which did not occur in other parts of the island. For example, there was the little blond-coloured horned viper, who would conceal himself just below the surface of the sand with only his eyes peeping out to keep a lookout. As these small vipers were not only poisonous but very aggressive, I had to watch where I was putting my feet. It was here, too, that I found two kinds of ant lion I had not come across elsewhere. A host of other insects were attracted by the great sheets of dune lilies, their snow-white lacy trumpet-shaped flowers trembling in the lightest breeze and their scent heavy and rich. Although in other parts of Corfu the olive groves grew right down to the shore, I was interested to see that the dunes acted as a sort of cross-roads between the groves and the sea. While this sandy middle zone had its own distinct inhabitants, it was also used by crabs from the seaward side and rats, mice, snakes and lizards from the olive groves on the landward side.

A hedgehog eats eggs of a ground-nesting bird

The Corfu dunes were difficult to get to, so they were particularly good from a naturalist's point of view. When you finally arrived you knew you could spend all day there and not see a soul. Walking to these dunes took two hours, and was especially exhausting if you had a lot of equipment to carry. What I used to do was walk to the nearest fishing village early in the morning and ask a friendly fisherman to take me out to the dunes in his boat, and then he would come back for me in the evening. When I was finally landed on the dunes with my nets and bottles and boxes (not forgetting a good supply of ginger beer, since it was hot and thirsty work), and the boat had disappeared along the shore, I had only my dogs for company. It was like being Robinson Crusoe.

A rabbit outside the entrance to its dune burrow

I was sitting among the towering dunes one day, finishing off my sandwiches and ginger beer, when suddenly on the skyline (which was the top of a tall dune) appeared a lone crab. He was a shore crab, and must have been about the circumference of a teacup. We were quite far from the sea, but I knew these crabs foraged some distance inland. They would tip-toe sideways across the sand, swivelling their telescopic eyes and picking up bits and pieces in their claws and testing them for edibility. I once came across about 30 of them feasting on a dead gull. This new crab, however, was behaving in a very curious manner. He rushed over the top of the dune, threw himself madly down the steep slope, missed his footing and rolled to the bottom in an avalanche of sand. As soon as he righted himself he started to dig frantically, presumably to conceal himself. At that moment, I saw why he was so panic-stricken. Over the top of the dune, like a huge blue and green dragon, came an enormous eyed lizard. He paused at the top, and I admired his glittering grass-green scales and the bright blue markings along his side, like the eyes on a peacock's tail. He lowered his head, flicking his tongue in and out and "licking" up the scent of the crab. Then, as skilfully as an expert skier, he slid down the sand dune and arrived at the bottom before the crab was fully concealed.

The crab did not really have a chance. The lizard, heavy-bodied and many times the size of his victim, knew all about how to hunt crabs. He

Sand dune

Visit the crowded beaches during the midsummer and the sand dunes, trampled by holidaymakers, seem virtually devoid of life. But return out of season, or visit a dune off the beaten track, as we did in west Jersey one fine June day, and you will be astonished at the profusion of wildlife.

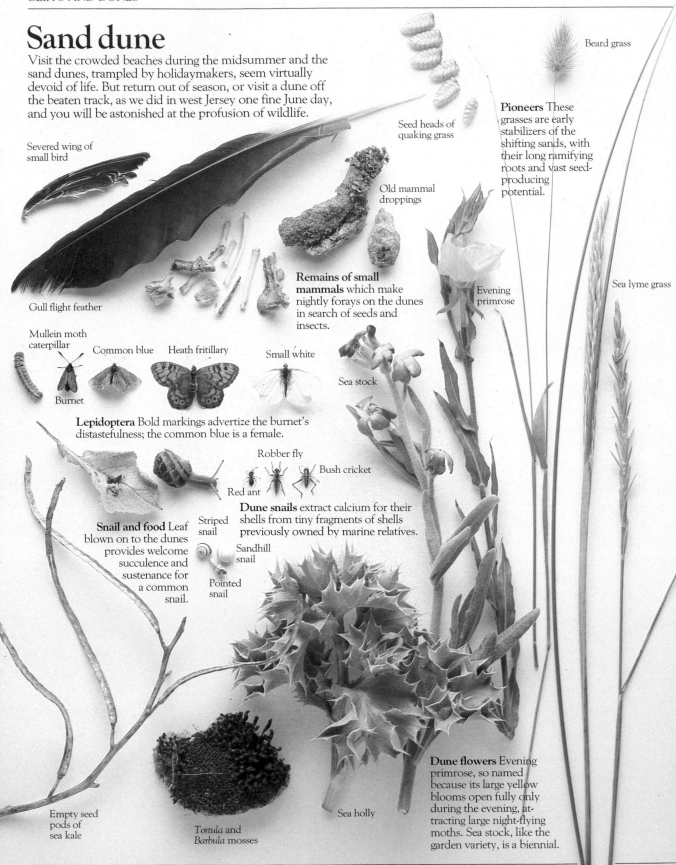

Beard grass

Seed heads of quaking grass

Old mammal droppings

Severed wing of small bird

Pioneers These grasses are early stabilizers of the shifting sands, with their long ramifying roots and vast seed-producing potential.

Gull flight feather

Remains of small mammals which make nightly forays on the dunes in search of seeds and insects.

Evening primrose

Sea lyme grass

Mullein moth caterpillar

Common blue

Heath fritillary

Small white

Sea stock

Burnet

Lepidoptera Bold markings advertize the burnet's distastefulness; the common blue is a female.

Robber fly

Bush cricket

Red ant

Snail and food Leaf blown on to the dunes provides welcome succulence and sustenance for a common snail.

Striped snail

Dune snails extract calcium for their shells from tiny fragments of shells previously owned by marine relatives.

Sandhill snail

Pointed snail

Empty seed pods of sea kale

Tortula and *Barbula* mosses

Sea holly

Dune flowers Evening primrose, so named because its large yellow blooms open fully only during the evening, attracting large night-flying moths. Sea stock, like the garden variety, is a biennial.

moved in with deceptive casualness, and the crab crouched there, his pincers up like a boxer. Suddenly the great reptile lunged, grabbed one of the crab's pincers in his mouth and then, with a sharp sideways jerk of his head, tore the pincer clean off. The crab scuttled a short distance away and then stood his ground gamely, waving his one remaining weapon of defence. The lizard spat out the claw and moved forward. Again there was a quick rush, a second shake of the head, and the crab had both pincers amputated. The lizard, now at his leisure, flipped the crab over on its back, tore off and ate the legs and then started scrunching his way through into the soft underbelly. Within half an hour there was nothing left of the poor crab except for a few bits of chitin, the carapace, and the two pincers. When I passed the spot again an hour later, hordes of ants had gathered greedily around these pathetic remains.

An eyed lizard (also called the ocellated lizard) on the trail of a shore crab

The fragility of the dune ecosystem

Dunes are frail things. Any trampling or uprooting of the binding web of plants by thoughtless holidaymakers and picnickers exposes the sand to the wind, which inevitably destroys the dune. But in some parts of the world, destruction is happening on a really grand scale.

When I was in Australia a few years ago, I was told by a friend of mine (an ardent conservationist) that I must visit some magnificent sand dunes on the coast of New South Wales—and see, moreover, what was being done to them. When I arrived, I was greeted by some of the largest dunes I have ever seen. Many were over 25 metres high, held together by forests of massive old eucalyptus trees. It was a magnificent sight: the bark was peeling off the trunks, as it does in the eucalyptus, and hanging in great fronds to reveal the pink-coloured trunk underneath, while in the branches of these mammoth trees were flocks of beautiful pink-and-grey galah cockatoos. The road wound along through the dunes and one could see how these great mountains of sand were protecting the inland areas from the powerful winds and waves that swept in towards the land from the Pacific Ocean.

Suddenly, the dunes came to an end. Ahead stretched absolutely flat sand, on which had been planted serried ranks of the Norfolk Island pine, which is not a native tree in Australia. I was horrified. Then I learned the reason for this destruction. A mineral company had discovered that in the dunes there was titanium, apparently a very valuable product for building such things as moon rockets. These beautiful old trees were being uprooted (in the process removing the homes of hundreds of animals) and then, with the aid of a huge machine, the dunes were being sucked up and the titanium extracted. The sand was spewed out to make a completely flat and featureless landscape, and planted with trees that were not even Australian! The firm which was responsible for this barbaric behaviour got wind of the fact that I had visited the area, and they wrote and said they wished they had known about my trip. They would have given me a conducted tour, because they would have liked to assure me that they were doing no harm to the environment; having extracted the titanium, they were then "replacing nature". It simply meant that this magnificent series of dunes, with all its plants and animals, has vanished for ever. On top of this, the coastline would probably become severely eroded in the future, now that the friendly dunes could no longer protect it against the wind and sea.

THE NATURALIST ON
SMOOTH SHORES

Most people think of a beach as being a great waste of sand, devoid of life and suitable only for sunbathing or the construction of sand castles. Of course, the naturalist knows this is not true. All you have to do is watch the hosts of sandpipers, plovers and other birds feeding fussily along the shoreline—they, like the naturalist, know that beneath the sand is a whole different world of creatures.

No sensible organism would live on top of the sand, being dried and blistered by the sun and wind at low tide and then rolled about like a marble when the tide comes in. You have to investigate beneath the sand if you want to find the creatures of the smooth shores, for it is here that you will find the many different kinds of burrowers and buriers and tube-dwellers. On the face of it, the sand of a smooth beach does not look a very attractive habitat, but it does have certain advantages. One of the reasons that such an abundance of life can be found here is the rather interesting fact that sand as an environment is surprisingly constant. A few centimetres below the surface, conditions are much the same whether the tide is in or out, or whether it is warm or cold, or sunny or raining. A film of water surrounds each sand grain, and this water acts as a sort of cement, sticking the grains together so that up to the high tide mark the sand is always moist. The temperature remains fairly constant all through the year, and the salinity of the water in the sand is unchanged even if there are heavy winter storms and rain.

Here, in this world beneath the sand, you find the hunters and the hunted exactly as you do in other communities. However, at first glance you may not readily appreciate what the hunted creatures feed on. In the darkness under all this sand how can the usual base of the food web—green plants—exist? The answer is, of course, that they cannot, and that the food for shore creatures is "imported" by the tides. Some sand-dwellers filter the incoming sea water through their bodies and extract food in the shape of suspended plankton and minute bits of detritus. Larger particles brought in by the tide fall to the bottom and provide sustenance for the creatures that creep along the submerged surface of the sand. Organic matter also gets mixed up in the sand itself, like a sort of soup, and you find that burrowing creatures eat the sand to extract the debris and its associated bacteria in much the same way that an earthworm eats soil.

Living between the tides
The lives and activities of shore creatures are naturally controlled by the rhythmic movements of the tides, which in their turn have their pulses manipulated by the sun and moon. Tides vary throughout the world, but in most places there are two high tides and two low ones each day. Then, twice a month, there are what are called spring tides—the water surges

Winter feeding
As the tide runs out, two knot and a sanderling take up their feeding stations. Sanderlings lightly pick at the surface for sandhoppers, running about like clockwork toys; knot tend to probe by shoving their bills in up to the hilt to extract crustaceans, worms and molluscs. Both birds breed in the Arctic tundra and turn up in large numbers along British coasts during autumn and winter.

Stick-in-the-muds
When the tide returns it covers the tubes of peacock worms (opposite) which unfurl their magnificent tentacles. Although they appear passive, minute "hairs" on the tentacles draw in water laden with food. The particles are then passed with mucus towards the mouth and sorted by size, the smallest being eaten.

SIEVING AND SHRIMP NETTING

Here are two useful methods for locating creatures hiding in sand. Sieving can be done with an ordinary fine-mesh garden sieve. Quickly dig up two or three spadefuls of sand from at or just below the water mark, and put them in the sieve. Then carefully sift away the sand at the water's surface in the calm shallows, to leave behind burrowing molluscs, worms, crustaceans and small fish. The other technique is shrimp netting, which grubs up shrimps, fish and similar half-buried creatures. You can make a reasonably sturdy shrimp net from a broom handle, half an old bicycle wheel rim, some coarse nylon netting and two short planks. Keep the base of the net a centimetre or two below the sand's surface as you push it slowly along. If sand piles up in the net, you can tip it into the sieve to uncover any interesting occupants.

Sieve away sand particles with a swirling circular motion

Use your arms and chest to push the shrimp net along

Sturdy shrimp net with reinforced base

very high up the beach and retreats very far as it goes out. Alternating with spring tides are neap tides, during which only a small portion of the middle beach is exposed between high and low water marks. Since the tides are so important to the lives of shore creatures, it is always advisable to get hold of your local tide table from a library or coastguard's office. This tells you the sea's behaviour in your area, and in consequence what the animal life is doing.

On any sandy beach, the area in which you actually find a particular animal species depends on how it obtains its food and oxygen. Fish, of course, must live in water, but other creatures such as the sandhopper can breathe air and so are able to live in the flotsam and jetsam of the strand line. (The strand is the cargo of weed, driftwood and shells collected and deposited by the tide as a long untidy necklace along the beach at high water mark.) But there are many creatures that live in the area between this strand necklace and the sea itself. These are called the intertidal animals. They get both oxygen and food from the water, so tides are as important to their survival as, say, shops and markets are for ours. Depending on the animal, it has to fit its "shopping" in with the tides. In order to do this, you find that different creatures live in different parts of the intertidal zone.

Suspension feeders like cockles must be covered with water or else they cannot feed. Some of the tiny sandhoppers, on the other hand, burrow into wet sand where they find food particles trapped between the grains (which must be the size of footballs to them). Therefore, while cockles have to live on the low shore in order to be under water most of the time, the more adaptable sandhoppers can live quite far up the beach. As the high tide drains away you may see hundreds of sandhoppers glistening like little fragments of glass as they scuttle about, madly digging themselves into the damp sand.

A little farther seaward you may find some bivalves such as the thin tellin and the banded wedge shell. These work over the sand like little hoovers, sucking up detritus from the bottom with their long siphons (unlike other bivalves, they are not suspension feeders). As the outgoing tide uncovers them, they dig quickly into the sand. They burrow very fast for a mollusc, because their slim flattened shells and their slimmer-still foot enable them to glide down into the sand like a knife blade.

The cockle has a heavily-ribbed shell which may help hold it in place against the pull of the tide. It has two short fused siphons, unlike a hoovering bivalve, and so cannot burrow very deeply. If you watch a cockle in shallow water you will see that when it wants to change its area of operations it actually leaps, pushing off from the sand with its foot. Not a very big leap, perhaps, since it just rolls over a few times, but it is a pretty good action for a mollusc and at least it allows the cockle to move about. But if the leaping abilities of the cockle are not very great, the athletic beach champions must surely be the razorshells. They are also suspension feeders with short siphons, coming out on the lower shore when the tide covers them, as cockles do. When the tide ebbs they plunge deep into the sand at an amazing speed. Their extraordinarily muscular foot and thin streamlined design allow them to dig in at half their length per second, wedging firmly into the sand.

If you want to catch a razorshell, there is a method of doing it which is rather sneaky and tricky, like tickling trout. You must approach very carefully over the sand because these molluscs, like many other shore-dwellers, are incredibly sensitive to vibrations and water currents. As you follow the retreating tide, look for the tip of the shell and a little spurt of water or a shallow depression left by the burrowing animal. Then sprinkle a handful of ordinary salt in the depression. As the salt slowly dissolves down into the damp sand it irritates the animal, who will start to back up. As soon as you see enough of the shell projecting above the sand, make a quick grab and pull it out. Do not hesitate or make a weak tug, since if the razorshell suspects you are after it then its foot will grip the sand like a suction cup as it prepares to dig back down. Once this happens you will only succeed in either cutting your hand or, what is worse, pulling the animal in two.

When the tide is out, you may see on the beach little mounds of sand in coils, like a sort of sand spaghetti, and just beside each mound will be a little dimple. The dimple is the entrance to the U-shaped mineshaft of the lugworm. Sand falls down the head end of the shaft, to be consumed by the worm. The animal digests the organic particles and then ejects the sand as coiled castings out of the tail shaft, much as earthworms make their casts on lawns. The lugworm burrow is strengthened by mucus, and body contractions pass from the tail to the head of the worm. These act like a pump to keep water flowing down into the tail shaft, across the red feathery gills of the worm and then out of the other end of the burrow. This flow of water also helps to loosen the grains, so that upper layers of sand are continually replaced as the worm eats them.

Muddy sand—a halfway world

In sheltered areas of the shore you sometimes find places where organic matter has accumulated. This brings about an increase in the various bacteria, which in turn use up the oxygen in the wet sand. So when digging in some areas you may come across zones in the sand that are black. This is due to the fact that, when the oxygen runs out, certain bacteria move in which can live without oxygen, and their activities discolour the sand. If the beach is in a very well-protected place, or is near an estuary or the mud flats of a marsh, you will probably find that there is a high organic content to the sand—that is to say, muddy sand.

A good mixture of mud and sand provides excellent anchorage for beds of cord grass, eel grass and sea lettuce, and some of the more sedentary bivalves. However, muddy sand has certain distinct drawbacks for the animals that live in it. Because of its stiffer composition, burrowing is made that much more difficult; also, because of the low oxygen content a lot of the creatures must rely on siphons or open-ended tubes for breathing. In muddy sand, therefore, one finds a whole range of animals which are different from the ones found on pure sand. For example, the gaper (called the soft-shelled clam in North America) has a massive siphon over 15 centimetres long, and is buried so deeply that it is extremely difficult to dig it out.

In this sort of habitat, see also if you can track down the beautiful peacock worms. These graceful creatures build long delicate tubes of fine mud; most of the tube, however, is buried and only a few centimetres protrude above the surface. When the tide comes in, from the end of the tube like a multicoloured fountain protrudes the worm itself. Its elegant feathery tentacles in brown, red and brilliant violet wave delicately about

A hasty retreat
Many marine invertebrates use the blood in their bodies as most efficient "hydro-skeletons". Watch *Ensis*, the razorshell, burrow down through sand. Its remarkably tenacious grip is achieved by the foot becoming gorged with blood to form a bulbous disc which acts like an anchor, while muscles attached to the shell pull the animal down. During the downward pull the shell valves are drawn closely together to aid the descent, but during the next phase, when the foot is once again filling with blood, the shells open slightly to stabilize the animal in its temporary position.

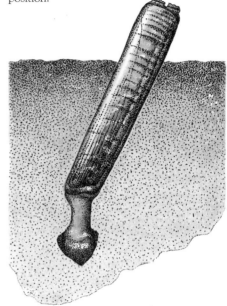

Sandy shore

One of the advantages of smooth stretches of sandy shore is that they are fairly easy to investigate. Our Jersey beach was backed by a series of high dunes; the day was warm and sunny and the sea was smooth. Like most beaches of this sort the strand line is a long necklace of flotsam and jetsam which can provide you with innumerable marine treasures. When the tide recedes the animals that are active at high water burrow deeply to prevent themselves drying out. Their hiding places can be detected by trails, casts or different shaped depressions. Tracking them down is comparatively easy; digging them out is another matter and you may well find that they can burrow faster than you can dig.

Lugworm signs Entrance hole and squiggly cast of the lugworm, which lives in sand much as the earthworm lives in soil.

Sea mouse washed up by a storm. When alive its back is brownish, but its side bristles shimmer in an iridescent multitude of colours.

Worm tubes Top portions of the tubes built by the sand mason worm, on the lower shore. The worm retreats to the bottom of its tube when disturbed.

Common whelk

Sea potato, the attractive name for the skeleton of a heart urchin shows holes for its tube feet.

Curved razor

Grooved razor

Razorshells The long thin streamlined shells allow the animals to burrow vertically through the sand at great speed.

Cast-off skin of burrowing prawn larva

The masked crab, looking like some strange creature from outer space, lies buried in the sand and draws in water through its long tubular antennae.

Prickly cockle

Pandora shell

Bivalve shells are often found in large numbers, but even so give little indication of the armies of these molluscs lying hidden in the sand.

Sandhoppers Move a piece of seaweed and numerous sandhoppers jump away.

Eggs of common whelk

Thin tellin

Common cockles

Banded wedge shell

The common necklace shell is an active predatory sea snail.

Eggs and egg cases come in a variety of colours and forms. The sea hare's sticky salmon-coloured egg ribbon has been dislodged from a nearby rock pool.

Mermaid's purse of common skate

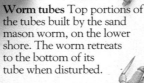

Sugar kelp

Crab bits and pieces The large carapace is from a spiny spider crab, the smaller one from a common shore crab.

Shipworms have drilled their burrows into this piece of driftwood.

Brown seaweeds torn from their rocks by rough waves provide a welcome source of food and shelter on the featureless sandy shore.

Knotted wrack

Breadcrumb sponge

Rocky shore refugees Odd bits of sponge and seaweeds have been broken from their rocky homes and cast ashore farther along the coast.

Pod flotation bladders of podweed

Gull feathers lodged in debris on the strand line.

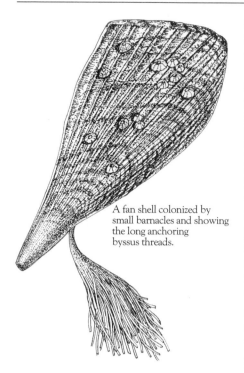

A fan shell colonized by small barnacles and showing the long anchoring byssus threads.

searching for minute suspended food particles. Should the worm sense the slightest vibration, however, it will immediately whisk itself back deep inside its protective tube.

In the days when I lived in Corfu, such things as aqualungs and masks and snorkels were unobtainable. In order to get the same effect that a mask gives you, I had made some lightweight wooden boxes with glass bottoms. When I held one of these boxes on the surface of the water it had the effect of smoothing out the ripples, so that in the clear Mediterranean I was able to view everything as though the water had suddenly and miraculously vanished. To test my first box I took it down to the mouth of a large bay where the mud and sand mingled and there was plenty of sea life. I was delighted with my new window on the undersea world. Wading waist-deep, with my back to the sun so that the glass was shadowed and I had no distracting reflections, I discovered a host of new creatures out and about on the sandy bed. Suddenly I came to a breathtaking sight—a mass of about 40 peacock worms, all close together. They looked like an enchanted flower bed, the colours shining with a marvellous iridescence as their delicate feathery "petals" moved to and fro. Entranced, I watched this beautiful sight for several minutes. As soon as I took a step closer, however, my entire magical flower bed vanished in the twinkling of an eye. The currents set up by my movements had warned the worms, and each one vanished into its tube. It was half an hour before the flower bed bloomed again, but it was a sight well worth waiting for.

In muddy sand you may also come across the fan shell, a really beautiful bivalve which is in fact the biggest found in Britain. It is shaped rather like a half-opened 18th century fan, and can measure up to 35 centimetres in length. The shell is a sort of orangey-amber in colour, flecked with little white protuberances over the broad end. Fan shells bury their pointed ends in the muddy sand, where they attach themselves to a buried rock or stone by what are called byssus threads. You can see similar sorts of threads on their distant cousins, the common mussels.

I had a very interesting experience in Corfu that involved the lovely fan shells. I knew a place where the sand and mud mixture was an ideal home for a great many of these molluscs and, because they were delicious to eat, the fishermen used to dive down and wrench them up from the sea bed. I was rowing along in my boat, *The Bootle Bumtrinket*, when I saw a fisherman friend of mine anchored out in the bay. I went alongside his boat and found that he had about 20 huge fan shells, which he was opening with his knife by cutting through the strong muscle that held the two halves of the shell locked together. To my astonishment, in each fan shell was a tiny greeny-white crab with a body the size of a large pea. They were in fact pea crabs, who use the fan shells as homes and share the food their hosts suck in. My fisherman friend told me that every fan shell he found in the area contained one of these little crabs. I kept a few for my aquarium and carefully dropped the others over the side, in the hope that they would find other shells to occupy since they had been so rudely deprived of their houses.

About 30 years later, I was making a film on Corfu and one of the sequences was to show diving for fan shells. The curious thing was that, of all the shells the fisherman opened, none contained a pea crab. Instead, inside each was a pair of small transparent shrimps that looked like mini-lobsters. This new fisherman told me that all the fan shells contained shrimps, yet 30 years earlier they had all held nothing but pea crabs. It was very interesting to find that one *commensal* animal should have completely replaced another.

Predators in sandy shallows

All the various bivalves, the sandhoppers, the lugworms and other creatures of the shore have their enemies, of course. Above them are the numerous shore birds such as oystercatchers and turnstones, plovers and sandpipers. Their enemies beneath the sand take different forms. The pretty necklace shell snail looks innocent enough, but is a deadly predator. It seeks out a bivalve and, using its exceptionally large foot, it grasps and immobilizes its victim. Then it bores a neat hole into the shell with an acid gland on its proboscis, like a burglar cutting a hole in a safe. Through the hole it inserts its proboscis and gradually sucks up the soft body of the bivalve, leaving only the empty shell, which you will find washed up on the strand line.

Some of the errant bristle worms, who move over and through the sand searching for their prey, constitute another line in predators. They are called errant because of their wandering habits. This type of worm has little black hooks on the end of its proboscis, which unrolls if you press it behind the head. One species is called the "white cat" because it is shiny and iridescent like mother-of-pearl. If you want to see these sand-dwelling predators you have to dig for them, but where to dig is the difficult thing, because few of them leave any surface signs of their activity. You may dig up one sort of bristle worm that is not a

Smooth shore syncopation

The bill shapes and sizes of wading birds show how, along one stretch of shore, birds feed at different depths on the rich harvest of crustaceans, molluscs and worms. The birds' actual behaviour while out on the sand flats also varies and helps with identification at a distance. The curlew's long curved bill gives it a striking silhouette as it feeds well out on the flats. Scattered parties and small flocks are common, walking at a steady pace. Bar-tailed godwits are birds of passage, seen during spring and autumn wading well out into the water on their long legs. They feed by shoving their bills down deep and moving them from side to side with a wedge-like motion. The oystercatcher's strong bill prises open cockles and mussels. These birds are tame and easily distinguished by their bold black and white plumage and long orange bill. Redshank both pick and probe, singly or in small groups. Knot tend to feed together in large numbers, probing the sand rapidly with a drill-like action. Their flocks form dense clouds which rise and fall in extraordinary precision. Small wary flocks of grey plover "touch down" on coastal flats and feed by running and picking. When at rest, they take up a hunched, almost dejected-looking stance. Turnstones methodically investigate isolated rocks or clumps of weed; often several will cooperate to prise up a particularly large object in order to get at the small crabs and molluscs lurking beneath.

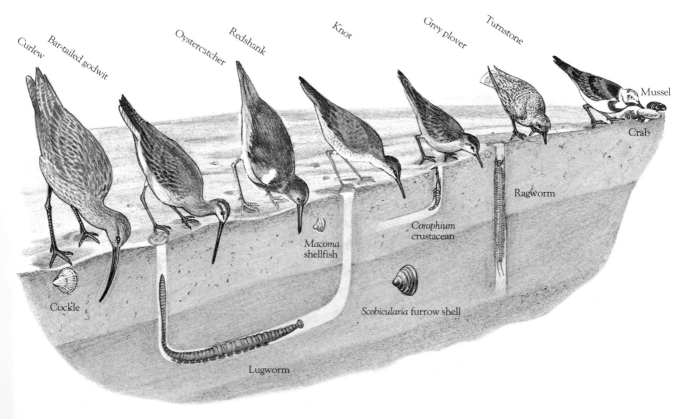

Curlew · Bar-tailed godwit · Oystercatcher · Redshank · Knot · Grey plover · Turnstone · Mussel · Crab · Ragworm · *Corophium* crustacean · *Macoma* shellfish · *Scobicularia* furrow shell · Cockle · Lugworm

predator, although it looks like one. This is the ragworm, who innocently shreds apart material with fierce black jaws. But beware—a large ragworm can give you a nasty nip.

Unexpected encounters in the shallows

One method I used to use for collecting the various sand-dwellers was to move along on all fours in shallow water, feeling in the sand with both my hands and feet as though fumbling for presents in a bran tub. Sometimes I used my glass-bottomed box and with this I could see sticking out of the sand the extended siphons of bivalves, the antennae of shrimps or even the extraordinary spike-like antennae which fit together to form the "snorkel" of the masked crab. These crustaceans were waiting patiently beneath the sand until night came, when they would come out and browse. However, this method of searching with your hands and feet is not always a very wise procedure.

In Corfu, I had just constructed a new saltwater aquarium and it needed some inhabitants. Taking my devoted band of mongrels and my collecting equipment, I made my way to a long sandy beach. I was going to try to capture a creature that I had long wanted to study, the heart urchin. These delightful little things, about the size of your palm, are covered with white spines like baby hedgehogs. At one end of the body the spines stick up like an Indian head-dress, and the body itself is a curious shape, as though someone had taken an apple and cut it in half so that one side is flat, the other side humped, while from above it looks like the classic Valentine card heart. On its flat side, this echinoderm has a sort of lip which acts as a shovel. As the urchin moves through the sand this lip scoops up material, the animal extracts the food and expels the wastes from a tiny hole at the end of the body.

The little heart urchins lived buried several centimetres down at the sea's edge. After I had traversed the length of the beach on all fours, I had collected some two dozen of them. Choosing two of the best, I released the others and then made my way up to the olive trees on top of a small cliff, to eat my sandwiches and admire my urchins trundling around in their glass jar. As I sat there, munching and admiring, I saw that the hitherto deserted beach had another occupant. A young fisherman had appeared, carrying the long-handled three-pronged trident used in Greece for spearing fish. He was moving slowly through the shallows, his trident poised and ready. I knew that he was hunting for flounder or plaice and I thought he might be successful, because in my pursuit of the heart urchins I had almost put my hand on several of these well-camouflaged flatfish. I watched him wading along, occasionally stabbing downward with the trident and then removing a wriggling fish from the prongs and putting it into his shoulder bag. I was just thinking what good luck he was having when he suddenly stopped, peered into the water and then jabbed his trident down with all his strength, just near his feet. At once, he uttered a yell of fright that echoed round the bay, leapt right out of the knee-deep water and then fell back in the shallows, writhing and trying to drag himself toward the beach. I rushed down to where he lay, white and sweating and shivering. He kept saying, "I thought it was a big plaice . . . I didn't realize . . . I thought it was a big plaice." When I wanted to wade out and retrieve his trident he stopped me, saying that he had speared a very bad fish, a very dangerous fish. Eventually, using

On warmer shores

Fiddler crabs (opposite, above) live on mud flats in the tropics and sub-tropics, digging and sifting the rich ooze. Male crabs have one claw particularly well-developed, used for signalling to females and to keep other males off their territory. The blue-eyed scallop (opposite, below left) from Australia is a filter feeder. It uses a crude form of jet propulsion to escape from predators; suddenly clamping together its valves, it lifts itself off the sea bed in a series of jerky jumps. A large tropical clam (opposite, below right) lying on the sandy sea bed has its fleshy mantle studded with single-celled plants called zooxanthellae.

long lengths of string and some waterweed hooks from my collecting equipment, we dragged the trident in. Impaled on the end, and still very much alive, was a large black torpedo fish—also called the electric ray. I knew these rays could give out a shock of up to 200 volts; and the handle of the trident was not made out of the usual wood, but was a piece of hollow metal tubing. The fisherman had received along his metal trident the full shock of the ray, which had literally knocked him off his feet. I wondered what would have happened to me if I had crawled over the ray during my urchin hunt.

We put the ray out of its misery and when we came to gut it we found it was a female with 25 embryos inside her (for these rays are *viviparous*— that is, they do not lay eggs but bear live young). The fisherman, grateful for my help, gave me the embryos for my home museum, and promised me the ray's cartilage skeleton. As I made my way home with my urchins I reckoned I had had a lucky day's collecting, in more ways than one.

Smooth sandy shores and the generally clear shallow water associated with them always look so limpid and innocent that one tends to forget there might be the occasional harmful creature. When Lee was about nine years old, she went on holiday to a place called Silver Beach in the panhandle of Florida. Her father wanted to take a picture of her with her net, and she waded into the water up to her knees to pose for the photograph. As she did so, she saw out of the corner of her eye a black shape swimming towards her. She urged her father to hurry up, as she did not know what this black shape might be. The photograph duly taken, she ran up to the shore and looked back, but the shape had disappeared. When the photograph was developed, however, there was Lee smiling gaily at the camera—and clearly visible in the transparent water, almost within reach, there lurked a very large and malevolent-looking sting ray.

In actual fact, rays are not as bad as they are made out to be. The electric ray, for example, only uses its shock to stun its prey, though of course if you step on one (or spear it with a metal trident) it will give you a shock in self-defence. Likewise, the sting ray uses its poison tail barb only as a means of defending itself.

Like their relatives the sharks, rays are true fish. But their skeletons are made up of cartilage or "gristle" (like the human ears and nose). Of the fish with bony skeletons that you can find near sandy shores, probably the most extraordinary are the flatfish. Young flatfish look like ordinary fish, but as they grow, an astonishing change takes place to prepare for a life lying flat on the sand. If a flatfish was shaped like an ordinary fish and it tried to lie flat (that is, sideways), one eye would be buried uselessly in the sand. As the young flatfish grows, one eye migrates, as it were, across the face until it reaches a position next to the other eye. Plaice, dab, sole and flounder lie on their left sides, so both their eyes are on the right side of the head. Turbots lie on their right sides, with their two eyes located on the left. It is one of the most extraordinary adaptations among fish, and if you manage to collect some young plaice you may be able to trace how the eye moves around the head.

Beachcombing on the strand line
As each tide comes in it brings a rich haul of things from the sea. When it ebbs it leaves a stranded straggling line which is a wonderful place for

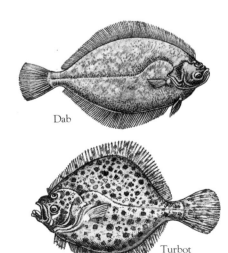

Dab

Turbot

beachcombing. There are few creatures living in this area of debris baking and drying in the sun and wind, but you will almost certainly find sandhoppers. They live in burrows they dig under the strand and then emerge in their millions in the evening, hopping about in search of bits of plant and animal matter cast up by the tide. You may also find a sea slater, the sea's equivalent of a woodlouse (which it closely resembles), though as a rule they seem to prefer rocky coasts to the strand line. There are also a number of different species of beetle that favour this area of the beach, and with them you can find the big bristly flesh-flies and some very small terrestrial worms.

Probably the greatest interest of the strand line to the naturalist is that it is a sort of graveyard, where the sea deposits all the remains that tell you what lives in and on the ocean. Storms wrench seaweed from the rocks along the shore, and you may find in the tangles of weed the corpse of a jellyfish or a crab. You might also come across pieces of wood with neat round holes and tunnels bored in them, which look as though they have been drilled by a carpenter with a brace and bit. These are produced by the so-called shipworms, which are not worms at all but bivalve molluscs, and by gribbles, which is a lovely name for small woodlouse-like crustaceans.

The strand line is really a marvellous place for a treasure hunt because you find the most unusual things. In south-west Wales, for example, hard shiny beans from tropical island plants in the Caribbean are frequently washed up, and these beans have been collected for teething babies to suck on. It is extraordinary to think of a tropical plant on the other side of the world shedding its seeds into the sea so that they can end up in the mouth of a baby in Wales. The strand line often contains the "tests" (shells) of sea urchins that look like some strange knobbly fruit, but are hollow as a drum, and also the flat pear-shaped "bones" of cuttlefish, white and crisp and brittle as a biscuit. Cuttlebone is the internal shell of this strange and lovely mollusc, just as a snail has its external shell. Then there are all the egg cases that you may come across tangled up in the weed. The so-called "mermaid's purses" are usually of a shark or ray, although the object that looks like a curved bathroom sponge is probably the egg case of the common whelk.

When you search along the strand line you will find innumerable trophies for your museum—shells of every shape and size and colour, skeletons of fish and birds, gull pellets, seaweeds from offshore areas, and many other things. Once, in Greece, I found the remains of a dead cow on the strand line; after a rather prolonged and smelly dissection its skull was added to my collection. On another occasion I found a turtle swept on to the shore by the tide. It was extremely dead, and I annoyed my family intensely by dissecting it on the veranda of our villa in order to add the shell and skeleton to my collection. One must remember that a naturalist is not always appreciated by his family.

While you are investigating the strand line, don't just concentrate on animal remains or shells. You might discover the most beautiful piece of wood which has been cleaned and softened and sculpted by the sea. I used to find, on the strand lines of beaches in Corfu, pieces of green bottle glass that had been sandpapered by the surf to look like giant frosty emeralds. The sea is a great painter and sculptor; use your eyes to discover the lovely works of art it gives you.

Seaweed scroungers
The crustacean *Idotea* lives well down the shore among seaweeds. It is one of the many small crustaceans which crawl about at night when the tide is out, scavenging on a mixture of plant and animal food. These relatives of the common woodlouse show a marked zonation down the shore, with relatively little overlap between species from strand line down to low water. The dune cockchafer is a large bumbling flying beetle which feeds on rotting seaweed along the strand line. Its fat larva lives underground on the roots of sand dune plants.

Idotea crustacean

Dune cockchafer

THE NATURALIST ON
ROCKY SHORES

Once, when I was staying in Cyprus with my brother, we used to go swimming at a small inlet near his home. The coast here was mostly rocky, full of tiny pools and caves, but this particular inlet had a small sandy beach. When we had finished swimming I would walk along the rocks to see what I could find. The area was very rich in sea life, with numerous rock pools full of beautiful sea anemones, starfish and spider crabs, and hermit crabs trundling about in their lovely shell homes. On one occasion I found a pool full of sea hares. There must have been two dozen of them, so I presumed it was a mating gathering.

It was on this strip of coast that I had a curious experience. I was scrambling about at the edge of the sea when I rounded a large boulder and came upon a small shallow rock pool. Squatting in the middle of it, like a bald-headed gentleman, was a young octopus. The moment I saw him I froze while he, as octopuses do in moments of stress, turned from grey and pink to green, then to purple and red, and then all the colours mixed in an iridescent sheen that seemed to float over his body. We stared at one another for a moment. I knew that he had come into the shallows to pursue crabs for his lunch, but he had not expected me. It was then that he astonished me by using a method of defence I did not know octopuses possessed. I must have made a slight incautious movement, because suddenly he pointed his funnel at me like a gun and shot a jet of water straight into my face. I was so surprised that I slipped and fell back against the rocks. This was all the time he needed. Quickly he humped and slithered his way out of the rock pool, plopped into the water and—using his funnel correctly, as a means of jet-propulsion—he shot off into deeper waters, trailing his tentacles behind him and belching forth great clouds of ink.

Of course, having water squirted at you by a worried octopus is a fairly rare occurrence. Nevertheless, most rocky shores are teeming with life and full of surprises for the naturalist. One of the first things to notice is that the plants and animals live in certain zones on the shore, depending on how often they need to be submerged by sea water. The plants and many of the animals attach themselves more or less permanently to the rock face, and even the more mobile creatures have a good grip on the rocky surface in order to stay in their own zones while being pushed and pulled by the tides.

On really exposed shores, where the waves are constantly crashing with great force, there is little life. The spores of seaweeds are unable to establish themselves, and it is only the tiny drifting larvae of acorn barnacles and limpets that can get a grip on the rocks in the surge of the waves. They mature into adults who have flat shapes and strong protective shells which almost become part of the rock itself. On the vast areas covered by barnacles, the larvae of the small periwinkles may settle,

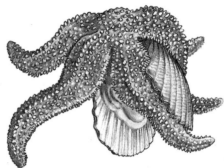

Armed combat
A cockle has little chance of escape once it is straddled and pinned down by a starfish. Here we have the final stage of the battle, with the starfish "extruding" its stomach to digest the bivalve's body tissues. Prior to this the starfish prises the two shell valves apart using its numerous tube feet in relays to tire its prey.

Rocky shore carnivore
A sea slug (opposite) browses on a hydroid colony. Instead of being stung by the hydroids' poison cells, as it eats it wraps them up with mucus, swallows them and passes them into the pink frills on its back, where they are stored as valuable ammunition. If the sea slug is molested the stinging cells are released into the water by tiny pores at the tip of each frill. As a desperate measure whole frills bursting with stinging cells may be dislodged against an enemy.

A rocky shore's rich hunting grounds
Between the boundaries of splash zone and low water mark lies an uneven terrain where one can never tire of the fund of wildlife. Most people tend to start searching well down the shore, but try looking first high up where the rocks receive only irregular dousings of sea spray. Here are thick black growths of lichen that harbour small sea slaters, while crevices are colonized by periwinkles eagerly sought after by a common shore bird, the rock pipit. Exposed shores usually have a good cover of encrusting barnacles, limpets and mussels; more sheltered shores have thick growths of seaweed. At low water mark, great brown seaweeds plastered well up against the rocks form a barrier that is difficult to penetrate.

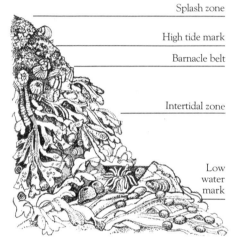

Splash zone

High tide mark

Barnacle belt

Intertidal zone

Low water mark

but as these snails grow they cannot take the pounding of the waves and so move to the relative safety of what is called the splash zone. This is the region where you get spray from the waves at high tide, and is quite extensive on exposed coasts.

The creatures that inhabit more sheltered shorelines are very numerous and of a tremendous variety, in spite of living in an environment which can be scorched by the sun and sometimes pounded by the waves. A rocky shore is similar to a forest in having many different microhabitats. In a forest you might find a whole world underneath a log or stone, or in a hole in a tree, or even on a single leaf; likewise on a rocky shore there are many different sorts of places among the rocks, each with its own special conditions and the organisms that have adapted to them. You will find a great carnival of creatures who lurk under boulders, or wedge themselves into crevices, or suspend themselves from overhangs of rock, or—with the skill of a carpenter's drill—actually bore themselves into the solid rock. Yet others lurk in great tangles of seaweed, much as creatures on land inhabit the undergrowth.

As the tide comes in, it carries adventuresome fish, octopuses and other creatures from the deep water. As it ebbs the fish sometimes find themselves stranded in rock pools and have to wait patiently for the next tide in order to escape. Busy naturalists absorbed in their work have the opposite problem, which is also suffered by some of the holidaymakers in Jersey, where our Zoo is. Instead of being released by the tide, they are trapped by it. Jersey has one of the largest tides in Europe—over ten metres—and almost every year we get people coming on holiday who go out to explore the enormous areas of rocks and pools left exposed by the retreating sea. Then they lie down to sunbathe, but meanwhile the tide has turned and is coming in almost at the speed of a galloping horse. The holidaymaker wakes up to find himself marooned on a rock, completely surrounded by the sea, and has to be rescued. The moral for the naturalist and holidaymaker alike is to be absorbed in your work but to keep one eye on the tide (and take some suntan lotion, just in case).

Plant zonation on rocky shores
It is a curious fact that on rocky shores the plants are not nearly as diverse as the animals. In the upper splash zone, for example, only patches of orange and grey lichens grow, with just below them a belt of black lichens interspersed with slimy blue-green algae which forms a rich grazing ground for the periwinkles. Below the splash zone on sheltered coasts, at the upper limit of the tide, is a small region bare of plants and occupied by barnacles and limpets. Next, slightly farther down, seaweeds and the pigmy lichen begin to grow again. The weeds are mostly brown and red algae, interspersed with a few green ones. Algae are simple plants, and the green ones represent the ancestors of the variety of higher plants that grow on land and in fresh water. But here in the sea, algae have retained their domain. The shore algae have graceful flexible bodies which enable them to bend with the waves like dancers bending with the rhythm of music. Many have gelatinous tissues which help to keep them from drying out, since at low tide they are exposed to the sun. Their jelly-like consistency also makes them slippery, which prevents them from being tangled and torn by the waves (in the same way that a long-distance swimmer coats his body with grease to enable him to move more

smoothly through the water). Although there are not very many species of shore algae, they coat the rocks and are very beautiful with their delicate pastel colouring.

Towards the lower intertidal zone you can see the leathery wracks and leaves of membraneous sea lettuce, which does in fact strangely resemble the lettuce that you buy at the greengrocer's. Then comes carragheen or "Irish moss", which is a red alga often eaten in Ireland (many species of seaweed are eaten in different parts of the world). Below this, at low tide mark, are vast beds of brown kelp, their slender blades like streamers rarely uncovered by the sea.

Clumps of seaweed provide marvellous shelter from the sun and wind and a cradle against the waves to any number of sea animals, their eggs and their young. Within the weed the small creatures who move on or through it each have a zone that they prefer, just like the seaweeds themselves, from the upper shore to the lower shore.

"Irish moss" (carragheen)

Creatures of the upper and middle shore

One roving inhabitant of the higher shore is the rough periwinkle who, very unusually for a snail, does not lay eggs but bears its young alive. Here at night you can find the sea slater, who looks like a sort of marine woodlouse, feeding on the weed and any debris trapped in it. Slaters are secretive creatures and very sensitive to light; they hide in crevices during the day, and even avoid going out in strong moonlight.

Grazing on the tufts of pigmy lichen and the grey clumps of flat wrack down to the middle shore there is a host of tiny life. It has been estimated that in this zone one could collect a quarter of a million animals per square metre—a myriad of young periwinkles, mites and even some insect larvae. Here you can find a tiny bivalve called *Lasaea rubra* which is only one millimetre long. Although it usually attaches itself to lichens, or even to empty barnacles, it can also move around quite well with its mobile foot that is longer than its shell.

Farther down the shore, where the sea covers the weed for an appreciable time twice each day, you begin to see many more creatures who live attached to the weed. There are several kinds of colonial animals—sponges, hydroids and bryozoans—which can be very much more beautiful and colourful than their freshwater relatives. (Bryozoans, the "moss animals" of lakes and ponds, are here called sea mats.) You will also come across coiled tubes that look like miniature rams' horns cemented to the weed. These are the homes of tubeworms, and if you watch closely under water you will see how they reach out from their tubes with lovely green tentacles to capture their food.

Among the more mobile creatures are the beautifully-patterned topshells, more periwinkles, a whole host of tiny crustaceans, and the wandering bristle worms. Relatives of the earthworm, the bristle worms come in a great variety of shapes and sizes and all of them are decorated with the most flamboyant lace-like fringes along the sides of their bodies.

One of the best collecting grounds in the weedy area is in the holdfasts of kelp. This seaweed clasps the rock, its holdfast acting like the roots of a tree, and from its trunk spread fronds in a great canopy. Growing on the sturdy roots are sea mats and hydroids, and under the roots are any number of different snails. If you are lucky you may find a lovely blue-spotted limpet, with its almost translucent shell streaked with a blue

Rocky shore

The strange randomness and disorder of shore life is deceptive at first sight. The experienced naturalist knows there is a great logic in why seashore plants and creatures exist where they do. There are three main factors: the position between the high and low water marks; the situation (exposed to the full force of breakers or in a sheltered cove); and the nature of the substrate (sand, rock or mud). The shore is never still, and on our calm and sunny June day in Jersey we saw continual action. Waves erode or break off rock fragments, leaving bare ground to be colonized, and every incoming tide brings with it fresh food in the shape of tiny planktonic organisms to be filtered from the water. On its return trip, the tide takes away the waste products and a cargo of eggs and larvae.

Red encrusting alga

Snakelocks anemone

Beadlet anemone

Sea hare

Slipper limpets live in chains of individuals piled on top of one another, up to nine high. The sex of each is determined by its position; the smaller upper ones are usually male, the lower larger ones tend to be female.

Plant-like animals Looking like a cross between a moss and a chrysanthemum, the anemone is a simple animal which lives anchored to rocks and catches current-borne food with its waving tentacles. The mouth is in the middle of the whirl of arms.

Coral weed

Common mussel with keelworm tubes

Desiccated brown seaweed

Barnacles on a limpet

Oyster drill

Cowrie

These sea snails, with the exception of the rough periwinkle, are all active predators. When alive the cowrie's shell is protected by a layer of tissue which retains its beautiful polished sheen.

Netted dogwhe

Banded dogwhelk

Skeleton of red alga

Sea mat on brown seaweed

Seaweed browsers Flat periwinkles browse on a piece of serrated (or toothed) wrack, rasping away with their file-like tongues called radulae.

Animal remains When alive the sea urchin has green spines with violet tips. The cuttle-bone is the internal shell of the cuttlefish; it is partly filled with air as a buoyancy aid.

Limpets The grey specimen was a victim of the whelk-like osyter drill.

Cushion star

Limpet

Dogwhelk

Painted topshell

Green urchin

Cuttlebone

Hair seaweed

Red seaweed

Blenny A common rock pool fish which stays immobile for long periods and feeds by rasping off barnacles and molluscs from rocks.

Hermit crab

Naked worm A sand mason removed from its tube, which was buried in a sandy cove.

Common shore crab

Sea lettuce

Spiny spider crab

Crabs in various guises The unadorned common shore crab shows the basic model. The hermit crab hides its soft unprotected body in a large whelk shell, while the spiny spider crab sprouts green algae which give good camouflage.

Juvenile

Female

Prawns The "berried" female carries a dark mass of up to 2,500 eggs.

Sea belt

Seaweeds The massive holdfast of furbellows is upturned to show numerous anchoring rootlets.

Furbellows

Colonized rock carries algae, sea mats and molluscs—a community in miniature.

Marine worms and their mouthparts
The bootlace worm can sometimes be found at low water mark, twined around seaweed fronds. Despite its thin fragile form it is a voracious predator—it shoots out its proboscis, spearing other worms which it swallows whole, like a snake. Ragworms are much more common, shredding up organic debris with powerful jaws. Slender paddleworms investigate the smallest of rock crevices, snaring prey with their spiky balloon-like proboscises.

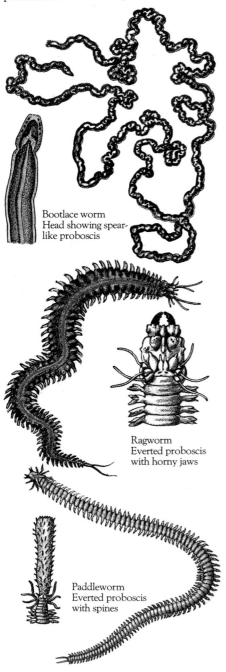

Bootlace worm
Head showing spear-like proboscis

Ragworm
Everted proboscis with horny jaws

Paddleworm
Everted proboscis with spines

that turns green when the light catches it. It is also here that you might come across the amazing bootlace worm. This worm can be up to five metres in length and is a sort of reddish-brown, looking very like the old-fashioned bootlace. It has numerous tiny eyes on its head with which it watches for its prey to approach. I remember the very first one I found was tangled up in the kelp holdfasts so intricately that it took me three-quarters of an hour to find the two ends; it took me another hour to unravel carefully the delicate folds of worm without breaking the fragile creature.

Colonizers of bare rock
Every rocky shore has patches where the seaweeds have been unable to obtain a foothold. If you examine these apparently "bare" areas more closely you will discover that they are not devoid of life, but covered with barnacles and limpets who seem to grow into and become part of the rock face. On lower rocks not covered by weeds there will probably be beds of mussels.

The various sedentary creatures who spend their time clinging to bare rock may appear on the face of it to be rather dull, but nothing could be more untrue. Look carefully at a barnacle, for instance, and you will find out what an extraordinary creature it is. There is a naturalists' saying about the barnacle, that "it welds its head to the rock and spends its life kicking food into its mouth with its legs". The legs extend from the shell like thin curved claws and wave in any prey that happen to be passing.

The common limpet is also a fascinating creature. It is really a snail that, during the course of evolution, has lost the lower spirals of its shell and become like a little cone. Unlike the barnacle, whose shell is cemented to the rock, the limpet moves snail-like around the rocks, browsing on the algae. It keeps down the growth of seaweed by doing this, but what is interesting is that a limpet browses in its own small territory exactly like a cow in a field. When the tide goes out it has to return to its "home", which is a little round depression it has made by grinding out the surface of the rock with the aid of its shell. Once a limpet has returned to its snug little depression it is almost impossible to dislodge it without damage, but if you are very cautious and patient you can wait until the limpet thinks that all is well, and with a sudden sharp push you can break its hold on the rock.

The common mussel, like many bivalves, is a suspension feeder, but instead of siphons it has evolved a very elaborate gill structure lined with tiny hairs called cilia. The cilia beat at the water, pulling it into the shell and dragging food particles to the mouth, and then expelling the strained water. Mussels are anchored to the rock by means of tough stringy threads called byssus threads.

One of the creatures that preys upon the mussel (which, after all, has no defence except to keep its shell tightly shut) is the starfish. This echinoderm with its five arms makes an easy job of eating mussels. The arms act like the tentacles of an octopus to steadily pull apart the two halves of the mussel's shell. Once the mussel gapes slightly, the starfish turns its stomach inside out and slips it down inside the mussel, to slowly dissolve and devour the contents.

Another enemy of both mussels and barnacles is the dogwhelk. This large powerful sea snail will sometimes eat limpets but the barnacles and mussels, which are incapable of moving away, form an easy prey for them. Very like the necklace shell of sandy shores, the dogwhelk has a tough drilling proboscis which either bores straight through the shell or else forces apart the valves on the barnacle. It is a curious fact that the colours of dogwhelks are variable, and it is thought that this depends on their diet. Yellow dogwhelks have usually been feeding on barnacles, whereas browny-black and mauve-pink ones have been eating mussels.

An upside-down life

On rocky shores when the tide is out, look for wide overhangs of rock. Here it is too shady for seaweeds to grow, but even when the sun shines outside, the overhang remains damp and cool and is a haven for many different animals. Various snails and worms hide away in the deep crevices, and if you look up at the ceilings in these half-caves you will find them decorated with a great variety of life. Plumose anemones dangle from the ceiling, their feathery tentacles retracted and their mouths closed. Their bodies are plump full of sea water, which they hold in to keep themselves moist until the tide returns. There are multicoloured tufts and hanging baskets of sponges and bryozoans and patches of colonial sea squirts. One of the most beautiful is the golden-starred sea squirt, which looks like a rather badly-made green sausage embroidered with little scarlet stars. In actual fact, the stars are groups of tiny sea squirt animals embedded in the gelatinous mass of green, blue, yellow or orange. The tiny sea squirts suck in water individually and, having fed off the contents, squirt it out from the centre of the star. These sea squirt colonies are favourite feeding grounds for little cowrie snails who move over them, browsing with their long proboscises licking up the tiny animals embedded in the jelly. Cowries even lay their egg capsules on the colony, and the eggs stick up like little orange vases between the stars.

Under rock overhangs you can also find solitary sea squirts, and if you examine their body shape you will see why they have the alternative name of tunicates. The body of each animal is contained in a stout bag, the tunic, and from this protrude two siphons. It is strange to think that these simple bag-like creatures could be the lowliest relatives of animals with backbones. The adult sea squirt, of course, has no backbone, but in its larval stage it closely resembles an early stage in the development of many vertebrate animals.

Under stones and boulders

As the tide retreats, you can look for creatures that hide either under or around the boulders on the shore. If you turn over a large rock, you will be astonished at the life that you will find, but be careful not to crush the creatures living on the top, and when you replace it (and *always* replace it) do so carefully to avoid damaging the creatures living beneath it.

The hard cases of tubeworms are cemented to the rock, and you can also find the lovely wiggly ribbon worms and ragworms, and the paddle-worms who swim so well by means of numerous lobes like paddles along their sides. There will probably be a common shore crab under any rock, but lower down on the shoreline you may find a velvet crab. Clad in a little furry suit, the velvet crab is most pugnacious and will threaten you with its pincers. Look out for (but do not tread on) the pinky-violet sea urchins that move along on their hydraulic tube feet and slowly browse

Getting a grip
A limpet viewed from underneath reveals its large sucker-like foot fringed with "marginal tentacles" which function like gills. Note also the rasping mouth flanked by two short tentacles. Limpets, unlike barnacles, move about on the rocks and eat young algal growths. By the time the tide has retreated they have returned to home base, a hollowed-out depression in the rock.

A closer look at some rock pool residents

A large rock pool acts like a magnet—the more visits you make, the more animals reveal themselves. Watch out for the velvet fiddler crab (opposite); for an animal with such an attractive name, it has a voracious disposition since it actively swims after small fish. An "empty" whelk shell may suddenly right itself and its occupant, a hermit crab, amble off. As it moves about its permanent passenger, a *Calliactis* anemone, sweeps the ground with its tentacles. A ghost shrimp, perfectly camouflaged against a cluster of hydroids, may suddenly lunge out to snatch a passing copepod. Chameleon prawns blend beautifully against their backgrounds. Sea spiders are difficult to spot as they feed on hydroids. The corkwing wrasse rasps crustaceans and molluscs off rocks with its horny jaws. Sea scorpions lurk in the shadows and dart out to engulf a passing prawn or fish with their large jaws.

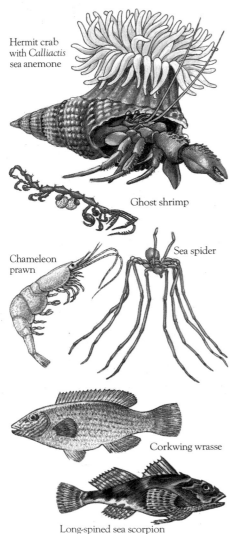

Hermit crab with *Calliactis* sea anemone

Ghost shrimp

Chameleon prawn

Sea spider

Corkwing wrasse

Long-spined sea scorpion

on the algae. The "test" of a long-dead sea urchin, which is often washed up on the strand line, is the shell which contained the soft parts of the urchin's body and which has been deprived of its spines by the action of the sea. If you very carefully file a cross-section on one of the plates of its skeleton, you will actually be able to tell the urchin's age by counting the growth rings. These are due to deposition of pigments during certain months of the year, rather than alternating periods of fast and slow growth as seen on fish scales and tree stumps.

An animal found in the water around boulders at the lowest spring tides, and one that may well surprise you, is the scallop. This handsome fan-shaped bivalve is capable of considerable movement, albeit in a rather clumsy way. What happens is this: a scallop lying quietly on the bottom slightly gapes its two halves, showing a line of tentacles between which are numerous eyes. Behind these is a sort of double curtain, one hanging from the upper valve and the other rising from the lower. By suddenly contracting the stout central muscle to snap the two valves shut, and at the same time closing certain parts of the two curtains, water is forced out between the open portions and jet-propels the scallop in a particular direction. The ability not only enables the scallop to move to new feeding grounds, but can also probably save its life when it is approached by an enemy, or it can even jet itself back into position if a wave suddenly takes it by surprise and turns it upside down.

The rock pool community

A rock pool has been likened to a miniature sea, but although it is always filled with water, the conditions in a rock pool are never as steady as those in the sea. The rock pool inhabitants must be able to withstand great swings in temperature, salinity and the oxygen and carbon dioxide content in the water. Nevertheless, from a naturalist's point of view there is nothing more rewarding than spending time investigating these lovely pools left by the sea.

At first glance, the average rock pool may yield only the lining of seaweed—pink tufts of red algae and green ones growing between them with little branches that look like Christmas trees, or with long fingers or membraneous blades. But as you look more closely things come into focus, as it were. You will notice that between the weeds are anemones feeding, their delicate tentacles waving gently about feeling for their prey. As soon as an unwary shrimp or something smaller touches one of these waving fronds, it is immediately entangled in the tentacles and drawn to the central mouth of the animal. The snakelocks anemone gets its name from its long snake-like arms and the fact that, unlike other anemones, it cannot retract completely when danger threatens. This curious anemone seeks out a position in the sunlight, because it has minute algae embedded in its tentacles. The function of this anemone-algae association has not yet been determined, and in any case such an animal-alga relationship is very rare on British coasts (although there are many instances of algae in partnership with corals or clams in the tropics).

On the thick column of the beadlet anemone you may be able to find one of the little sea spiders, which is only very distantly related to land spiders. The sea spider has no abdomen to speak of, and its stomach extends down into its legs. The eggs develop in the legs, too, but only in the male sea spider—the female lays them in the male's legs and leaves

them for him to hatch out. The larvae which hatch out of the eggs can either remain attached to the male spider or become parasites on a sea fir.

One creature that you may confuse with the sea anemone is the cup coral, the only type of coral that lives on British shores. They live individually, not colonially like the typical reef-building corals, and beneath their waving tentacles is a hard skeleton. You sometimes find these corals with a small barnacle which grows on the edge of the cup, the barnacle being only about a millimetre across.

Most rock pools have a thriving shrimp and prawn population. One of the most extraordinary is the ghost shrimp, which always reminds me of an aquatic praying mantis. It crouches still and almost invisible in its forest of weed, but with powerful forelegs ready to grasp any passing prey. Then there is the curious chameleon prawn, which has red, yellow and blue pigment cells scattered in its skin. Like a painter with his palette, by judiciously mixing these colours the chameleon prawn can change colour to merge with its surroundings. But perhaps chameleon is not quite the right name, since the true chameleon can change colour in a matter of minutes whereas the chameleon prawn takes about a week. In addition to these creatures, you will also find the ubiquitous common prawn, which comes to rock pools to spawn in the summer.

Some of the most beautiful and interesting creatures of the rocky shore pools are the sea slugs. They come decked out in a wide variety of colours with all sorts of wonderful frills and trimmings on their bodies—lobes, scalloped edges, spikes and flying buttresses. The best thing to do with sea slugs (if you are not lucky enough to have a saltwater aquarium) is simply to sketch and paint them or else take photographs— they are almost impossible to preserve in spirits, because they lose their colours and become pathetic blobs.

One of the commonest sea slugs is the sea lemon, a pale caramel beast with brown blotches and a sort of bustle of little tubercles at one end of the body. If you ever find one laying its eggs, it is really a beautiful sight. The eggs come out like pale gauzy streamers so that the whole thing looks like some strange Christmas decoration. The sea lemon is one of the few creatures that enjoy grazing on the spiky sponges. Then there is the grey sea slug, whose whole body is covered with pale whitish finger-like protuberances which make it look rather like a Merino sheep without legs. This sea slug eats anemones, but by some unexplained process manages to avoid triggering the anemone's nematocysts (stinging cells) and these cells end up incorporated into the protuberances on the slug's back. The grey sea slug can then use these plundered weapons for its own defence.

In the early summer you can find another sea slug, the soft-brown or olive-green sea hare. This creature is well named because it has two long ear-like tentacles which make it look very much like a leveret (baby hare). They come into rock pools in order to strew about their bright orange spawn and to browse, hare-like, on the sea lettuce.

Fish that spend most of their lives in rock pools are well adapted for life there. Most are stealthy hunters, cunningly camouflaged and shaped so that they can get about in among the rocks and thick weed. You will find sticklebacks, blennies and gobies, and a number of fish with suckers that allow them to hold fast to the rocks when the tide is draining from the pools. The sea "scorpions", which are fish related to the freshwater

See-through sea squirt
A solitary sea squirt, enclosed in a translucent tunic, allows one to glimpse something of its internal structure—part of the gut and a perforated pharynx. With the body wall removed all is revealed. A living sea squirt sits on the rocks and draws water into its body by the large top siphon, food particles being trapped by mucus which streams over the pharynx and is transferred to the gut. The water then passes through perforations in the pharynx, giving up oxygen, and is then forcibly ejected through the small side siphon. Colonial sea squirts form beautiful gelatinous encrustations on rocks of the lower shore.

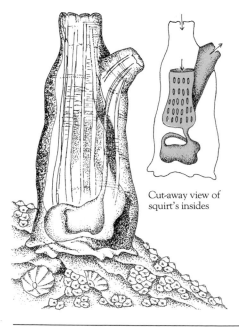

Cut-away view of squirt's insides

miller's thumb, have very squat bodies and spiny gill covers that they raise as a sort of armour should anything attack them. In some of the deeper rock pools of the lower shore you may find the beautifully coloured wrasses who, strange to say, sleep on their sides.

Hiding in the weed in a rock pool are the pipefishes, which look rather like sea horses extended into a tube. Their mouthparts are very similar to those of the sea horse. They also live on the plankton, and they carefully examine every morsel of food with their big eyes before sucking it in. Like the sea horses, the male great pipefish has a pouch in which the female deposits the eggs, so it is father who does the job of looking after the eggs and rearing the young. These babies stay very close to their father, and should danger threaten they hurriedly swim back to the safety of his pouch.

There is constant to-ing and fro-ing of small life between the rock pools. Worms and snails travel in their search for food or else move from one patch of weed to another as rapidly as possible, for not only do they have their underwater predators to guard against, but also sharp-eyed gulls look fondly upon the rock pools as marvellous hunting grounds. Some of the most active creatures you will see are the young hermit crabs who live in the deserted shells of periwinkles, topshells or dogwhelks. The hermit crab is not constructed like other crustaceans. Only its forequarters have a hard chitinous shell, whereas the hind part of the body is soft and almost snail-like. This crab can retreat into its shell just as effectively as a snail, and it draws in its big front claws to seal the opening against intruders. Of course, as the hermits grow they must each find a larger home. They are very fussy about this, rolling the potential new residence about, turning it over and over and carefully tapping it with claws and antennae to make sure it is the right shape and size. Once they are satisfied, they quickly extract themselves from the old shell and dive into the new home, turning round and grasping the inside central column with little sickle-shaped appendages on their abdomens.

If one thinks of a rock pool as a miniature sea, then a hermit crab can almost be called a miniature rock pool. Sometimes they have sea anemones attached to their shells, who hitch a ride and catch with their tentacles the small things that the crab stirs up as it trundles about the pool. The anemone may also help the crab by protecting it from predators with its stinging cells. Squeezed up inside the shell beside the crab there is usually another guest, a ragworm. It lies just in the path of the incoming currents of water that the crab sweeps past its gills, which is rich in bits and pieces from the crab's meal. It is thought that the ragworm, in repayment for this hospitality, keeps the interior of the host's shell nice and clean.

Not all the guests of the hermit crab are so thoughtful. On the outside of the shell there are often heavy encrustations of barnacles, tubeworms, hydroids, sponges and bryozoans. Two sorts of parasitic barnacle burrow into the crab's abdomen, and if you find a parasitized hermit crab you can see the barnacle's reproductive organs hanging down as yellowish lumps when the crab leaves its shell.

Rock pools are magnificent, bizarre and fascinating worlds. They are like miniature tropical forests in their colour and diversity, and the naturalist can spend many happy hours engrossed in watching their inhabitants, working out their food webs and studying their ecology.

The incomparable urchin

A living sea urchin bristles with many moveable spines and long waving tube feet. These two types of projections work together, allowing the animal to move along the sea bed and even walk up vertical rock faces. The tube feet also have a sensory and respiratory function. When the animal dies the spines fall off, revealing the hard calcareous "test" with five rows of pores through which the tube feet are extruded. The urchin's feeding apparatus consists of five chisel-shaped teeth worked by a complex arrangement of muscles and rods, the whole structure being known by the wonderful name of "Aristotle's lantern".

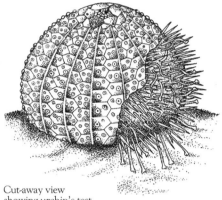

Cut-away view showing urchin's test

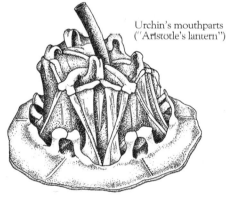

Urchin's mouthparts ("Aristotle's lantern")

THE NATURALIST ON
SEAS AND OCEANS

The ocean is the biggest ecosystem in the world. All the seas are interconnected and there is a continuous circulation of water in the currents and upwellings, but different parts of this vast ecosystem have produced their own sorts of habitat. The Mediterranean, for example, has practically no tide, whereas the tides in some parts of the Atlantic are measured in tens of metres. The ocean averages about 3,600 metres in depth with certain parts, like the Mariana Trench off the Philippines, so deep that you could submerge Mount Everest in them. But every part of this vast volume of water is inhabited by life of one kind or another. The weight of all the marine plants and animals (called their *biomass*) far exceeds that of all land and freshwater organisms combined. Again, the ocean is the most variable ecosystem in terms of energy produced by photosynthesizers. The waters of a coral reef, for example, can be as productive and as rich as a tropical rain forest, and yet in the waters of the open ocean there is by and large so little life that it might be compared to a desert. Although the open oceans have 20 times the area of the rain forests of the world, their total production is not even double that of the rain forests. The idea of "farming the sea", which was at one time a very popular conception, is therefore now known to be quite impractical. In any case, I think it is preferable to be an explorer unravelling the secrets of the ocean, rather than disturbing the balance of this vast ecosystem by attempting to farm it.

The food web in the open ocean is based on a tremendous assortment of phytoplankton. This is made up of such things as the beautiful diatoms in their microscopic glass houses, and dinoflagellates which can actually become illuminated and glow. Next, there are the tiny creatures that feed on the phytoplankton. Most of these are copepods (a kind of crustacean), but there are also tiny shrimp-like krill which are the basic food for some of the whales, the largest mammals on the planet. Dead plankton and the bodies of bigger animals and plants drift down to the huge depths of the ocean beds, where weird fish live in perpetual darkness, strung with "lights" like a ship's portholes at night and rigged with luminescent "bait" to attract prey. Contrast these with a sunlit coral reef, where you get the fantastic multicoloured host of fish that revolve about the reef like some brilliant carnival. Around the reef, the base of the food web is not only the plankton but also various algae that grow in the tissues of living coral and flourish on the outside and inside of the coral skeletons.

The colourful complex reef

A coral reef supports a teeming mass of organisms that play the same roles as creatures on land. I think I only realized the parallels between coral communities and land ones when I visited a wonderful reef that

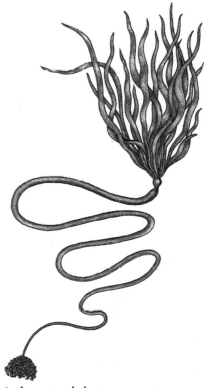

A phenomenal plant
"Sea otter's cabbage" is a brown seaweed which grows to over 40 metres in one season. It is found off the coast of California, in deep water with strong tidal currents. This region is also famous for the even larger giant kelp *Macrocystis*, which grows to 65 metres.

Coral reef partnership
Tropical clownfish (opposite) associate with large sea anemones, scurrying unharmed into their poisonous tentacles when danger threatens. The anemone probably benefits from the association by picking up left-overs from the clown's meal that drop on to its tentacles. The colourful fish show a strange sex-change during their life history; young ones are male, while older larger individuals are always female.

fringes the island of Mauritius, in the Indian Ocean. Floating in water that was so salty one literally could not sink, I felt the way a hawk must feel as it floats over forests and meadows, except that here the forests and meadows were made up of branches of coral and clumps of weed. Slowly I came to look at these great coral beds in the same way as I looked at a tropical forest, and I began to see similarities. There were the equivalents of the herbivores, in the shape of molluscs who graze on algae, and the omnivores, in the shape of multicoloured parrot fish browsing off algae and coral polyps—you could hear their parrot-like beaks scraping away quite clearly through the water. And then there were the predators— slim swift fish that hid in the blue branches of the stag's horn coral and suddenly slid out in order to attack. I found that very soon I could divide up the great coral beds as one would do a forest, seeing fish that lived on the "ground", i.e. the sea bed, and others that made their lives in the upper branches of the coral jungle. Some were predators cruising over the coral like eagles, and some were harmless and colourful like the hummingbirds or parrots in the rain forest. I spent six weeks in Mauritius and every morning I went out as soon as it got light and spent an hour fish-watching. During the whole of that six weeks, I never came back to the hotel without having seen at least one, and sometimes several, species that I had never seen before. This might give you some inkling of the complexity of life on the reef.

Offshore life

I have been lucky, during the course of my life, to have travelled on and in many cases, to have swum in most of the oceans of the world. Being brought up on an island in the Mediterranean, during the summer months I used to spend more time in the sea than on land. I had my own small boat, in which I rowed up and down the coast collecting specimens of all sorts, but I couldn't get any deep-water specimens until I made friends with the local fishermen. They would tell me when they were going to pull their nets in, and I would go along to help. In addition to the fish in the nets there was a mass of other things which, though useless to the fishermen, were like gold dust to me. There were huge creamy-white crabs decorated with scarlet blotches; there were sea horses like delicate chessmen; there were corals and rocks covered with seaweed, that when shaken out into a container of water proved to be a veritable zoological garden of life forms.

Of course, today one can have a seawater aquarium with electric pumps to aerate and freshen the water, and one can even make artificial sea water with special packets of salts. In those days, however, living in the wilds of Corfu, there was no electricity and so my aquarium was quite small, and in hot weather I had to lug buckets of fresh sea water up the hill to the villa four or five times each day. I could only keep my specimens for a couple of days before releasing them, since in spite of my efforts the water soon became foul. Nevertheless, my primitive aquarium did enable me to observe some splendid things. I saw crabs changing their skins and then, with the new carapace as soft as damp paper, they would lurk shyly in the rocks, trying to avoid predators. I saw a male sea horse "give birth" to eight babies and I watched an octopus the size of the end of my thumb hatch from an egg. A spider crab, from whose carapace I had removed the disguise of weed that he had planted on his back,

Common octopus

Male short-nosed sea horse showing brood pouch

A Mediterranean crab in defensive posture

redecorated himself with the things I had given him—seaweed, bits of brightly coloured coral, and even a living sea anemone.

I remember one night in Corfu we went for a moonlight picnic. It was midsummer and very hot, and the water was full of "phosphorescence". This is due to the seasonal bloom of certain species of dinoflagellates in the plankton. They have the ability to change chemical energy into light energy, and when the water is disturbed they glow with a weird light. Wading into the sea was like marching through a golden-green fire. I swam out into the bay and was lying on my back contemplating the moon when suddenly, with great snorts and gurgles, a school of dolphins arose all around me and startled me out of my wits. They plunged and played around me for several minutes, churning up the water so that it glowed brightly, and then sped out to sea leaving a burning trail of light behind them.

The open sea

It is, I think, a great pity that nowadays most people travel by air, for they miss so much. Travelling by sea, as I have done for most of my life, is very rewarding. I have seen six sperm whales rolling majestically in the water like giant tar barrels, letting out fountains of spray from their blowholes. I have watched the effortless glide of a giant albatross which followed the ship day after day, never varying its position and never, while I observed it, flapping its wings even once. Off the West African coast I once saw a huge concourse of Portuguese men-o'-war, those graceful but dangerous stinging jellyfish, magenta-pink in colour and trailing their tentacles like ribbons on a bonnet. The jellyfish herd stretched as far as the eye could see and took an hour to drift past us.

The manta ray is one of the oddest-looking of all fish to be seen in the open sea. It is shaped in a distorted triangle, with "wings" like an ultra-modern aircraft. The manta uses these great wings, six metres across, to "fly" through the water, flapping like a bird. The first time I saw these creatures was off the coast of Sri Lanka. Six huge manta rays suddenly appeared. Their great wings would come clear of the water and curl over at the tips. As they headed farther out to sea, one of them suddenly leapt out of the water and crashed back on to the surface with a terrific report and a splash. Immediately the others followed suit, so that the air was filled with the sound of the manta rays leaping and crashing back into the sea, with a noise like a heavy gun bombardment. This behaviour is thought to be a method that not only mantas, but also other big fish and even whales use to try and get rid of irritating parasites on their skin. But as the mantas I saw were all leaping after one another, like a sort of ballet, I can only conclude that they were simply enjoying themselves, or else perhaps one of them had seen an enemy and so the whole school was trying to frighten it off with the noise of their bodies crashing on the surface of the water.

Sailing through the Sargasso Sea, the vast area of floating seaweed in the Caribbean, was an exciting experience. It was thought if a ship sailed into this almost solid mass of weed it would get trapped. Unable to free themselves, the crew would die of starvation, and the ship would gently rot away in the arms of the weed. It is of course a fable, even though at one time everyone was convinced that the Sargasso Sea was a short of ship's graveyard. Nevertheless, leaving the legend aside, the

The manta's misnomers
Devil fish and devil ray are two common names for the giant manta ray, but although it looks terrifying, it is completely harmless, cruising through the surface waters by rhythmic undulations of its great pectoral fins. This ray subsists on a diet of plankton funnelled into the mouth by the small anterior fins.

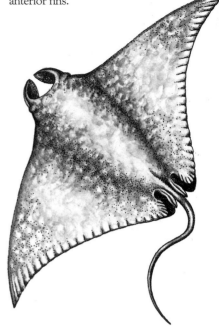

The countless krill

As insects swarm on land, so do crustaceans swarm the seas. Small shrimp-like animals are especially prolific in the surface waters of the polar seas. They are known collectively by the Norwegian sailors' term krill, and form the staple diet of many large whales.

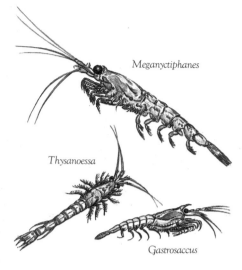

Meganyctiphanes

Thysanoessa

Gastrosaccus

Two great marine mammals

The humpback whale (opposite, above) is an individual among baleen whales in having extraordinarily long flippers and bosses on its snout. It often rests at the surface with one flipper extended like a sail, which it brings down with a great resounding smack. It also has the habit of leaping clear out of the water in a wonderful backward arching flip. Like many of the great whales it feeds in cold waters and migrates to tropical seas to breed. The dusky dolphin (opposite, below) is small by dolphin standards, reaching less than two metres. A fast and energetic swimmer, it is confined to the colder waters of the Southern Hemisphere. Schools 20 strong of these inquisitive creatures are common; they will come to the aid of any of their fellows who are wounded or have become stranded in fishing nets.

Sargasso seaweed itself is very distinctive and a number of creatures, ranging from shrimps to fish, have adapted themselves perfectly to resemble the weed. They are so well camouflaged that they are extremely hard to see among the fronds, even if you know that they are there. The first time I went to South America, the ship I was travelling in passed through the Sargasso Sea and I was all prepared. I had borrowed several plastic buckets from the kitchen and with the help of the Chippy (the ship's carpenter) I had bent some stout wire into a series of hooks and tied them to a long length of nylon line. As soon as we started to pass through the great mats of yellow-green weed, I flung my hooks overboard and soon pulled up quantities of weed, and with it its attendant fauna. I had a wonderful day, shaking these great wigs of weed out into my buckets of sea water and watching what emerged. It was a world of its own. Tiny fish were shaped so that they looked like the deeply lobed pieces of weed; shrimps were so transparent that they took on the colour of the weed; and tiny shells stuck to the fronds and were so similar to the floats of the seaweed that you couldn't tell the difference.

The oceans may still have lurking in their depths many creatures which are unknown to man. Comparatively recently, for example, the coelacanth was discovered swimming happily off the Comoro Islands in the Indian Ocean. Everyone had thought that this lobe-finned fish, a relative of the first land vertebrates, had been extinct for millions of years. Not so very long ago, a Danish ship trawling in deep waters in the Atlantic pulled in an enormously long elver (young eel). It was so long that, if it was in proportion to the elver of the common eel, it meant that deep in the ocean there are eels nearly 30 metres long. Sharing these gloomy depths with giant eels are giant squid, some of which attain enormous size. One of the biggest was discovered aground on a sandbank in a place with the charming name of Thimble Tickle in North America. The tide was ebbing, and so to make sure the huge creature did not escape, the fishermen threw their grapnels at it. The barbed hooks punctured the body, and the men secured the grapnels with ropes tied to a tree. The poor giant squid, unable to escape, writhed about in its death agonies. When it was finally dead, its body was measured at over six metres in length, and one of its tentacles was over ten metres long.

In the early 1800s, which is only a moment ago in the time that life has existed on Earth, Africa and South America were just beginning to be explored. Africa was known as the "Dark Continent". The ocean is still today virtually unknown to us—it is our dark continent. The seas contain species we have never seen and creatures with extraordinary behaviour patterns that we know nothing about. How, for example, do eels from streams in Europe and North America find their way thousands of miles to the Sargasso Sea, where they spawn and die? And how do the new elvers find their way back to the same streams their parents came from? We know that spiny lobsters perform mass migrations across the sea bed, but we don't know why they do it, or where they are going, or where they came from. Recently we have discovered that whales sing to each other in the depths of the ocean. Why they sing and what they say to each other are still a mystery. For a naturalist today, the sea can provide the interest and the fabulous excitement that must have infected the early naturalists in their explorations of the great newly-discovered continents.

THE NATURALIST
AT HOME

Although observing and experimenting in the field is of course vitally important to all naturalists, there are many observations and experiments which are much more easily done when you get home. It is also essential to make a considered and permanent record of the work that you have done in the field, because even the best memories are fallible. The only way you will be able to compare a specimen with the details of one you saw months, or even days ago, is to keep a clear and organised record of all the observations that you make. The naturalist at home is in many ways just as important as the naturalist in the field. Most of the great naturalists did a lot of work at home and people such as Darwin and Fabre were lucky enough to have very large rooms in which they kept not only preserved but also live specimens, so that they could conduct their experiments. Lee and I are lucky because we live in two places—in the zoo in Jersey and in our house in France—and so we can keep and preserve our specimens in two places.

But you do not have to have a lot of space or elaborate equipment. It is possible to carry out most of the techniques described in this section on a shoestring budget, adapting everyday items as containers for your specimens, and purchasing a few inexpensive pieces of equipment. It is true that a few experiments require special equipment which is certainly not cheap. But this is rarely essential. For the naturalist, the limits to activities both outdoors and at home are usually ingenuity and imagination, while the most important items of equipment, at home as well as in the field, are your ears and eyes.

When you have finished field collecting and you have taken your specimens home, the first thing to do is to sort them out carefully. Make sure that the animals are correctly housed and fed, and the plants watered and adequately supported. Then decide which specimens you want to keep permanently, and which you will keep for a time before releasing them into the wild. Always release them into their own habitat—only here will they be able to find the right conditions.

With your permanent specimens you can work out what we call special project boxes. That is to say groupings of your various specimens combining photographs, drawings and the different preservation techniques that you have to use. For example, one project might be a dissection of a hedgerow or a cross-section of a forest; another might be the life history of a butterfly or a beetle. The room in which you work and keep your specimens should be like a miniature museum and should display all the interest you take in the natural world and its conservation. Always try to use your work as a naturalist in a worthwhile way, and always remember that it is part of your role as a naturalist to educate others and help them to observe, appreciate, and therefore conserve the natural world around them.

Unpacking the naturalist's rucksack
Returning from a springtime field trip to a hedgerow and a pond, the rucksack has just been emptied on the table (opposite). Notice that all the items have been carefully labelled in the field, to avoid confusion. Already sorted out, to the top right of the picture, are plants from the hedgerow, for preservation with the flower press which is leaning against the wall at the top. These specimens have already been put into glasses of water, to keep them fresh. The fluffy seed heads of the colts-foot, the serrated leaves of the white dead-nettle and the narrow pointed leaves of the stitchwort are all clearly visible. There is also a brimstone butterfly, just removed from its protective paper envelope. In the middle, in a plastic box, are stinging nettle leaves curled over to form homes for red admiral caterpillars. These larvae will be reared, keeping one adult for the collection and releasing others into the wild. At the bottom are galls and fungi for preservation, and a selection of soil-living creatures for study with the hand lens. There is also a jar of tadpoles, collected in order to study the fascinating meta-morphosis from tadpole to adult frog. Still in the rucksack are several plant specimens carefully wrapped in plastic bags, to preserve humidity and keep the plants fresh. A twig from a sweet chestnut tree shows the young spring leaves that have emerged from the tree's "sticky buds".

TECHNIQUES AND EQUIPMENT

SETTING UP A WORKROOM

The naturalist's workroom is a most important place and it must be efficiently laid out, since it is here you keep your specimens and conduct experiments that would be difficult or impossible to carry out in the field. It is best, if you are lucky enough, to have a special room for your materials. Failing this luxury, a table and some wall space in a fairly quiet part of the house—perhaps your bedroom—will do. Remember that wherever you choose the room must be dry, because many specimens rot in damp air.

Try to get the biggest work table you can find, with drawers in it for storing equipment. You can often find old-fashioned kitchen tables in second-hand shops, and these are ideal. Place the table near a window, since daylight is best for doing fine work, but hang curtains at the window so that you can draw them to prevent sunlight fading your specimens. If you do need artificial light use an adjustable table lamp (fluorescent light tends to strain your eyes). Make sure the window can be opened for ventilation when you are working with ether or formalin.

Ideally, your room should have a sink with running water, but failing this some shallow plastic containers will suffice. Cover the wall space with cork tiles or pegboard on to which you can fix charts, notes, drawings and (the most important) a big year-planner. The planner is excellent for long-term observations.

The workroom really cannot have too many shelves—not only for books and magazines but also for your museum specimens and your live creatures in their cages or aquaria. Remember also to leave a space in one corner for the storage of things which you buy cheaply in bulk, such as cottonwool and surgical spirit. This space can also be used for the free equipment a naturalist can get in his own home—matchboxes, cardboard boxes, jam-jars and any number of similar containers.

The dissecting kit

The dissecting kit is the naturalist's prize possession, and is used not only for dissection but also for the general handling of tiny things. A good naturalist uses his instruments like an artist uses his brushes—gently and carefully, and always washing and drying them when finished. Unlike an artist, however, you cannot paint over your mistakes so the golden rule is: Go carefully. Use a drawing (available in most biology books) to guide you in the dissection and identification of internal parts. As you dissect, be sure to keep your specimens from drying out by dampening them with a little water or saline.

The purpose-made workroom

A naturalist who is lucky enough to have his own workroom might lay it out as shown here. A large well-lit table, plenty of shelves and cupboards and some wall space are most important. A sink with running water is useful but not essential.

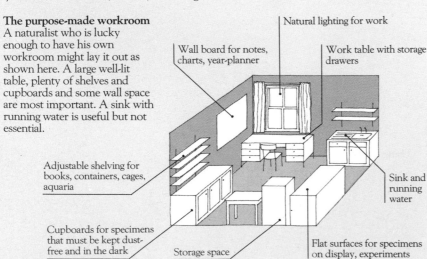

Natural lighting for work

Wall board for notes, charts, year-planner

Work table with storage drawers

Adjustable shelving for books, containers, cages, aquaria

Cupboards for specimens that must be kept dust-free and in the dark

Storage space

Flat surfaces for specimens on display, experiments

Sink and running water

DISSECTING INSTRUMENTS AND THEIR USES

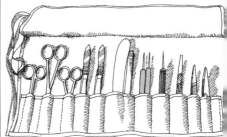

Complete kit

Ready-made kits usually come in a cloth roll with pockets for the instruments. If you buy instruments separately, wrap them in any soft cloth and keep them in a biscuit tin or, better still, a pencil case.

Scalpels

One-piece scalpels are still obtainable but are difficult to sharpen. A medium-sized disposable-blade type with a range of blade shapes is best for most work. Follow instructions on blade-changing carefully, for the blades are very sharp.

Straight   Half-curved   Curved   Concave

Scissors

Try to obtain two pairs of scissors: a small pointed pair for very fine work and a round-bladed pair (so as not to puncture the organs) for cutting through thick skin or small bones.

Probes (mounted needles)
These are used for separating different parts of the specimen and lifting delicate things. Have one blunt and one sharp, with wooden handles.

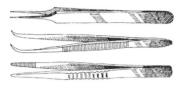

Forceps (tweezers)
A small pointed pair, perhaps with curved tips, for holding tiny items delicately, and a stronger pair for gripping and pulling tougher bits and pieces.

Paintbrushes
These are used for dampening dry areas, smoothing hairs and lifting up very delicate specimens.

Dropper (pipette)
An everyday item—an ordinary eye dropper from a medicine cupboard is fine —which is useful for putting drops of water under the microscope, etc.

Dissecting dos and don'ts
1 Do wash your hands thoroughly before and after all dissecting work. If you can, wear thin rubber "surgeon's" gloves.
2 Don't dissect an animal that was really sick. Mammals and birds in particular can pass on diseases or parasites to humans.
3 Don't dissect if you have cuts or sores on your hands.
4 Do be careful not to cut yourself. If this happens, wash the wound with soap and water and apply disinfectant, and then check with your doctor. It is possible (especially with mammals) to contract a disease such as tetanus, anthrax or even rabies—so beware.
5 Do thoroughly clean all your instruments and work surfaces with disinfectant when you have finished. Dispose of remains hygienically, in sealed plastic bags or by burning them.

USING THE HAND LENS AND MICROSCOPE
Nowadays the equipment needed to reveal the world of small animals and plants around you is not too costly. A hand lens or a microscope will introduce you to new creatures and will reveal the detail in parts of the larger ones, from a bird's feather to a fly's eye.

The hand lens
For a naturalist, the hand lens is indispensable. You will need one for field work anyway, so this can be used at your work table as well when you are doing dissections or arranging small specimens. The best all-round hand lens to get is ×8 or ×10. Hold the specimen in a good light or you will not be able to make out the really interesting fine detail. Reminder: Tie some brightly-coloured wool to your lens in case you drop it outside.

The binocular microscope
Often called a dissecting microscope, the binocular microscope is an expensive piece of equipment but a must for the serious naturalist. It gives you a clear view for examination and dissection of very small specimens. "Binocular" means two eyepieces giving three-dimensional vision, unlike a hand lens or a light microscope. Some models have a rotating carousel with three lenses, for example ×10, ×20 and ×50.

For dissection under the binocular microscope you need very bright lights on the specimen (some models have special built-in lamps). Practise on a few unimportant objects at first, for it is quite an art. Until you get used

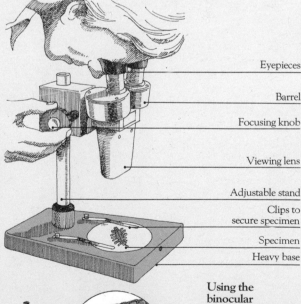

Eyepieces

Barrel

Focusing knob

Viewing lens

Adjustable stand

Clips to secure specimen

Specimen

Heavy base

Using the hand lens
Hold the lens about 5 cm from your eye. Bring the specimen to about the same distance from the lens, then fine-focus—it is sometimes easier to move the specimen rather than the lens.

What you can see
The almost invisible flea magnified 50 times is revealed in great detail.

Using the binocular microscope
The microscope is designed to focus about 20 cm above the specimen. This allows room for your hands and dissecting instruments beneath the lens.

to it, at ×50 magnification your fine-pointed forceps will look and feel like a pair of giant bolt-cutters! The binocular microscope is ideal for dissecting small mammals such as mice and shrews.

The light microscope
In the same way that binoculars can bring distant things near to you or an aqualung can allow you to see the world of underwater life, this wonderful instrument enlarges the naturalist's world immeasurably. It brings to life the tiny world that swarms all around you in a drop of pond water or in the water from your butt or gutter. The microscope allows you to examine the composition of a bird's feather or the delicate structure of a fly's wing. It makes you realize for the first time that you are really a Gulliver in Lilliput. With this type of microscope, you can only see semi-transparent things. To prepare a specimen, you must cut a very thin slice of it (a section), add a few drops of a biological dye (called a stain), and mount the section on a glass slide.

Microscope illumination
The microscope shown here has a mirror which reflects light from an adjustable desk lamp through the specimen and up the barrel. An alternative design has a built-in bulb to provide illumination for viewing. The higher the magnification you use, the more light will be necessary to illuminate the specimen.

MAGNIFYING SPECIMENS

The hand lens gives magnifications of ×5 to ×20 (×10 is the commonest magnification). It will reveal details of small animals and plants, e.g. the mouthparts of a butterfly or the flower of a buttercup.

The binocular microscope provides magnifications of ×10 to ×50. With it you can see details of very small animals and plants, e.g. the root hairs on a plant or the scales on a butterfly wing.

The light microscope allows subject magnifications of ×20 to ×500. At ×100 you can see virtually invisible worms and algae in soil and water. At ×500 you can see a single cell.

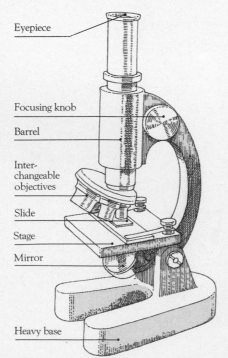

Eyepiece

Focusing knob

Barrel

Interchangeable objectives

Slide

Stage

Mirror

Heavy base

NOTEBOOKS

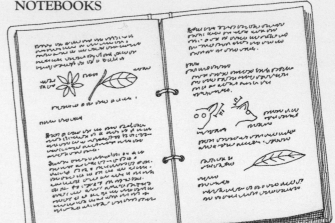

The notebook is one of the most important of tools for the naturalist. You should always have two: a stoutly-covered one for notes in the field (quick drawings, notes of colour, etc.) and a more complex one for your study, into which you can transfer and elaborate your field notes. The field notes, taken on the spot, are invaluable because your memory can play tricks. At the end of an exciting day, unless you use your field notebook, you might forget whether a bird was purple, green or blue, or whether a fish was shaped like a dart or a kite. The home notebook is your final record. Keep it safe in your study. Try and get a large loose-leaf one so that in places you can insert sheets of drawing paper on which you can sketch the creatures you have observed. A loose-leaf notebook has other advantages. You can keep it up to date easily, inserting new material about particular species, so that it is easy to find. It also allows you to include photographs of the plants and creatures you have observed. You can also illustrate animal behaviour patterns, dissections, and other stages of the experiments you do. Remember the essentials to write in your field notebook and transfer to your home notebook—date, day, site, time of day, the weather, surroundings, detailed description of your subject, observations of its activities, what happened to it while you were watching it.

Using index cards
In addition to the detailed records in your notebook, it is helpful to use index cards. Using one card for each specimen, write down its name, where and when you got it, and any other relevant information. Number each specimen and each corresponding card. It is easier to have your collection on file like this than to have to search your notebooks for the information. Remember that specimens without data are useless.

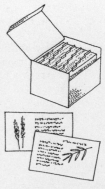

HOME PHOTOGRAPHY

If you are collecting with a camera, always try to photograph the specimens in the field rather than bringing them home (see page 196). A camera is also very useful in the workroom. One of the best cameras for an amateur naturalist to start with (and one that offers excellent value) is the adaptable "SLR" (single lens reflex) camera, which uses 35 mm film. A normal camera lens of 50 mm will give you excellent pictures of objects the size of a large butterfly or a rose, but if you want really good detailed pictures of small things then you will have to invest in specialized close-up equipment. The simplest accessories are a reversing ring, allowing you to use the standard lens back-to-front, and extension tubes, which you fit between lens and camera. Both of these allow you to focus objects closer to the camera than with a standard lens alone. Extension tubes will give images up to life size with a standard lens. You can also use supplementary close-up lenses.

For still higher magnifications and better image quality a special macro lens is ideal. You can use this mounted directly on the camera, or with extension tubes or bellows to give progressively higher image magnifications. Used in conjunction with a bellows unit a 55 mm macro lens will provide magnifications up to five times larger than life size. Whatever equipment you use, the amount of the subject that is in focus is always small when you work close up, so focus carefully.

Remember, if you are taking a close-up picture of a plant or a creature, take a photograph of the habitat around it as well as a reminder of where you saw or found it. A subject like a butterfly can be "slowed down" for photography without damage by putting it in the refrigerator for a time.

There is now a wide range of highly sophisticated (and expensive) photographic equipment on the market, but whether your equipment is expensive or cheap, remember that the vital rule of the nature photographer is: the picture is of less importance than the welfare of the subject.

Close-up equipment

An SLR camera and standard lens will take good close-up pictures with attachments such as a supplementary lens or reversing ring. A special macro lens is best. You can extend its range with a set of extension tubes or, better still, a bellows unit.

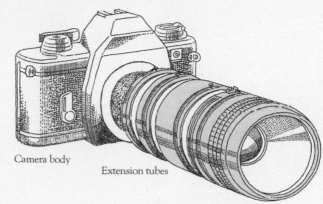

Camera body

Extension tubes

Macro lens

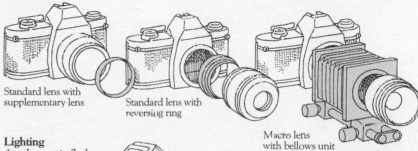

Standard lens with supplementary lens

Standard lens with reversing ring

Macro lens with bellows unit

Lighting

An electronic flash with an adjustable head is best. It enables you to "bounce" illumination off ceiling or walls, to avoid the harshness of frontal lighting. A ringflash gives even illumination ideal for close-ups.

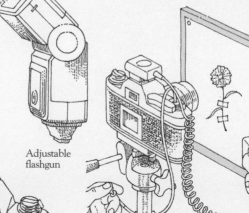

Adjustable flashgun

Home studio setup
Your camera must always be steady, so ideally you should have a tripod and cable release. If you take a specimen home, try and take a "prop" with it—the branch on which you found it, or something similar. Choose a backdrop of a contrasting colour to your specimen, so that the subject stands out.

Ringflash

PRESERVING METHODS

Before adding something to your collection you should always ask yourself whether it is going to be of value to you for future reference or whether you are just collecting for the sake of it. The Victorian period, when amateur naturalists took hundreds of specimens, is mercifully gone. Today it is essential to collect with a great sense of responsibility (see page 320).

There are two ways of preserving specimens—the dry method, for example the pressing of flowers or taxidermy, and the wet method, which involves preserving in spirits. You have to choose the method of preservation which best suits the specimen so that its features remain intact and its colour stays as natural as possible. Take great care with the preservation of your specimens—it is easy to destroy items needlessly.

Wet preservation

Preserve your specimens in either alcohol or formalin, both of which can be obtained from your chemist or a biological supplier. They preserve soft tissues well, but you will need strong glass or plastic containers with tight-fitting tops. An important advantage of this method of preservation is that there is no risk of your specimens being damaged by mould or pests, provided that you pickle them correctly. But wet preservation has the disadvantage that it does not allow internal examination of the specimen.

For preservation in alcohol, you buy pure (or "absolute") alcohol. To make the 70 per cent solution required for preserving, dilute seven parts alcohol in three parts distilled water. Alternatively, you can buy ready-diluted surgical spirit. But if you do this, you must check the label to see if the dilution is correct. To produce formalin, you usually buy formaldehyde in a 40 per cent solution. This needs further dilution—add one part 40 per cent formaldehyde to three parts distilled water to make the ten per cent formalin that is best for

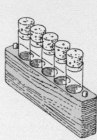

Containers
For specimens preserved in spirits, all types of jars are useful, provided that you use plastic washers to protect the lids from corrosion. The best are Kilner jars or corked test tubes.

PRESERVATION TECHNIQUES

	Wet	Dry
Technique	Pickle the specimen in preserving "spirit" (alcohol or formalin) in clear containers	Leave the specimen set out in a dry atmosphere, or press it between layers of absorbent material
Containers	Clear glass jars with sealable lids, eg Kilner jars—ensure that metal lids are protected from corrosion by a rubber or plastic seal; test-tubes with tightly fitting plastic caps	Dust-proof boxes and cabinets; for insects use a material that will not exude harmful oils or resins, eg oak or mahogany. Otherwise use sealed plastic boxes. For shells, eggs, and bones, glass-topped boxes are ideal
Suitable for	All aquatic specimens; whole animals (especially soft-bodied invertebrates such as worms and caterpillars); fish, reptiles and amphibians; dissections	Most plants (except for specimens that are very succulent); shells and coral; birds' eggs and feathers; mammal and bird skins; skulls, teeth and bones; hard-bodied invertebrates (such as insects and spiders)
Precautions	Store at low temperature (6°C is ideal); ensure that containers seal well; check jars regularly for evaporation and top up if necessary	Store in dark, very dry place; ensure containers are dust-proof; use insect repellers; check regularly for signs of pest damage; don't handle specimens too often
Problems	Alcohol causes slight shrinkage; formalin causes slight hardening; colour change; fire risk	Colours fade slightly; risk of pest damage; fire risk

preserving. It is also possible to buy formaldehyde in crystal form and make a ten per cent solution yourself.

Store your jars in a cool place to prevent evaporation and minimize the risk of fire. Put a label on the outside of the jar, but because these tend to fall off, put another inside. Write all the details in Indian ink on thick paper, otherwise the preservative will make your writing vanish.

Dry preservation

If you can preserve your specimens without alcohol, it is much cheaper and more satisfactory, but with dry specimens watch out for mould and insect pests. Try to keep specimens in a dry and dust-proof place. Tins are useful, but they tend to "sweat", so you should put muslin bags full of silica gel in them to absorb the moisture. Ideally you should have a proper specimen cabinet, but new ones are very expensive, though you can get second-hand ones more cheaply. But wherever you keep your specimens, protect them against pest damage. You can do this with either old-fashioned mothballs, flaked naphthalene or a blob of creosote in a cup. A piece of cottonwool dipped in a concentrated solution of phenol, and allowed to dry, will work in the same way. Both creosote and naphthalene need replacing as they evaporate

slowly. If your dried plants become attacked by insects, dip them in a solution of pentachlorophenol and then re-dry them. All these chemicals are available both from chemists' shops and from specialist biological suppliers.

Everyday containers

The resourceful naturalist can adapt containers that are thrown away in the average kitchen. Yoghurt and margarine pots are good for dry things such as birds' eggs or groups of beetles. Plastic containers with tightly sealed lids are good for larger specimens. A lot of jars and pots that once contained jam or honey are useful for "pickled" specimens. Clear glass ones are best, since you don't have to keep opening them to see what is inside.

Long-term storage

For long-term storage of a lot of specimens, you can use things such as cardboard or wooden boxes. An old chest of drawers or even an office filing cabinet is very useful if you make it light-proof and dust proof with the aid of plastic draught excluders.

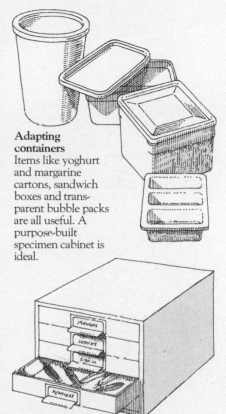

Adapting containers
Items like yoghurt and margarine cartons, sandwich boxes and transparent bubble packs are all useful. A purpose-built specimen cabinet is ideal.

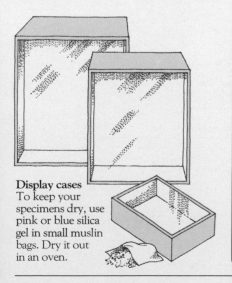

Display cases
To keep your specimens dry, use pink or blue silica gel in small muslin bags. Dry it out in an oven.

INFESTATION

Check your collection every month to make sure that your dried specimens have not been attacked by insects or mould, and that there has been no loss of fluid by evaporation in specimens preserved in spirit. Mould on dried specimens shows as a furry or powdery covering. Soak the specimen in 10 per cent formalin for 30 minutes, to kill the mould. But even after this treatment a regular check of specimens is vital.

Never allow insect pests to establish themselves in your collection. Holes in your specimens and a fine dust or flaky bits underneath them provide signs of infestation. It is best to destroy affected specimens, but if they are too valuable fumigate them with ethylene dichloride or ethyl formate. Follow the instructions carefully since these chemicals can be dangerous if not used correctly.

Common pests
The clothes moth, and museum and carpet beetles are shown here with their larvae. Lice and mites are also destructive.

Clothes moth

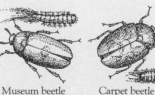

Museum beetle   Carpet beetle

STUDYING AND PRESERVING PLANTS AND FUNGI

THE STRUCTURE OF PLANTS

Like animals, plants and fungi have their anatomies. A study of these can teach you how the different parts of a plant have adapted to do different jobs. The roots, for example, not only grasp the soil and anchor the plant, but also act as pipelines, sucking up water and nutrients to feed the plant. The stiff stem holds the plant aloft (if possible above its neighbours), since most plants sky rocket towards the sun. The leaves act like solar energy panels, catching and trapping as much of the light energy as possible. The spores and flowers are designed so that the plant can reproduce effectively. And then inside the plant, like the veins in a human body, are minute tubes and pipes which carry the "blood"—water and dissolved nutrients—to the different parts of the plant's body. If you carefully dissect a plant and study it under a hand lens or microscope, you will see specialized structures and formations

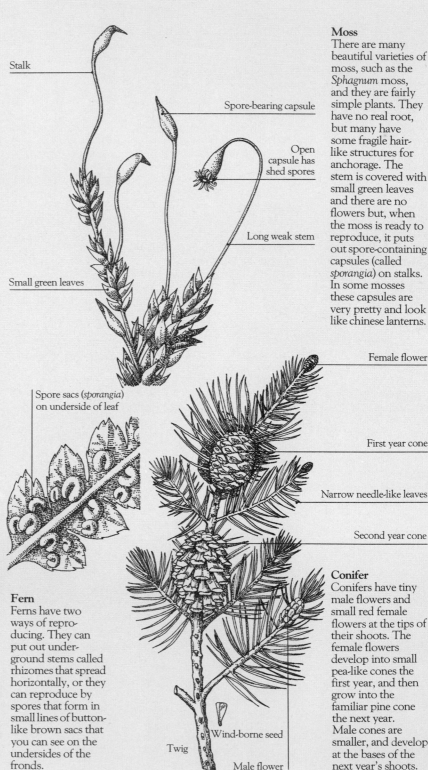

Stalk

Spore-bearing capsule

Open capsule has shed spores

Long weak stem

Small green leaves

Moss

There are many beautiful varieties of moss, such as the *Sphagnum* moss, and they are fairly simple plants. They have no real root, but many have some fragile hair-like structures for anchorage. The stem is covered with small green leaves and there are no flowers but, when the moss is ready to reproduce, it puts out spore-containing capsules (called *sporangia*) on stalks. In some mosses these capsules are very pretty and look like chinese lanterns.

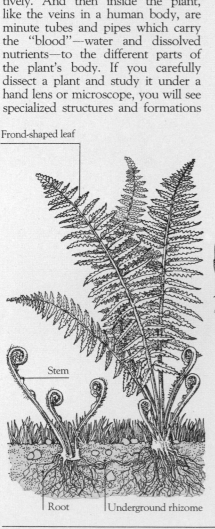

Frond-shaped leaf

Stem

Root

Underground rhizome

Spore sacs (*sporangia*) on underside of leaf

Fern

Ferns have two ways of reproducing. They can put out underground stems called rhizomes that spread horizontally, or they can reproduce by spores that form in small lines of button-like brown sacs that you can see on the undersides of the fronds.

Female flower

First year cone

Narrow needle-like leaves

Second year cone

Twig

Wind-borne seed

Male flower

Conifer

Conifers have tiny male flowers and small red female flowers at the tips of their shoots. The female flowers develop into small pea-like cones the first year, and then grow into the familiar pine cone the next year. Male cones are smaller, and develop at the bases of the next year's shoots.

of cells that are as intricate and fascinating as the make-up of any animal.

Plants vary greatly in their complexity. Algae, which include seaweeds and the green growths that cover tree trunks and damp walls, are the simplest of plants. They have hardly any specialized parts, consisting only of a flattened body (or *thallus*). Some also have a root-like holdfast to anchor them to a firm object. Mosses are slightly more complex, having leaflets and spore-producing capsules on the stalks. Ferns, with their intricately shaped leaves, stems, and roots, are more sophisticated still. Ferns are also the simplest plants to have vascular tissue—specialized tubes for carrying water and nutrients. Flowering plants are far more sophisticated and specialized—even an everyday plant such as the common dandelion can have highly complex flowers.

The flower is one of the most important parts of a plant's anatomy. It is worth learning how to dissect one, because the details of the number and arrangement of petals, sepals, and stamens is important in plant identification. To dissect most plants all you need is a sharp scalpel, a probe, some fine forceps and a hand lens. One way to dissect a flower is to cut through it vertically so that the two halves reveal the inner parts. Another method, particularly useful with more complex flowers, is to dissect it carefully from the outside, working inwards.

Buttercup

The buttercup is a good example of a typical flower. The five sepals enclose the bud, and within the flower's yellow petals is the male part (the stamen) that consists of the pollen contained in the anther which is held up by a filament. The female part (the carpel) consists of the stigma supported on the style, and the ovule (egg). It is an unspecialized flower. It can be pollinated by any creature crawling over it, transferring pollen from anthers to stigmas.

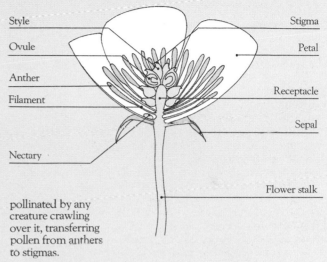

White deadnettle

The white deadnettle flower has the same parts as the buttercup, but has a completely different, much more specialized shape. It relies on bees or similar insects to pollinate it—the pollen from the anthers is placed so that it will brush off on the back of the bee. The bee will then transfer the pollen to another deadnettle flower.

Dandelion

The common dandelion as well as being a lovely flower is very complex. It is a member of the *Compositae* family. Members of this family have flowers made up of numerous florets, each one like a miniature flower with male and female parts.

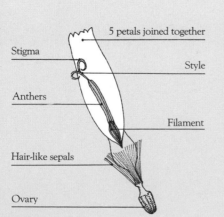

Grasses

All the grasses have flowers but they are quite different from the flowers of the buttercup. Instead of petals and sepals grasses have numbers of leaf-like bracts. Inside these all the various parts of the flower are enclosed.

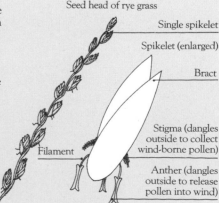

PLANT PARTS

It is fascinating to discover how plants are constructed. With a hand lens or a low-power microscope, together with your basic dissecting tools, you can examine the intricate architecture of any plant. Some houseplants are good subjects for examining, because their large parts are easy to dissect.

Making a section—a thin slice of a specimen—allows you to study it in greater detail. It is as though you could cut a huge building in half and see all the rooms and offices that make up the whole structure. A section is so thin that light will pass through it. This allows you to look at it with the aid of a light microscope. It is possible to buy a special tool that is called a microtome for cutting extremely thin sections, like a hand-operated slicer for sausages in a delicatessen. You can take sections across different parts of a plant—a longitudinal section will reveal different aspects of the specimen compared to a cross-section.

Leaves, stems and buds

Leaves vary greatly in size and shape, but most of them have the parts shown below. Some leaves, such as those on heather and the pine tree's "needles", are highly modified, and it can be difficult to recognize all the parts. Try dissecting various leaves and identifying all their comparable parts.

Looking at the stem enables you to study the "veins" of the plant. Cut a cross-section and you will see the thin fragile sap-carrying tubes called phloem, and also the xylem, which are thicker-walled water-carrying tubes.

The bud is a sort of deflated plant. When conditions are right, usually in spring, the bud expands, and its tiny leaves are ready to function. Good examples for dissection are the "sticky buds" of the horse chestnut tree and the common brussel sprout.

Roots

The roots of a plant are generally hidden, but they are among the most interesting parts of a plant's anatomy.

How to make a section

Whenever you are cutting soft tissues, it is advisable to use a support. A material such as cut-out pith, dense foam rubber, balsa wood or expanded polystyrene is ideal.

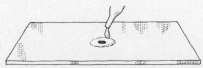

1 Using a scalpel or razor blade, slice your sections as thinly as possible, holding the specimen carefully.

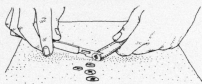

2 Take the thinnest of your sections and place it on a microscope slide. Add a drop of biological stain to colour it.

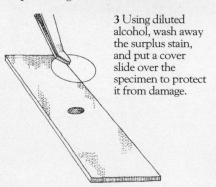

3 Using diluted alcohol, wash away the surplus stain, and put a cover slide over the specimen to protect it from damage.

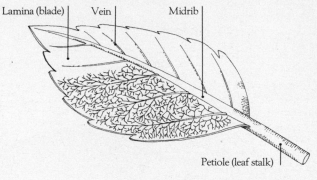

Lamina (blade) Vein Midrib

Petiole (leaf stalk)

Examining a leaf

Leaves work like sun traps. They are designed to obtain the maximum light energy from the sun. A broad flat shape is most efficient. The ribs and veins support the leaf, as well as carrying nutrients. A section shows the sap- and water-carrying vessels.

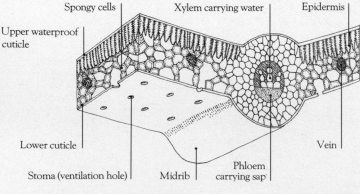

Spongy cells Xylem carrying water Epidermis

Upper waterproof cuticle

Lower cuticle

Stoma (ventilation hole) Midrib Phloem carrying sap Vein

Stem cross-section

The white dead-nettle provides a good example of a stem, since its cross-section shows clearly the separate bundles of vessels inside. With its hollow centre, the whole stem forms a tube, with strength-ening bulges around the outside. This combines strength with lightness.

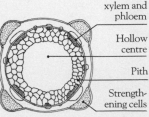

Bundles of xylem and phloem

Hollow centre

Pith

Strength-ening cells

They are also vital providers of mineral foods and water. The root tips are covered with tiny hairs that give them a bigger water-gathering surface. Some plants, like ivy, will grow their own roots if you break a piece off and put it into water. You can watch these develop and measure their growth.

The tip of the root is the only part that grows. Cut off the tip of a root (an onion root is best) and squash it gently on a glass slide. With a microscope you will be able to see the cells dividing. By using an acetic orcein stain, you will also be able to see the chromosomes in each cell.

Seeds, fruits and berries

The fruit or berry contains the seed, which in turn contains the embryo plant. A number of seeds contain stored food that is meant for the embryo plant. But a lot of animals (including humans) like to eat this. By the way, the greengrocer's distinction between fruit and vegetables is not a biological one, since things like runner beans, marrows and tomatoes are all fruits because they contain seeds.

Seeds, fruits and vegetables differ widely in size, shape and colour. For example, the tamblacocque seed from Mauritius looks like a little Chinese face, while the seed of the common dandelion is shaped like a parachute, to help with wind dispersal.

Fruit formation

The diagrams on the right show how the strawberry fruit develops from the flower. A section through the flower itself shows the features of the fruit already present. After fertilization, the petals of the flower wither away, while the seeds grow and accumulate food reserves to nourish the embryo plants inside. Animals eating this part help the seed's dispersal.

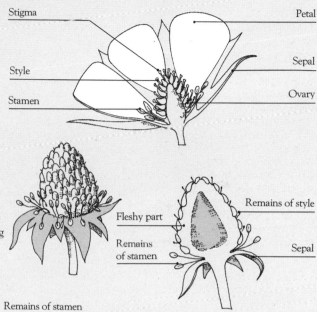

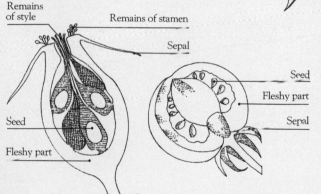

Types of fruit

The fruit of the tomato, shown left, and the rose hip, far left, show the variety that you can find in these "baby plants". As well as the seeds and the remains of the flower, these fruits have a fleshy portion containing food.

Studying roots

It is possible to cut a cross-section through a growing root tip, but it is often more revealing to halve a root lengthwise. This allows you to see not only the xylem and phloem at the centre of the root, but also the tip. This is the portion that pushes through the earth as the root grows. It is protected by the hard root cap.

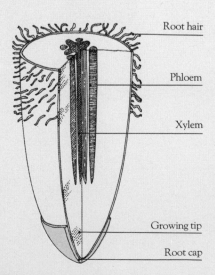

Bud and shoot

Inside a bud are tightly folded leaves and the condensed shoot, ready to burst forth as soon as conditions are favourable. The buds that grow on the ends of twigs are known as terminal or "apical" buds, those growing along the twig are called lateral or "axillary". Terminal buds continue the growth of the length of the twig, lateral buds create branches.

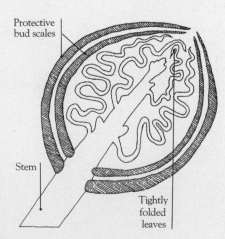

DRYING AND PRESSING FLOWERS

It is usually best to store flowers and other plant material dry unless you intend to examine them later with a microscope, in which case specimens should be pickled. The colours will fade, so you should take photographs or make colour sketches of the flowers when they are fresh. Most important, do not forget to label the specimens throughout the period of drying and pressing. An alternative method of preservation is the hot sand method, often used for fungi (see page 254). You will find that this will help the plants keep their shape, but they will lose a great deal of their colour.

Preparation

To get successful results from pressing it is important to pick and then store the plants properly while you are still in the field. Keep them as uncrushed and moist as possible, and allow them some ventilation. You must not allow them to dry up, so deal with them as soon as you return to your work-room. If you cannot attend to them immediately, wash the specimens clean as soon as you get back and then cleanly cut off the ends of the stems. Place these in fresh water in a tall vase. This will support them until you can press the plants.

The flower press

Your most important item of equipment is a flower press. With it you can preserve not only flowers but also leaves, stems and even roots. The faster all these specimens dry, the better the final colour retention, but there will always be a certain amount of shrinkage and warping. The basic principle is to sandwich the flowers between absorbent sheets of paper, apply sufficient pressure, and allow ventilation to dry them. The size of the press you use depends on the dimensions of the plants you want to press. Standard museum herbarium sheets are 45 × 30 cm, and this is about the right size for most flowers and herbs. It is always best, before you start collecting seriously, to practise

Preparing the press

Plants are very delicate. You should therefore be sure to carry specimens home in a suitable container such as a vasculum (the botanist's purpose-made carrying case). You must keep the plants fresh until you press them.

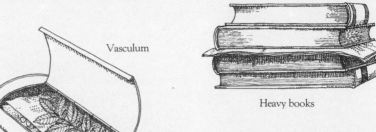

Vasculum

Rigid polythene container

Specimens preserved in water

Types of press

If you haven't got a professional plant press, you can improvize with books, bricks, or even boards slipped under the leg of a bed.

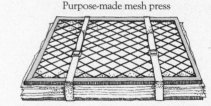

Heavy books

Improvised tennis racquet press with boards

Purpose-made mesh press

REHYDRATION FOR STUDY

It is possible to rehydrate dried specimens. In other words, you can put water back into them, to make them fairly life-like when you need to study their anatomy. It is even possible to examine rehydrated specimens under a microscope.

To rehydrate a plant, you must immerse it in water and simmer it gently on the stove until it is soft. This can take up to 20 or 30 minutes, depending, of course, on the type of plant. If it was originally a soft and rather wet plant, such as one of the mosses, it will probably

rehydrate after about five minutes in warm water. But tougher drier plants, such as groundsel, may require 30 minutes of simmering before they rehydrate.

You must be very careful when handling a rehydrated plant. Specimens become soft, flexible and very easily torn.

How to press a flower

It is essential that the plant is clean and dry before you press it. Soak the plant in clean water, and dab it dry with soft tissues. This helps to remove any insects or other creatures that may be attached to the specimen.

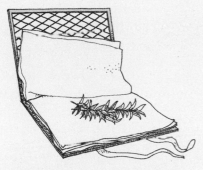

1 Arrange the specimen on two sheets of absorbent material such as blotting paper or newspaper. This will show off all the aspects of the flowers, buds and leaves (both upper and lower surfaces) to the best advantage.

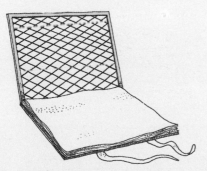

2 Put three sheets of paper on the plant, add another specimen and three more sheets, and so on, up to ten items.

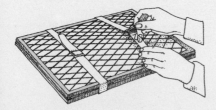

3 Close the press and leave in a warm dry room, changing sheets every two or three days. Plants will be dry in two to four weeks.

first on some common garden plants, so that you can perfect your technique.

The number of sheets of paper you use and how frequently you change them varies with the type of plant you are pressing. A fleshy specimen requires a lot of paper and frequent changes in order to dry it out. It is best to lay out really succulent and floppy plants using the flotation method (see page 255). Remember that the finished dried plant will be very brittle, so make sure you arrange it carefully from the start. Some naturalists are currently experimenting with microwave ovens for drying plants. You put the plant in a small press and then place this in the microwave oven for about two minutes on an average setting. Experiment with unwanted material first, and beware of fire.

Storage for pressed flowers

Store your flowers on clean white sheets of parchment or thin card, securing them with clear plastic tape, gummed paper or clear glue. Putting your dried plants in the deep freeze for three days before storage will kill any pests. When you have done this, label each sheet carefully. Your label should show clearly the plant's common and

scientific names. It is also worth compiling a brief description of when, where and in what type of habitat you found it. In addition, you should describe the plant from which you took the specimen, noting its size, type and the number of leaves and flowers.

Store the plants dry and flat in a way that protects them from damp and breakage. You can use cardboard boxes, drawers, plastic bags, or a loose-leaf file or album (see page 246).

Awkward customers

Some plants present particular problems of pressing and storage. These include bulky items such as stems and fruit, as well as some flowers. These are easier to store if you first cut them in two to aid drying. While drying them, be sure to pad them out with wads of paper to protect them from getting crushed. With specimens that you have cut in half, display one half with the cut surface uppermost, and the other uncut-side-up, so that you show both the inside and outside.

Large plants, such as bushes and trees, seem to present daunting problems. But it is usually possible to select parts of the plant and press these in the normal way.

A FLOWER AND ITS VISITORS

You can do a very interesting investigation to reveal all the different animals that depend on a particular plant. Select a common species, such as a sunflower, and collect and press the stems, buds, flowers and seeds. Observe which insects and other creatures visit or live on the various parts of the plant. Gather and preserve one of each. Note the different visitors at different times of day. You can sketch or photograph the creatures that you cannot collect, such as birds, or alternatively find pictures of them in books or magazines. Then make a display that will reveal the food web linking up the various creatures.

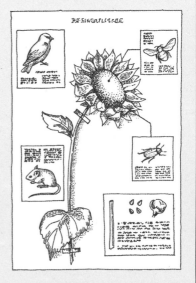

STUDYING FUNGI AND LICHENS

It is usually best to "collect" fungi and lichens by photographing or drawing them, because in this way you can record their colours accurately. It is possible to preserve them, but the colour is generally lost as a result. You can pickle specimens whole, but otherwise it is best to use a method of dry preservation. Flat crusty (or "crustose") lichens can be dried attached to the things on which they grow, for example rocks or sticks. Two to three days in a warm room will be sufficient to dry them. The fungi and soft branchy (or "foliose") lichens are best preserved by hot sand drying.

The hot sand bath

This method preserves the shape of fungi, but will cause some of the colour to fade. You need a metal or oven-proof dish and some clean fine sand. Bake the sand in the oven first, then put a layer in the tray and arrange the specimens on it. Next carefully and slowly pour the hot sand in from all sides to cover the fungi. Leave this to dry for 24 hours, and tip the sand out gently. If you find that the specimen is not completely dry, you should repeat the process.

The hot sand method is a difficult process to control, and it is advisable to use it only when you have spare specimens. It is easy to reduce your fungi to pieces of charcoal.

When the specimen is dry, clean off the sand carefully with a paintbrush. Remember that the dried specimen will be very delicate. You must handle it carefully by the stalk. When it is ready, you should store it in a rigid container such as a cardboard box.

Spore prints

Fungi reproduce not by seeds, but by minute spores that are released from the underside of the mushroom's or toadstool's "cap". You can collect these spores as they are released, and make a spore print.

This print is rather like a fingerprint. Select young umbrella-shaped fungi.

You will obtain the best prints from a specimen with an even rim around the cap. Fungi with uneven caps will not give good results. Cut off the stalks and place the remaining portion underside down on a sheet of clean paper. It is essential to work in a draught-free place, otherwise the spores will be blown off the paper. In order to make sure that your specimen is protected from draughts while you are making the print, cover it with a bowl or similar container. Leave the fungus overnight and then carefully remove the cap. You will find that a pattern of spores will show up on the paper. The spokes of the radiating gills on the underside of the toadstool's cap will be clearly visible.

In order to preserve the print, spray it with hair lacquer or the clear fixative that artists use for their paintings.

You must do this carefully to avoid smudging the spores. The best method is to spray horizontally towards the print from about 60 centimetres away, and about the same distance above. This will allow the spray to settle gently on the spores. Spray the print three times, after which it should be fully fixed. For an attractive contrast effect, try using papers of colours different from the underside of the fungus cap. Use yellow paper for brown spores, and black for specimens with a white underside.

The structure of lichens

Lichens consist of two organisms that live together—a fungus and an alga (a very simple plant). Both partners gain from this relationship. The alga is provided with a sheltered environment and a supply of raw nutrients, while

The hot sand method
After you have baked the sand in an oven, pour it over the specimen in a tray. You should wear a glove for this process as both the sand and the container will be hot.

Cleaning the specimen
When the fungus has cooled, remove the sand gently with a paintbrush.

How to make spore prints
Spores are distributed by the wind. To make a good print you must take every precaution to avoid draughts.

1 Using a scalpel or razor blade, carefully cut the stalk from the cap, as close to the top as possible.

2 Place the fungus on paper and cover it with a box or bowl while you are making the print.

3 Spray fixative on to the print from above and to one side, to avoid disturbing the spores.

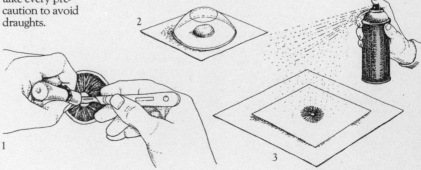

the fungus benefits from the energy-containing nutrients that the alga produces by photosynthesis. In order to examine the construction of lichens, look at them under a microscope.

Although they are incredibly hardy, growing in arctic conditions and in areas of drought, lichens are extremely susceptible to impurities in the air. They are therefore very good indicators of pollution, and few species grow in towns and cities.

Lichens grow very slowly. Many of the small crustose types increase only about one millimetre in size per year, while even the larger foliose lichens grow no faster than one centimetre per year. Do not remove too large a patch (greater than, say, ten per cent).

Lichens under the microscope

The two basic lichen shapes (crustose and foliose) are shown on the right, but there are numerous variations on these, each with a particular combination of fungus and alga. You can see the partnership by looking at the lichen under a light microscope at a magnification of ×250. Powdery green lichen found on tree trunks is a good example.

Crustose lichen

Foliose lichen

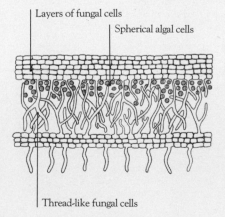

Layers of fungal cells

Spherical algal cells

Thread-like fungal cells

MOSSES AND LIVERWORTS

These are simple plants that do not have flowers, but reproduce by spores in the same way as the fungi opposite. In mosses the spores are often carried in small capsules held above the leaves by stalks that look like swans' necks.

Mosses require damp conditions in order to reproduce and they usually grow in damp habitats. Mosses and liverworts contain a lot of water, and they require special care when you are preserving them.

Methods of preservation

A very simple preservation method is to hang the moss or liverwort in a muslin bag, and place this in warm dry air. This preserves the specimen's shape very well. In addition, the method allows you to store the dried moss or liverwort in a small box. Try to find some specimens of moss with the spore capsules intact, and preserve them in this way.

You can also use a plant press similar to the device used for flowers (see page 252). For mosses, the process requires less pressure and more changes of paper, so that it takes longer than for flowers. When using this method press both whole clumps of moss and individual leaves.

Flotation is a preservation method commonly used for fragile succulent plants such as mosses and seaweeds. The technique captures the natural wavy shape of seaweeds, which it is difficult to see when you lay the leaves out on dry paper.

Using a muslin bag

The muslin bag provides a very simple method of preservation. To dry the specimen thoroughly, place it in the bag and hang it in a warm dry airstream. An airing cupboard is perfect for this. At ordinary room temperature, the specimen should be dry in seven to ten days.

How to preserve by flotation

If you want to preserve specimens of seaweed that you have collected, flotation offers a reliable method.

1 Soak the seaweed in plenty of fresh water overnight. This will remove most of the salt from the specimen.

Seaweed preserved with the salt left in it will eventually rot.

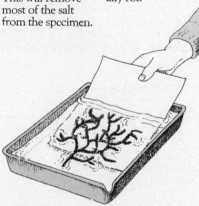

2 Float the plant in a shallow tray filled with clean water, and slide a strong piece of paper (such as parchment or even cardboard) slowly under it. Then lift the specimen out of the water carefully, to preserve its shape.

3 Allow the plant to dry in the air for an hour, before pressing it very lightly in a flower press. Using extra sheets of paper in the press, and keeping to a light pressure will ensure that the specimen maintains its shape.

BARK AND LEAVES

If you want to make a comprehensive study of an entire bush or tree, you should try and preserve all the different parts of it in some way. Besides the blossom, which you can press in the same way as you would any other flower (see page 252), you should keep a record of the seeds and fruit, and the bark and leaves.

Recording tree bark

If you do not make a record of bark by rubbing in the field (see page 119), it is possible to produce a plaster cast. You can make these casts of many things, but bark gives especially good results. You can make a mould out of clay or plasticine and then produce the cast itself from plaster of Paris or modelling resin when you get home. To do this you will have to build clay end walls for the mould and pour in the plaster. When this has dried, you remove the clay and paint the cast in realistic colours.

Leaves

There are several methods of preserving and recording the leaves you find. Whichever technique you use, do not remove too many leaves from trees. It is best to take only hard autumnal leaves that are just starting to change colour, since at this time the tree no longer needs them and they are almost ready to drop to the ground.

The simplest method of making a record of leaves is to press them, as you do flowers. As well as preserving individual specimens, you can make attractive composite pictures of the different leaves in your garden or local wood. Experimenting with the beautiful autumnal colours can be particularly rewarding.

You can record the surface features of a leaf by rubbing, in the same way that you would rub a piece of bark. To make a simpler representation of the leaf's outline, it is possible to make a silhouette. You hold the leaf flat on a piece of cardboard or paper, and spray or brush paint over it, so that the outline appears on the paper. When applying the paint, it is essential to

A FERN'S LIFE CYCLE

Ferns have an interesting life cycle that is divided into two parts. The familiar fern with its large leafy fronds forms one part of the cycle. On the undersides of some of these fronds are the small brown buttons that contain spores. When these spores are released and land in a damp place they develop into the other part of the cycle. This is the prothallus, which is small, green and heart-shaped, with tiny roots. On the underside are the sex organs, tiny spherical male ones that contain sperm, and small cup-shaped female ones that each contain an egg. The sperm swims through the surface film of water, fertilizes the egg, and a new fern then grows.

Arrange a display to show this process. Press large items such as the fern fronds, and examine smaller things such as the spore capsules, and make drawings.

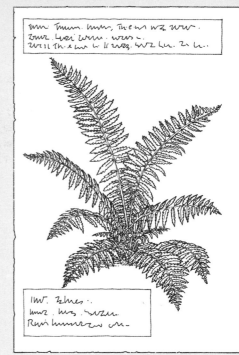

Making a cast
1 Put clay on the bark, and tap it into the grooves using a board and hammer.

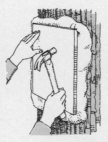

2 Curve the clay slightly and add ends to make a mould that will take the plaster. Support it with sand.

3 Pour a thick mixture of plaster of Paris and water into your mould until you have filled it. Leave it to set.

4 When the plaster has set, remove the clay carefully. Clean the cast with water and a small stiff brush.

use outward strokes, so that it does not get under the edge of the leaf and spoil the outline effect.

There are two techniques that enable you to show very clearly the leaf's structure of veins. The first, the leaf print, is made in a similar way to the operation of a printing press. Leaf skeletons are made by removing the flesh from the leaf, leaving the veins.

Making a leaf skeleton
Simmer the leaf in water for an hour and leave it for a week in the same water until the flesh is soft. Rinse, gently remove the flesh with a paintbrush, and bleach.

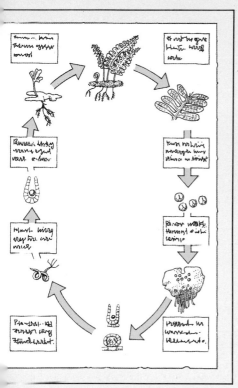

SEEDS, FRUITS AND GALLS

The best way to preserve these is to keep them dried. If they get moist they will start to germinate. Most cones, seeds and nuts will be quite dry when you find them. However it is best to ensure their preservation by leaving them on a hard flat surface for a few days in a warm dry room. Label and store them with other parts of the same plant. Fleshy fruits go mouldy if you do not dry them quickly. A hot sand bath (see page 254) provides a good method of preserving fruits such as berries and currants. Cut them in half first, so you can display both the inside and the outside. In addition, you can press small fruit.

When you are collecting material from a tree, you should also look out for galls—the small round objects caused by creatures such as gall wasps and gall mites (see page 121). If you collect galls during the autumn and keep them in a cool place, the gall-making creatures and others that associate with them will hatch out. Hard dry galls such as the oak apple will keep quite well if you let them dry, but you should press galls attached to leaves. To dry soft fleshy galls, a hot sand bath is best.

Inside the gall
Cutting open a gall reveals the chambers inside. The number of exits shows how many insects inhabited it.

Printing with leaves
You first spread oil, paint or shoe polish on to the leaf. You must apply this thinly and evenly to produce a good result. Place the leaf wet side down on a sheet of clean paper, cover this with blotting paper, and press and rub the leaf. Let the print dry and afterwards spray it with fixative.

THE ANATOMY OF A TREE

Choose a local tree to study. It may be one in your garden, or in a nearby park or wood. Measure its height and canopy area (see page 110), and make a scale drawing of the tree as the centrepiece.

By collecting and preserving things such as twigs and over-wintering buds you can gradually build up a study of the tree's anatomy. You can also show the difference between the spring, summer and autumn leaves. Preserve the tree's blossom, its seed or fruit, and the galls you find on it. Finally, make bark rubbings or take plaster casts of the trunk or branches. By dissecting and drawing the specimens you can build up the whole picture.

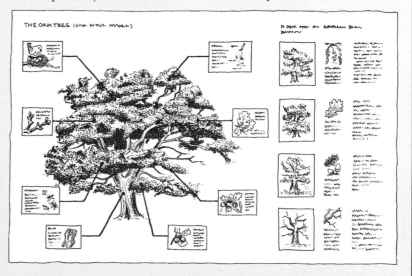

KEEPING LIVING PLANTS

THE HOME GREENHOUSE

Growing plants can be great fun, as well as being very instructive for the naturalist. Plants grow and develop just as animals do. They also respond to their environment, though they do so much more slowly than animals. The nervous system of an animal allows it to react swiftly to seek out a sunny spot or to find water, whereas a plant, which has no nervous system, responds to light and moisture by growing towards them.

You do not need much special equipment for growing plants and doing experiments on them, and the long-term rewards are great. Cultivating further specimens is especially easy, since once your plants are growing you can produce more from their seeds or by propagation—your greenhouse will be self-sufficient.

Use your wall chart or year-planner for recording dates and notes on planting, germination and growth measurement.

It is best to buy seeds in packets rather than damaging natural habitats by gathering your own. Peas, beans, sunflowers, wheat and cress are very useful for experiments, because they grow quickly, and you can eat them after you have finished observing them. But you can take some seeds from common countryside plants without causing harm.

Conditions for growth

Most seeds need four things in order to germinate and grow into seedlings —light, moisture, warmth and air. But, of course, different plants require different amounts. There are four experiments which you can do that demonstrate these necessities. But many plants require variations on these themes. For example, some seeds will germinate only in the dark, because they must be sure that they are in the soil. However, their leaves need light once the plant starts to grow. You can devise experiments of your own to show which plants react in this way. Some seeds, such as apple pips, will only germinate after a cold period. If you collect some in the

STUDYING CONDITIONS FOR PLANT GROWTH

Light

This experiment is to show that light is necessary for plant growth. Fill two plant pots with damp soil, and put in each container three or four seeds that have been soaked overnight. Place both pots on the window ledge. Cover one with a cardboard tube, closed at one end so that light does not enter. Punch a few holes in the tube to allow air to circulate. This will also permit some light to enter, so you should put this specimen on the shady side of the window, to minimize this effect. Water both pots every day. Both specimens will then receive ample warmth, air, and moisture, but the covered plant will have hardly any light.

Result The seeds in the uncovered pot should germinate and grow well. Those in the dark pot may start to grow, and produce long pale stems and leaves, but these will soon die off.

Moisture

With this experiment you will see how necessary moisture is for both germination and growth. Take two plant pots and fill one with ordinary damp soil and one with dry soil. To make this soil really dry it is best to bake it in the oven. Soak some seeds overnight and put three or four in each pot. Keep both the pots on a window ledge, where they will have a good supply of warmth and light. Water the pot with the damp soil every day, but do not water the other.

Result The seeds in the pot that you water should germinate. Those in the dry pot probably will not germinate. If they do, they will soon die.

Warmth

It is best to do this experiment in cold weather, preferably during a frost. These conditions will help you demonstrate that a warm environment is also necessary for the processes of germination and growth. Prepare two pots with seeds, as in the previous experiments. Use damp soil and make sure that both pots are watered regularly. Put one on the inside window sill of a warm room, the other on the outside ledge of the same window.

Result The plant inside will grow well, but the specimen outside, although it may germinate, will be killed by the cold.

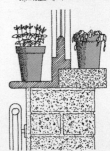

Oxygen

This experiment shows that air is necessary for the plant. Set up two plant pots with seeds as in the other experiments. Put one in a clear plastic bag, the top of which you attach tightly to a length of round tubing. This will allow you to water the plant, though do this sparingly and close the tube with a rubber stopper. Set up both the pots on a window ledge, and water both every day, always remembering to replace the stopper on the specimen inside the plastic bag.

Result The seeds in the open pot will germinate and grow, but those in the pot covered by the bag will suffocate and eventually go mouldy.

Comparing light sources

Sunlight, and illumination from artificial sources such as light bulbs and fluorescent tubes, is composed of different wavelengths, and plants will respond to these in different ways. Prepare three pots of seedlings. Place one on a sunny window ledge, one near a fluorescent lamp and one near a desk lamp with a normal bulb. Do not put the plant too near the bulb, or it will be affected by the heat. Draw a graph of the growth rates, as shown below.

How seedlings respond to light

All plants grow towards the light. You can see this by placing any plant near a window and watching its growth. You can do an experiment with seedlings to show which part of the plant responds to the light in this way. Grow about 20 seedlings in a tray near a window. Once they have germinated, cover the growing tips of some of the plants with small caps of aluminium foil. The plants with the metal caps should grow straight up, while the others bend to the light. It is the tips that respond.

Raising plants from mud

1 After a country walk, scrape mud from your boots and take the dust out of your pockets. Mix this with water to a semi-liquid consistency, and leave overnight to soak.
2 Sterilize a tray of potting compost by baking it in the oven. This will ensure that it does not contain any viable seeds that will germinate and spoil the experiment.
3 Add your mixture of mud and water to the compost, cover the container with a sheet of glass, and leave it in a warm place. Keep it moist but not wet.
4 After two or three weeks several seedlings will appear.

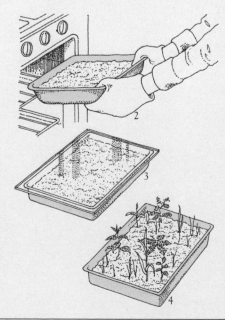

autumn and try to grow a few in the ordinary way, they probably will not germinate. But try keeping some in the refrigerator for two weeks. This will simulate the cold weather, and so the seeds should germinate. In nature this strategy of the seeds stops them growing in the autumn. Consequently, the young plants are not killed off by frosts and snow in the winter.

Does the amount of light matter?

All plants need light in order to carry out photosynthesis—the process by which they produce food. But how much light is needed for a plant to grow effectively? And does the nature of the light matter? Do plants need sunlight, or will they grow under domestic light bulbs? To investigate these questions, take two established seedlings, keep them on a window ledge for a few days and measure their growth. Then reduce or extend the light period given to one of the plants, measure the growth, and make a graph of the growth rate. Plot the growth rate of the "control" seedling that you have left on the window ledge as well. You can then compare the results easily. You can also compare how the plant grows when given more or less light than the normal daylight hours, whether the plant grows as well in artificial light as daylight, and whether fluorescent light is as good as ordinary domestic light bulbs.

Plants from seeds and berries

In addition to the seeds and berries that you take directly from plants, many others will have already been dispersed by the wind or animals. You can see how common this process is by simply scraping the mud off your shoes after an autumn walk in the woods and fields. Plant the mud in a pot, providing all the conditions necessary for growth—light, warmth, water, and air. You will be amazed at the variety of plants that spring up.

Which foods do plants need?

New seedlings can live for a while on the food reserves stored in their seeds. But eventually they need a new supply of mineral nutrients (as well as energy from sunlight) for growth. You can do some experiments to find out which nutrients are the most important for young plants. Good seeds to use are wheat, peas or runner beans—all are widely obtainable and will germinate easily. You also need jars or test-tubes.

In order to compare how well the plants will grow with different nutrients, you will need to prepare a basic plant nutrient solution. You can do this using the ingredients listed below. If you cannot obtain these, biological suppliers sell kits for the whole experiment. In addition, you will require some distilled water. You can buy this from a garage or car accessory shop. Alternatively, melted ice from the walls of your refrigerator will do,

although it will not be completely pure. Dilute one part of the basic solution in four parts distilled water. You can then make different growing solutions and compare the results.

Cuttings

Almost any part of a plant, if you cut it off and give it the right care, will develop into a new plant like its parent. The best parts to take are the growing shoots or twigs, or the side shoots of a plant. Plant them in damp seed compost and put them in a damp shady place to allow them to take root. It is very helpful to add a little hormone rooting powder to the cut surface of the piece of the plant before putting it in the compost. This is a chemical preparation that encourages the growth of roots, enabling your cuttings to establish themselves quickly. Many plants, especially house plants, root more easily if you leave the cut-

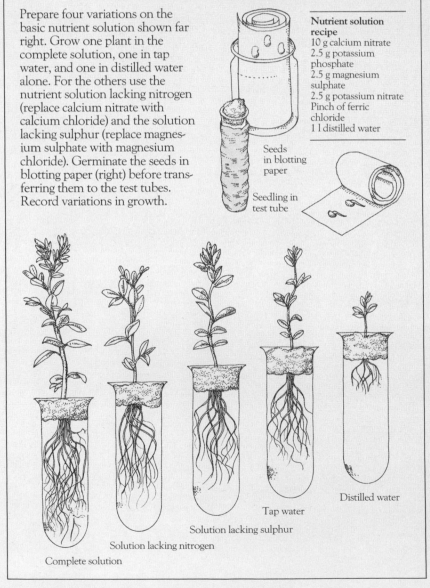

COMPARING PLANT FOODS

Prepare four variations on the basic nutrient solution shown far right. Grow one plant in the complete solution, one in tap water, and one in distilled water alone. For the others use the nutrient solution lacking nitrogen (replace calcium nitrate with calcium chloride) and the solution lacking sulphur (replace magnesium sulphate with magnesium chloride). Germinate the seeds in blotting paper (right) before transferring them to the test tubes. Record variations in growth.

Nutrient solution recipe
10 g calcium nitrate
2.5 g potassium phosphate
2.5 g magnesium sulphate
2.5 g potassium nitrate
Pinch of ferric chloride
1 l distilled water

Seeds in blotting paper

Seedling in test tube

Complete solution

Solution lacking nitrogen

Solution lacking sulphur

Tap water

Distilled water

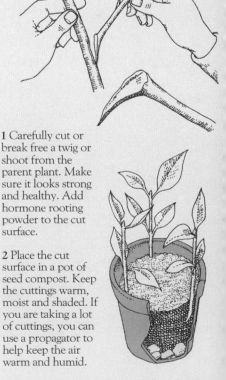

Taking cuttings

1 Carefully cut or break free a twig or shoot from the parent plant. Make sure it looks strong and healthy. Add hormone rooting powder to the cut surface.

2 Place the cut surface in a pot of seed compost. Keep the cuttings warm, moist and shaded. If you are taking a lot of cuttings, you can use a propagator to help keep the air warm and humid.

ting in water for about ten days. You can then transfer them to the compost, but do not damage the roots.

Not all your cuttings will survive, so to start with limit yourself to common house or garden plants. Try different parts of the same plant—for example a piece of stem, a section of the root, and a piece of leaf.

Grafting

Grafting works almost like a transplant. You use the roots and stem of an established plant (which is called the stock) and transplant on to it the growing stem and buds of another, very closely related, plant (known as the scion). What you are actually doing is putting another branch on to your plant. The joint where the new branch meets the plant must fit exactly. Press them very close together, and then secure them with damp raffia, making sure that you give adequate support to the upper parts, so that your graft does not droop towards the ground.

Grafting is usually done to shrubs such as rhododendrons and azaleas, as well as fruit trees, but it is possible to graft many other species. But whatever type of plant you want to graft, you must choose specimens that are closely related. Horticulturalists often graft one rose on to another, or a fruit tree on to a similar species. The methods are shown on the right.

Time-lapse photography

This is an excellent way of making records of plant growth. Always set up the camera in exactly the same position. If possible, leave it there, supported by a tripod. A large piece of graph paper fixed to a board and supported upright makes a useful background. It will allow you to measure off the growth on the final pictures.

You must keep the lighting consistent. An automatic flashgun will give the correct illumination for photography even if the normal lighting round the subject varies according to the time of day. For a plant that grows quickly, take photographs every one or two days.

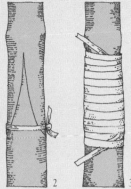

Saddle grafting
1 Choose stock and scion of similar size, and cut them into V-shapes.
2 Fit the stock and
scion together and bind them with damp raffia.
3 When fully bound, leave in a warm place.

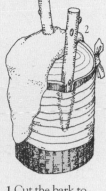

Crown grafting
You can use this method with many fruit trees, to join several scions to a single stock.

1 Cut the bark to accept the scions, and insert these.
2 Bind the graft with raffia, and apply grafting wax.

Budding
1 Using a sharp penknife, slice a bud, plus a piece of bark, from the parent plant.

2 Carefully make a T-shaped cut in the stock plant.
3 Insert the bud into the cut, and bind the two together.

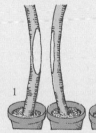

Making an approach graft
1 Remove a strip of bark from both stock and scion, so that they match.

2 Bind the cut surfaces together.
3 Cut away the upper part of the stock and the lower part of the scion.

Taking picture sequences
You can produce sequences of photographs taken
at regular intervals to record plant growth. Use a graph paper background so that it is easy to
compare the sizes of the subject as it grows. File prints together to show the growth rate.

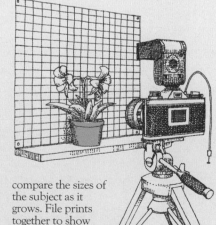

FOOD FOR ANIMALS

Many herbivorous animals, and certain other species such as mice, enjoy fresh plant food. You will find that they thrive on barley. It is best to buy the seeds in bulk. Soak about 20 seeds overnight in water, and then sow them in a yoghurt pot filled with soil. Keep the pot warm, moist and well lit, and within ten days the barley will have grown to a good size.

If you have a number of creatures that need barley, it is a good idea to set up a "production line" of pots, sowing seeds every two or three days so that you have a constant supply of fresh green food. You can then simply place one or two pots in the cage each day. You can re-use the soil four or five times, but renew it after this.

Keeping a bottle garden

A self-contained bottle or tank garden is ideal for growing ferns, mosses, liverworts and other small plants, especially those that do not need very much light. Once stocked it will continually recycle—all you need to do is water it occasionally. It can provide material for study that is difficult to find in the field. Mosses and ferns are best adapted to damp shady conditions, so don't keep your bottle garden too warm or too bright.

Making a bottle garden

Start by putting in a layer of charcoal pieces mixed with pebbles, for drainage purposes. This should be about 3 cm deep. You can then add potting compost, to a depth of about 5 to 10 cm. Use long, flexible sticks to introduce the plants into the narrow-necked carboy bottle. It is easiest to start at the edges and work inwards. Keep the bottle garden in good, but not bright, light.

New shoots

Half-grown plants

Fully-grown plants

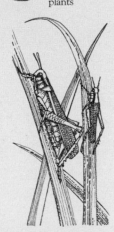

The food supply
Home-grown wheat or barley makes excellent food for herbivores such as grasshoppers (shown on the right), locusts and similar insects, as well as animals such as mice and voles. If you sow seeds at two- or three-day intervals, you will have a constant supply. The food will be fresh and wastage is low.

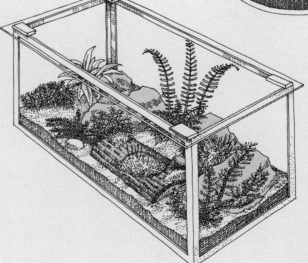

Keeping plants in an aquarium
A tank makes a good plant container. It provides a less costly alternative to the traditional carboy bottle. In order to grow mosses, you should place stones in the bottom of the aquarium. Pieces of wood will also provide an acceptable habitat for mosses. Covering the tank with glass keeps it humid.

GROWING STONES AND PIPS

With care and patience, you can culti-vate many of the seeds from fruits you eat. Apart from a few basic requirements—constant water, light, draught-free conditions and a regular temperature—most are easy to grow.

Avocado

Probably the largest fruit seed you will find is the avocado. This can grow into an attractive house plant. The most popular method of germinating is to support the seed in water, in the same way that you would grow a hyacinth bulb. Make sure the stone's more rounded end is covered with water and put it in a dark place where a temperature of 21°C can be main-tained—an airing cupboard is ideal. The stone swells slowly, and in a period of ten days to six weeks roots and a yellow-green shoot start to appear. At this stage the plant must be transferred to the light. When the roots begin to clog the container, you should transfer the plant to a pot about 12 centimetres in diameter. After about four months, the plant will require a 30 centimetre pot.

Citrus fruits

One seed from a citrus fruit will usually produce a number of seed-lings. You must rear all the seedlings from one pip, since not all of them will be capable of producing fertile plants.

Citrus pips need a temperature of 16–21°C. Lemon pips are the most tolerant of lower temperatures. Plant the pips with a shallow covering of compost and place them in a sunny position. Most citrus fruits are Medi-terranean in origin, and will tolerate a cool spell during the winter.

Lemons are generally the easiest to cultivate, while Seville oranges, man-darins and tangerines will all grow well. All citrus fruits produce attrac-tive plants with glossy dark green foliage, and white scented flowers.

Small trees

Many trees will produce attractive and interesting plants, but you must re-member that most can withstand a hard winter period. You should therefore consign them to a cool room during the winter months.

Among the trees you can cultivate in this way are the apple and the chestnut. For the apple, use pips from very ripe fruit. Plant them about a centimetre deep in potting compost. They will take any time from three to eight weeks to germinate.

You can plant chestnuts in the garden, but it is easier to locate speci-mens if you place the seeds in tins with holes punched in the bottoms for drainage. Plant in the autumn, cover-ing them with a thin layer of sand. Bring them indoors in February,

giving them a little gentle heat, to stimulate germination. When estab-lished, put the plants in the garden.

Pineapple

It is quite easy to grow a pineapple plant from the top of the fruit. Cut off the top, leaving a disc of flesh about one centimetre thick. Leave this for 48 hours to dry. Before planting, moisten the flesh slightly, and apply some root-ing powder. Place in a container of well-drained compost, keeping the leaves clear of the soil. The pot should be large enough to allow a 2.5 centi-metre border, for easy watering. Keep the pineapple in a light warm place.

How to grow an avocado
1 Support the stone in a jar of water using pins or match-sticks. Leave in the dark until germin-ation.
2 Transfer to the light and leave it in the jar until the roots get too large.
3 Place the plant in a pot 12 cm across.

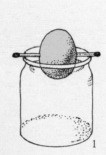

Propagating lemon pips
Sow pips in a propagator with a clear lid. This will keep in the mois-ture. A jam jar or polythene bag over the pot will also do.

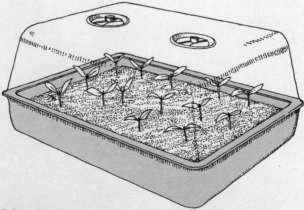

Cultivating pine-apples
Cut off the top to leave a disc about 1 cm deep. When it is dry apply rooting powder and plant. A polythene bag acts like a propa-gator.

STUDYING AND PRESERVING SMALL ANIMALS

FINDING SMALL ANIMALS

Small animals range from insects such as butterflies, beetles, flies and grasshoppers to small invertebrates like spiders, worms, slugs and snails. These are all fascinating creatures to examine and study, and one of their advantages is that you can catch and keep them quite easily. You can therefore do experiments on them in your garden or workroom. Most of these small animals are fairly easy to house and do not require the elaborate cages that bigger creatures often need.

You can find these small creatures virtually anywhere. For example, if you simply examine all the nooks and crannies in your own house, you will be surprised at the number of small animals you will find. You should also look on the leaves, flowers and stems of both house and garden plants. When you turn them over, piles of stones or logs will also reveal any number of creatures. But if you do this, be sure to replace the stones or logs in the position in which you found them. You can find further sources of small animals by sifting through leaf litter or beachcombing along the shore if you live near the sea. Puddles of rainwater, a rainwater butt or a clogged gutter will reward you with any number of fascinating microscopic creatures. Similarly, rock pools will provide a rich source of microscopic aquatic beings.

In addition to all these places, you must remember to look on the larger creatures that you find for the tiny parasites that many of them carry.

As well as collecting creatures yourself, you can, of course, catch things by the numerous methods described earlier in this book—the pitfall trap, the sweep net, smoking out, the light trap, pond netting and tray beating. If you collect some leaf litter in a plastic bag, you can use a device known as a Tullgren funnel to extract the creatures living in it. Alternatively you can sieve the litter on to a clean flat surface and pick up any animals that drop out with the aid of a pooter. A good method for sifting soil is to shake it up with a strong solution of salt water. Any small animal in it will then float to the surface.

Birds' nests are a good source—not only of parasites from the birds themselves, but also of creatures that feed on the nest material. If you put some mothballs in a nest, and place it in a large polythene bag, the creatures will emerge, to escape from the smell. Pick them up with a pooter.

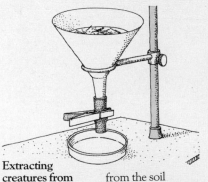

Extracting creatures from the soil

A Baermann funnel has a spout that you can block. You use it for extracting nematodes (water-living roundworms) from the soil water. Place soil or litter in a muslin bag and put water in the funnel while the bottom is blocked. Nematodes sink into the spout.

The Tullgren funnel

Pack leaf litter into a funnel (a plastic one is best as it causes least condensation). Wedge the funnel into a jar and lay blotting paper in the base of the jar. Place a table lamp above it. The light and heat will drive the creatures down to the paper.

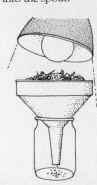

Flotation

You can remove small creatures from the soil by means of flotation. Shake up one part of soil with three parts of strong salt solution. Put the mixture in a wide jug or bowl, and leave it. The animals float to the top.

Creatures float to top of water

Soil sinks to bottom

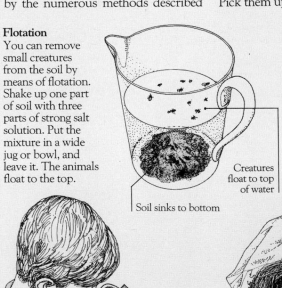

Animals in a nest

Old nests can often provide a rich source of small animals. You can extract these from the nest by using mothballs. Place mothballs in the nest and place it inside a large clear polythene bag. Quite soon the creatures will crawl out of the nest, trying to escape the smell.

Insects coming out of nest

Using the pooter

The pooter enables you to suck up the small and delicate creatures found in soil and leaf litter. You use it very much like a vacuum cleaner for invertebrates. Sieve the collected material on to a clean white surface. You can then suck up the creatures into the pooter.

PRESERVING SOFT-BODIED CREATURES

Animals such as worms, nematodes and caterpillars, after being placed in the killing jar, should be preserved with care. The best way is to pickle them in alcohol. In order to maintain the shape of the creature, you must treat the specimen first. You do this by putting it in alcohol that is diluted in the same volume of water. Leave it there for an hour, and then transfer it to normal preserving alcohol. Keep the specimens individually or in groups in small test tubes.

Creatures like snails and slugs pose other problems when you try to preserve them. This is because they secrete a great deal of mucus as soon as you put them in the preservative. It is therefore necessary to clean them up with blotting paper or a paintbrush before putting them into the final preservative solution. When you have done this, block the specimen tubes with cottonwool, and place them in a jar full of preservative. The liquid can penetrate the cottonwool, keeping the tubes topped up.

Microscopic animals

Most minute water-living creatures move very rapidly. In order to study them properly under a light microscope, you must slow down their movements. A drop of glycerine or iodine will retard them sufficiently.

When you put them under the microscope, start on low magnification, so that you can see a large area of the subject and locate the interesting parts of its body that you want to study more closely. You can then turn to a higher power.

Common microscopic creatures

These are some of the interesting and common organisms that you can collect and study under your microscope. In sea water, for example, there are countless tiny larvae which (if not eaten by other creatures) will grow into large shrimps or prawns. In addition, there are many small aquatic adult organisms in both sea water and fresh water.

Nematode This semi-transparent worm, 1 mm long, lives in water

Rotifer Tubby relatives of nematodes, they measure about 0.2 mm.

Daphnia The well-known "water flea" is really a tiny crustacean 1 mm long.

Euglena A single-celled organism, about 0.05 mm long, that swims by lashing its tail.

Paramecium This organism moves by waving body hairs. It measures 0.25 mm.

Nauplius This is the larva of another tiny crustacean. It is only 0.5 mm long.

Zoea This larva, 1 mm long, grows into a small shrimp.

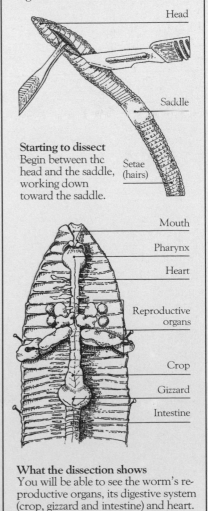

INSIDE THE EARTHWORM

A worm seems the same along its entire length, but you will be surprised how much the internal organs differ. For dissection, choose the largest earthworm you can find and put it in a killing jar. Lay it on its back in a dissecting tray and cover with salt water. (You can tell when a worm is on its back because the tiny hairs point upwards.) Cut delicately down the front of the worm using a sharp scalpel. Spread the skin open and pin it back to reveal the internal organs of the worm.

Head

Saddle

Starting to dissect Begin between the head and the saddle, working down toward the saddle.

Setae (hairs)

Mouth

Pharynx

Heart

Reproductive organs

Crop

Gizzard

Intestine

What the dissection shows You will be able to see the worm's reproductive organs, its digestive system (crop, gizzard and intestine) and heart.

INSECTS AND SIMILAR CREATURES

Creatures such as butterflies, dragonflies and beetles are best preserved dry. You do this in three stages. First, most of your killed specimens will stiffen if they have been in the anaesthetic jar for any length of time. They need to be relaxed so that you can move the limbs or wings around. In a fresh specimen, of course, this is easier. The second stage is setting, which involves putting the creature into the correct pose and position, so that it is possible to examine easily all parts of its anatomy and so that it looks attractive. When you have done this, you let the specimen dry and stiffen. The third stage is mounting—fixing the creature in position for storage and viewing.

For setting and mounting you should always use pins that are thin, very sharp and rustproof. Stainless-steel entomological pins are best for this. Always handle the insects and pins carefully with forceps.

Relaxing the specimen

In order to do this you must use a relaxing fluid. You can buy this from a biological supplier. Alternatively, you can use plain boiled water with a few drops of disinfectant. You also require a sealable container, such as a small tin. Clean and scald this to sterilize it, allow it to dry, and place an

How to set a butterfly

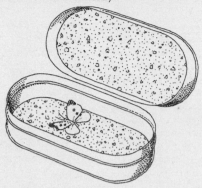

1 Make sure the butterfly is completely relaxed by leaving it in a sealed tin containing relaxing fluid. You will need to leave it like this for at least a day. After this, test the specimen for flexibility.

absorbent pad on the bottom. Add the fluid and specimen, and close the lid tightly. After one or two days the specimen should be relaxed. You can test this by moving one of the insect's legs slightly. Leave the specimen in the tin until it is pliable.

Setting

This is especially important for specimens with delicate wings such as moths and butterflies, or grasshoppers and beetles. If you want to show the wings, use a setting board. You can either buy one of these or make it

2 For this stage you will require a cork setting board with a groove that will accommodate the insect's body. Pin the insect to the board. Its wings will probably stick up, so rearrange them.

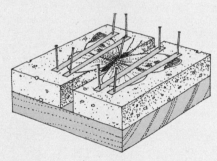

3 When you have placed the butterfly in the position that you require, secure its wings to the board with strips of grease-proof paper. Drying time is about two weeks, but depends on size.

INSIDE THE COCKROACH

The cockroach is a good example of an insect to dissect because it is not specialized, and its body has most of the basic organs you will find in other insects. Cockroaches are also common and you can buy them from biological suppliers.

After you have put it in the killing jar, arrange the cockroach on its back in the dissecting tray. Hold its legs out with pins and cover it with salt water. Cut carefully up the midline, pull the body casing back and pin it. Lift out the gizzard and unravel the intestines to see the nerve running down the back.

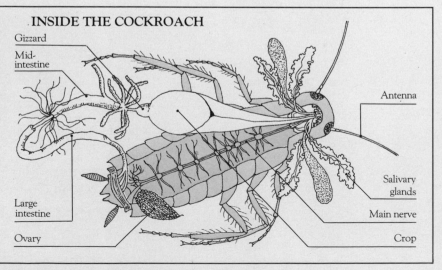

Gizzard

Mid-intestine

Large intestine

Ovary

Antenna

Salivary glands

Main nerve

Crop

yourself using thick cork tiles. You adjust the gap to fit the specimen's body. When you have set the insect in the correct position, cover the top of the wings with strips of greaseproof paper. Before adding it to your collection, you should leave it like this in a warm dry place out of reach of pests for a month. This allows it to dry.

Mounting

After the insect has dried out, remove all the pins that keep the legs or wings in position. Using the pin through the specimen's thorax, fix the insect in place in your permanent case or tray. For creatures where no setting is required (flies, wasps and spiders, for example) soften the body to relax it, and pin through the thorax slightly to one side of the midline. If you are mounting a beetle, it is best to insert the pin through one wing case. For long-term storage, you must protect specimens with insecticides.

SNAILS AND OTHER SHELLED CREATURES

You can find snails, crabs, shellfish and sea urchins in many different places. Rock pools on the sea shore provide a good source, as do ponds and wetlands. Chalk grasslands will also yield many specimens.

To preserve both snail and shell, put it in the killing jar, then into preserving alcohol, and gently remove the body with the aid of forceps and probes. This will enable you to see what the creature itself looks like. If you are only interested in the shell, kill the creature humanely by putting it in boiling water and leaving it there for 15 minutes. Clean older empty shells with a wooden pointed stick and wash with white spirits or turpentine.

To preserve the shells dry, put them on cotton wool in a clear plastic box, or glue them to a card with labels below. A coat of clear varnish will improve the look of many shells.

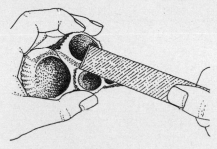

Inside a snail's shell
Use a coarse metal file, sandpaper block or fine wood rasp to file away part of a shell. This will reveal the internal structure with curved walls and central column.

Cleaning shells
Clean old shells carefully using a wooden pointed instrument which will not crack or scour the shell.

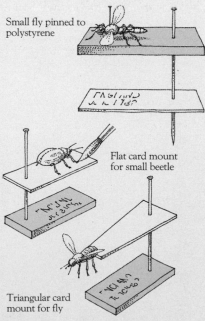

Small fly pinned to polystyrene

Flat card mount for small beetle

Triangular card mount for fly

Methods of mounting
A pin through the insect's thorax enables you to remove it for examination. Glueing the body to card also allows you to look at the underside—often vital for identification. With tiny beetles, glue the legs to a card.

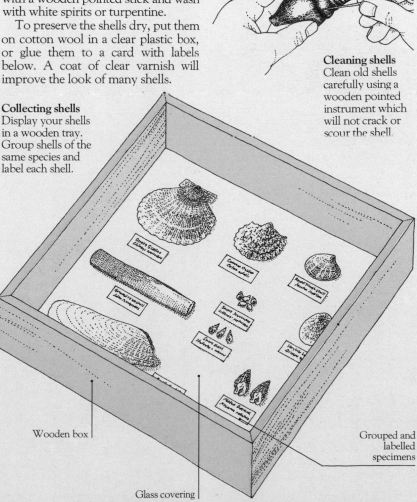

Collecting shells
Display your shells in a wooden tray. Group shells of the same species and label each shell.

Wooden box

Glass covering

Grouped and labelled specimens

STUDYING AND PRESERVING LARGER ANIMALS

COLLECTING AND HANDLING

The larger animals include a wide range of creatures—fish, mammals, birds, reptiles and amphibians. You can preserve these animals in spirits, and study them whole, but many of them contain interesting organs that it is possible to dissect. In addition, you can preserve their skeletons and skins.

You should never, of course, go out and kill animals merely to add to your collection. But there are plenty of alternative ways of finding specimens to study. There are numerous dead creatures to be found in the countryside or on the roads. Sometimes you can find one that has fallen into an

Protecting your hands Wear tight-fitting rubber gloves when handling animals.

empty bottle or a can, and has consequently died. Even your cat will sometimes supply you with specimens, and you are bound to have deaths among your own creatures.

Whenever you dissect any of these creatures, you must always guard against the risk of diseases and parasites. Handle specimens with the greatest care, keep everything clean, follow the rules for dissection (see page 243), and wear rubber gloves.

Trapped in a bottle
Thoughtless litterbugs may provide you with material.

When collecting and preserving the animals proves difficult, you can often build up a profile of a particular species without collecting the creature itself. This is especially effective with birds, as you can gather feathers, pellets, broken eggshells, and even nests in midwinter when you are sure they have been abandoned.

FISH, AMPHIBIANS AND REPTILES

These animals are extremely difficult to skin, so it is better to pickle them whole in alcohol, or preserve their skeletons. It is best to sketch or photograph specimens first, because the preservatives will make their colours fade. Leave the specimen for two days in alcohol diluted with an equal quantity of water, then transfer it to normal preserving alcohol. If the

Preserving fish
Pin the specimen on a cork tile using stainless-steel entomological pins, so that it keeps its shape. After about a week, the fish will be rigid and you can remove the tile.

animal is quite big (a large goldfish, for example), make a slit in its belly so that the preservatives will penetrate. Many animals will curl up and stiffen, so you should pin them to a cork tile beforehand. You can remove the pins and tiles after about two weeks.

The frog's development
If you raise frogs from spawn you can preserve specimens at each stage.

INSIDE A FROG

Because they have been collected for dissection and research in vast quantities, as well as being eaten widely, frogs are now quite rare. Do not kill one unless you have bred it yourself.

For dissection, pin the frog on its back on the dissecting tray. Carefully lift the thin belly skin clear of the inner organs and cut it with scissors or a scalpel, then pin this skin back. As you dissect,

keep all the organs moist with salt water from an eye dropper. Be careful not to burst any blood vessels or digestive tubes. If you do so the body fluids will flow everywhere—wash the specimen thoroughly with salt water.

Your dissection will reveal the liver at the top of the frog's body, and the coiled intestines towards the bottom. In a female, the lumpy spotted ovaries will also be visible.

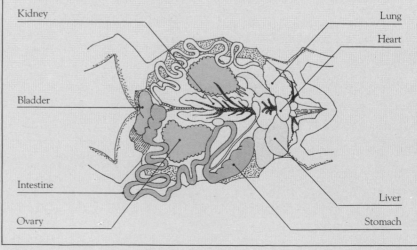

Kidney

Bladder

Intestine

Ovary

Lung

Heart

Liver

Stomach

BIRDS

It is possible to preserve many parts of birds. You can keep eggs and feathers dry. If you keep watch on the birds in your garden you will be able to collect feathers, eggshells and nests (when the birds have finished with them), and so build up a picture of the bird life in your garden.

Skin, skull and skeleton

Preparing a bird skeleton is quite difficult, because birds have light bones that are very fragile. (These light bones are specially adapted to allow the bird to fly.) You should treat the bird as you would a mammal (see page 273), but proceed very carefully. A skull is simpler to preserve, and will still make an informative display, since the beak will tell you what type of bird it came from. If you find only a wing you can open it out to dry and then store it flat. To skin a bird, follow the same principles as for skinning mammals (see page 272). You keep the beak attached, and the wings intact. As you skin the body it may be greasy, so it is a good idea to use sawdust as an absorbent. When you have finished, mount the specimen on a frame of cotton wool wrapped round a stick or piece of wire.

Feathers

A collection of feathers can be fascinating. By examining one under the hand lens you can see the barbs and interlocking barbules of which the feather is made up. You may also find lice and other parasites on very fresh feathers. It is interesting to study these under the microscope.

There are two main types of feathers —the soft "down" feathers, which act as insulation material, and the stronger flight feathers, which make flying possible. You can collect feathers anywhere, and the best times are spring and summer, when birds moult.

It is difficult to identify birds from one or two feathers, unless you can find very obvious specimens, such as the blue wing feather of the jay. If you find a whole dead bird, it is worth taking a selection of sample feathers

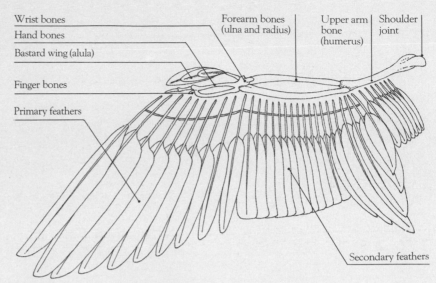

Wrist bones
Hand bones
Bastard wing (alula)
Finger bones
Primary feathers
Forearm bones (ulna and radius)
Upper arm bone (humerus)
Shoulder joint
Secondary feathers

The structure of a bird's wing
Make a detailed sketch if you find a wing, to show the position of the feathers. Then remove these carefully. Scrape all flesh from the bones and reassemble the wing.

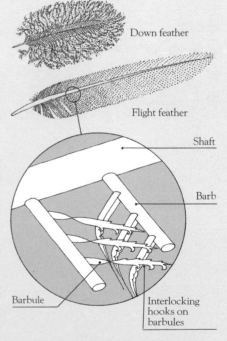

Down feather
Flight feather
Shaft
Barb
Barbule
Interlocking hooks on barbules

Preserving skins
You can skin a bird as you would a mammal. Take a small strip of wood and wrap it in cottonwool, binding in some mothballs, until you have a cylindrical shape. Insert this into the bird's skin.

for reference. It may also help you to identify others that you already have.

Keep the feathers that you find in a loose-leaf file. You can either fix them to the pages or protect them with clear plastic wallets. Keep them clean by using the same method as the birds themselves—preen them in the direction of the barbs. You can do this effectively with a stiff paintbrush.

Eggshells

You should never collect eggs from a nest. Collect only broken eggshells that you find scattered on the ground;

Examining feathers
The fluffy down is designed to trap air close to the bird's body, to keep the creature warm. The flight feathers are quite different in appearance. They are long and flat, and consist of a structure of interlocking barbules. These hook into each other to provide a continuous surface, smoothed by preening.

if any shells that you collect are dirty, wash them very gently in warm soapy water with a little disinfectant. Use a small paintbrush to remove any traces of debris on the shells. You can mount on card the shells you have cleaned, fixing them with a small spot of glue. But it is best to store them in a small box, protected with cottonwool. This method of storage allows you to remove each item in your collection and examine it, but you must always handle the shells with great care since it is all too easy to crush them.

Nests

Go nest collecting only in late autumn or winter. The majority of nests will have been used by the birds by this time. Since most birds build new ones each year, you can collect the old ones. In addition, winter, when the leaves have fallen from the trees, is the easiest time to see the nests. Even in the late autumn you must watch the nest to make quite sure that it is not being used. In addition, certain birds do not abandon their nests each year. As a general rule, it is the small birds that nest in hedgerows or low shrub who build a new nest for each brood that they rear. Larger species, that nest high in trees, in holes, or in the eaves of houses, generally re-use their nests year after year. They add to the structure of the nest each breeding season. You must never take the nests of these types of birds.

When you bring a nest that you have collected back to your workroom, put it in a polythene bag for a few days with a strong insecticide. You can then take it out and dissect it with the aid of your forceps. It is best to do this on a white surface (a large sheet of cardboard or something similar is best). Start from the inside. Sort out all the items that the birds used to build the nest. You will be able to group these components and mount them on a card. Alongside them you can mount any of the small insects you have found in the nest (see page 264). Alternatively you can keep the nest whole, spraying it with fixative or lacquer so that it hardens slightly and keeps its shape.

Examining pellets

Many species of birds cough up (or "regurgitate") indigestible material from their gizzards. These come up as compact masses known as pellets. These are very interesting to dissect,

Nest from suburban garden

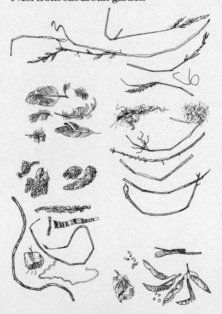

Nest from rural wood

Complete nests before dissection

Making an eggshell collection
The fragments of eggshell that you collect will be very fragile. Keep them in a sturdy box, protected with cottonwool. A glass cover is best, so that you can see the contents of the box easily. Group the eggshells from the same species, to aid identification of other specimens that you will find.

Nesting materials
Birds are amazingly resourceful in finding and using a wide variety of materials for nest building. Most birds improvize, using twine, string or other man-made things in addition to grass and twigs. Try to compare all the component parts of a nest from a hedge in a town garden with the same kind of nest in an area remote from human beings. Notice the difference in the materials used. You can make a very interesting display of nest components, grouping them according to their origins. Town birds use materials like cloth and string, in addition to the items (twigs, feathers, seed pods and moss) favoured by birds building their nests in the country.

because they can tell you a lot about what the bird feeds on.

If you find an owl roost, for example, collect the pellets from below it throughout the year. By dissecting them you can find out how the diet changes with the fluctuation of prey. It is best to break the pellets into two or three pieces and leave them to soak in water for a time. Then you can use forceps to tease them apart gently while they are still in the water. Any bones that you find can be put in household bleach for a few minutes. Then dry them and mount on cards.

Skull fragments

Fur

Jaws

Incisors

Vertebrae

Ribs

Limbs

Foot

Shoulder joint

Pellet dissection
A typical owl pellet will contain well-preserved remains of the bird's prey.

This example includes the bones and fur of a small mammal (that was probably a vole).

IDENTIFYING PELLETS

The pellets left behind by birds such as owls, crows and hawks vary considerably in size and shape. In addition, they differ in surface appearance. Rook and crow pellets, for example, have a loose surface texture, while the material in owl pellets is bound together much more closely into a tight mass.

	Where found	Appearance	Contents
Hawk (e.g. kestrel)	Near deserted buildings, under large trees, especially in open ground	25 × 15 mm Smooth, pointed at one end	No bones. Fur, feathers, claws, beaks
Barn owl	Near quiet buildings, under large trees	50 × 25 mm Black and shiny when fresh. Very round ends, may be spherical	Bones and fur of small mammals. Sometimes bones, feathers and beaks of small birds e.g. sparrow
Tawny owl	Under derelict buildings or large trees, especially conifers	60 × 50 mm Cylindrical, pointed ends, lumpy with bones at surface	Bones and fur of small mammals. Bones, feathers and beaks of small birds
Gull	Anywhere, especially under cliff breeding sites	Up to 40 × 20 mm Loose mass of material	Plant debris, some bones including fish bones, sea shells
Crow, rook	Under nest high in tree tops	About 25 mm across. Loose mass of material	Lots of plant fragments, stems, small pebbles

MAMMALS

Although it is possible to preserve mammals whole in alcohol or other spirits, this does not produce a very attractive result. The animal looks as if it is drowning in the liquid. In addition, although you can see the creature's external appearance, it is impossible to examine it inside.

You can get much more information from a specimen if you preserve the skin and skeleton separately. Both will be valuable additions to your reference collection. An advantage of keeping a mammal skeleton is that you can examine the skull and teeth. These provide probably the most important clues to identification.

Skinning and preserving skins

When skinning a mammal, try to use a specimen that has recently died. It will be easier to skin, and the finished result is much less likely to develop mould than the skin of a creature that has been dead for over two days.

With small mammals it is quite easy to remove the skin and preserve it mounted on a piece of stiff card. The actual art of taxidermy—stuffing a skin to give a three-dimensional life-like appearance—is very difficult indeed. The card-mounting method, producing what is known as a museum skin, is much easier, and adequate for a reference collection.

You should measure the mammal before you start to skin it, so that you can prepare a piece of card of the correct size. This card should be slightly greater than the length of the mammal. It is usually best to choose a width at least half the length. Cut the card to a tapering shape at the head end, and leave the rest with straight sides. Remember when you prepare the card that skin shrinks as it dries.

You can then remove the skin. As you proceed, and when you have finished skinning, it is advisable to rub a mixture of powdered borax and alum on to the skin. This will remove the grease from the skin's surface, and help preserve it. When the skin is completely removed, roll it on to the card as shown on the left. You should also insert something into the specimen's tail, so that this keeps its shape. A cocktail stick is ideal for this.

When you have finished, you should label the creature as fully as you can and attach the label to the card. Include as much information as you can on the label. It is worth recording the animal's names, size and the date and place of collection.

Cleaning bones and teeth

Keeping mammal skeletons is useful in two ways. They show clearly how the mammal is constructed, and they

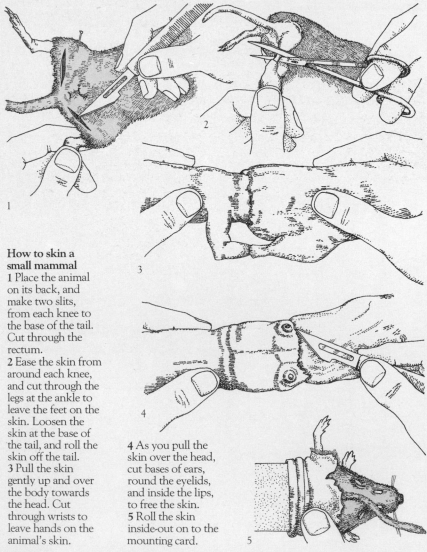

How to skin a small mammal
1 Place the animal on its back, and make two slits, from each knee to the base of the tail. Cut through the rectum.
2 Ease the skin from around each knee, and cut through the legs at the ankle to leave the feet on the skin. Loosen the skin at the base of the tail, and roll the skin off the tail.
3 Pull the skin gently up and over the body towards the head. Cut through wrists to leave hands on the animal's skin.
4 As you pull the skin over the head, cut bases of ears, round the eyelids, and inside the lips, to free the skin.
5 Roll the skin inside-out on to the mounting card.

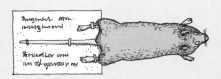

Displaying the skin
Make a label and attach this to the card on which you have mounted the skin. This should give full details for future reference.

INSIDE THE MOUSE

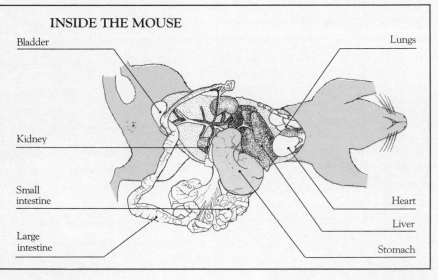

Bladder

Kidney

Small
intestine

Large
intestine

Lungs

Heart

Liver

Stomach

To dissect a mouse, cut down the middle of the animal's underside, in the same way that you cut a frog (see page 268). Pin back the skin to reveal the internal organs. If you unravel the coiled intestine and carefully lift it away, you will be able to see the blood vessels and kidneys near the backbone. Older specimens have a lot of fatty material in and around the organs. You have to clear this away in order to see the internal details of the specimen.

are important in identification of the other specimens that you find.

There are several ways of separating and cleaning up the bones and teeth of mammals and other bony creatures such as fish, amphibians, reptiles and birds. The simplest method is simply to leave the animal in a bucket of rainwater outdoors. This allows the flesh to "macerate". In other words, it becomes soft and gradually falls to pieces. This process takes time (usually several weeks), and produces a very unpleasant smell, but works well.

The second method is to simmer the animal in a macerating fluid. A good substance to use is a weak solution of sodium hydroxide (washing soda). This is highly corrosive, so you should wear gloves while using it, and be careful not to spill the solution.

As a third alternative you can use a biological method of cleaning skeletons. The larvae of dermestid beetles or blowfly maggots will eat away the flesh, leaving only the bones. Ants will also do this job for you.

Whichever method you employ, it will be necessary to finish off the task with a sharp piece of wood (for example, a cocktail stick) to ensure that all the flesh is removed. Next bleach the bones (a normal household bleach solution will do for this) and they will be ready to set and dry.

Cleaning in water
Leaving the bones in a bucket of rainwater makes the flesh disintegrate. This process gives probably the best results of all the methods, but the natural decay produces a bad smell.

Simmering on the stove
Using macerating fluid will speed up the cleaning process. Simmer the bones in an old saucepan. After about ten minutes the flesh will begin to separate from the bones.

Using insects
Some of the insects that you have in your home zoo can be useful for cleaning mammal skeletons. Keep the beetles and the carcass together in a sealed container. Dermestid larvae are ideal, but you can also use slower carpet bettle larvae.

Using ants
Place the carcass on a baseboard under a bucket or bowl near an ants' nest. The bucket will protect it from larger predators. A crack in the bucket allows the ants to enter and clean the specimen with their tiny jaws.

MOUNTING SKULLS AND SKELETONS

After you have removed the flesh from the bones, rub off any tiny pieces of meat still on them with forceps or a cocktail stick. Soak the bones in a solution of household bleach for two or three hours. Finally, hang them in an ammonia solution or petrol. When you do this, suspend the bones so that they are well above the bottom of the container, to allow the grease to fall clear. Be particularly careful if using highly inflammable petrol.

The skeleton will then be ready for setting. To replace the ligaments that normally hold the bones together, use either small spots of glue or wire supports. You may also require a wire frame to support the skeleton while you are joining the bones together.

Removing grease

Thoroughly removing flesh from a skeleton is not enough. You must also get rid of the grease. Hang the bones in a jar of degreasing fluid such as petrol, car degreasing agent or ammonia solution. The grease falls to the bottom.

Teeth and jaws

These help in mammal identification. Insectivores such as shrews, hedgehogs and moles have fairly unspecialized teeth. Cheek teeth are often pointed, and canines small. Bats have similar teeth, but larger canines. Rodents, such as mice, rats and squirrels, have large incisors at the front of each jaw.

Lower jaw of shrew

Lower jaw of bat

Lower jaw of rat

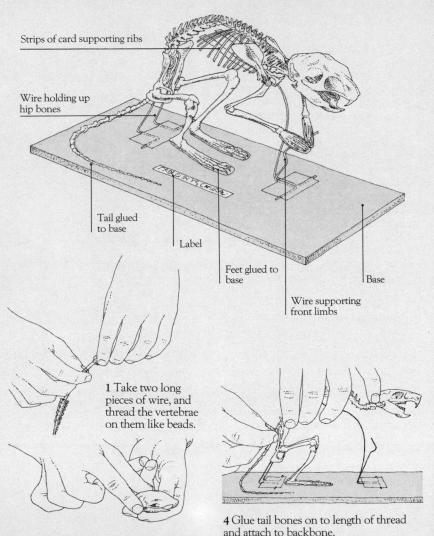

Strips of card supporting ribs

Wire holding up hip bones

Tail glued to base

Label

Feet glued to base

Base

Wire supporting front limbs

1 Take two long pieces of wire, and thread the vertebrae on them like beads.

2 Glue together the upper and lower jaws, and leave them to set. Attach the skull to the end of the wire. Assemble limbs, sticking them together with glue.

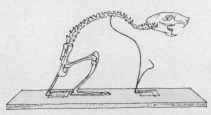

3 Glue pelvic bones and back legs to backbone. Support the skeleton with the ends of the wires, and add a wire from the baseboard to the centre of the backbone.

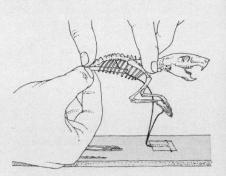

4 Glue tail bones on to length of thread and attach to backbone.

5 Glue ribs to vertebrae, supporting them with strips of card. Stick front limbs securely to rib cage.

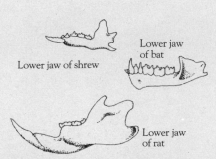

TAXIDERMY

When done well, taxidermy produces a much more lifelike record of the original specimen than a museum skin mounted on a card. But taxidermy is not an easy technique. It is most important to learn as much as you can about the species you are preserving. You must know the specimen's exact shape, so that you stuff the skin in the right way. In addition, it is vital to know the animal's typical postures, so that you mount it in a lifelike position.

First skin the bird or small mammal you want to preserve (see page 272). Cleaning the skin and removing any traces of grease are particularly important. You can wash off blood with cold water, while to remove the grease it is best to clean the specimen carefully with borax or even sawdust.

Mammal taxidermy

To prepare a small mammal skin for mounting, you must first construct an artificial body, as near in size and shape to the original as possible. To do this make a wire frame, and wrap this round with wood wool to build up the correct shape. Leave the ends of the wire protruding at the front, so that you can attach the skull. This can be either the mammal's own skull, cleaned thoroughly, or a plaster cast. Mark the positions of shoulders and limbs on the body.

Next cut four wires, with sharp ends, each about twice the length of the specimen's limbs. Push each wire up through the mammal's foot, following the line of the bones. Stuff the front limbs with tow or wood wool and join the wires to the body. Before tackling the hind legs, insert a wire to strengthen the tail, filling this with tow if necessary.

It is advisable to examine the body carefully at this stage, comparing it with photographs and drawings of a live specimen. Look for any hollow areas. The shoulders will probably need filling with tow.

After drawing the skin fully over the body and stitching it up, it is necessary to pose the animal naturally and attach it to a baseboard. Once again, you must consult drawings and photographs of the live mammal, because even a well-stuffed specimen will look unnatural if positioned wrongly. Glue the feet to the base, or pass the end of the leg wire through the board to secure the mounted specimen. Carefully add glass eyes of the correct colour. Push them well into the skull.

Mounting birds

You can use the mammal technique for bird taxidermy, but birds pose additional problems. To stop the wings from spreading, tie them together while stuffing the skin. Keep them in a natural position, and secure them by tying together the two humerus bones.

An additional difficulty is that the bird's feathers are likely to be ruffled when you have finished. You can remedy this by smoothing the feathers and then binding the bird with soft cotton or wool until it is fully dry. A piece of cardboard around the tail will also help return feathers to their natural place as they dry.

How to mount a small mammal

1 After measuring the skin, make an artificial body of the right size. Twist wood wool round a wire frame, and leave the wires sticking out.

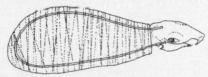

2 Clean the skull, or make a cast, and attach this to the protruding wires on the body.

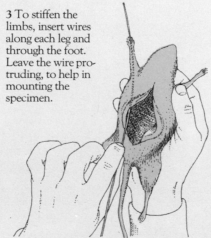

3 To stiffen the limbs, insert wires along each leg and through the foot. Leave the wire protruding, to help in mounting the specimen.

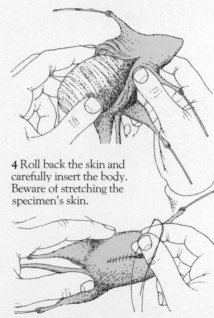

4 Roll back the skin and carefully insert the body. Beware of stretching the specimen's skin.

5 Push each leg wire into the body, so that it is firm. Sew up the skin.

6 Choose a natural pose, and attach the finished mammal to a baseboard using the protruding wires.

KEEPING LIVE ANIMALS

KEEPING LIVING THINGS

I have been very lucky throughout my life in being able to keep numerous wild creatures. When I was a boy in Corfu I kept everything from eagle owls to sea horses, and from snakes to butterflies. Later on, when I collected animals for zoos, I met and studied animals from Africa, Asia and South America. Thousands of different creatures must have lived with me at one time or another. In fact in my own zoo in Jersey I live in the middle of over a thousand creatures.

As I know from experience, the keeping and caring for any live creatures—whether they are worms or white mice, rhinoceroses or elephants —is a responsibility that you must not undertake lightly. Once you have an animal in captivity it is entirely dependent on you for its well-being. Make it a rule that the animal's welfare must always come first. A lot of legislation exists to protect animals, and there are many organizations, such as the RSPCA, which exist to endorse these laws. So before taking any creature into your care, it is essential to make sure that you can give it a comfortable home with an adequate food supply, and that once you have finished studying it you can release it back into its natural environment.

Freshwater aquarium

Install a filter to remove impurities from the water. Place rocks nearby to keep it firmly in position. Use an air pump to operate the filter. Stock the tank with plants and a selection of freshwater creatures.

AQUARIA

An aquarium is probably one of the most satisfying things for a naturalist to set up. Aquaria are generally easy to keep, and provide you with an endless source of interest. The creatures that are kept in them are also often easy to breed. An aquarium forms a miniature pond-like ecosystem. The plants grow and provide oxygen and food for the animals. The animals' waste products and their remains provide food for the decomposers, which pass food back to the plants again. A perfectly balanced aquarium in which this process takes place needs little attention.

Your aquarium may be cold or tropical, and have either salt or fresh water, so altogether there are four separate types of aquarium. The easiest for most naturalists to set up is the cold freshwater aquarium. This allows you to study the aquatic life of your area. But even if you do not live near the sea, it is still possible to create a seawater aquarium by using marine salts bought from a local supplier. This will turn fresh water into sea water, but the process is complicated, and the aquarium will need much more attention than a freshwater type. If you live near the sea you can obtain fresh sea water every few days.

Some hints on keeping aquaria

When designing your aquarium, try to select a range of plants and animals that will fill up the various niches in the tank. You must remember the different roles that the various inhabitants will fulfil. Plants will produce food and oxygen, most fish will live in the "mid-water" area, while shrimps, catfish and some snails live near the bottom, keeping the sand clean. Some filter-feeders such as freshwater mussels will help keep the water clear.

If you are intending to use your aquarium for serious long-term study, it is advisable to buy plants from a specialized stockist. Pond plants often carry diseases and parasites. You must

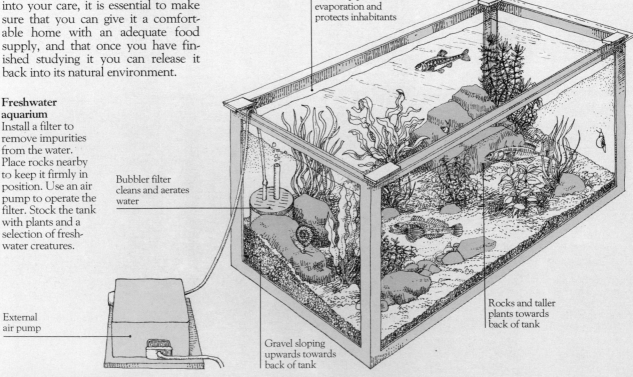

Glass top prevents evaporation and protects inhabitants

Bubbler filter cleans and aerates water

External air pump

Gravel sloping upwards towards back of tank

Rocks and taller plants towards back of tank

keep your tank in a well-lit position, but not in bright sun, which will make it too hot. If you have to keep it in a fairly dark place, you should consider fitting a light. You can build this into the lid of the aquarium, and control it with a time switch. This will help to provide the vital light energy which the tank will need.

There may be problems with debris accumulating on the tank bottom. If this happens, clean it out with a siphon tube. But don't be too fussy—any tank needs some natural refuse.

Setting up an aquarium

1 Clean the aquarium and scour it with paper. Rinse it thoroughly with clean water. If you have an undergravel filter, fit it in place.

2 Collect some gravel from a local stream. You should boil this for about 20 minutes to sterilize it, rinsing it until it is very clean.

3 Place the gravel in the tank so that it slopes down from back to front. You can then install any other filters, and other equipment.

4 Next insert the plants (small ones at the front, larger ones behind); heaping gravel over the roots to keep them in place.

5 Then add water by pouring it slowly on to a sheet of paper in the tank bottom. This will protect gravel and plants from disturbance. Leave the aquarium for two to three weeks to "settle".

THE GARDEN POND

If your garden does not have a pond, it is possible to build one. There are many different methods, such as lining the pond with concrete or glass fibre, but the easiest way is to line the pond with butyl rubber. This is specially made as a pond liner, but thick pvc sheeting will also do. Ordinary plastic sheeting, though, is not adequate.

Choose a suitable site in the garden. A place that is not exposed to the sun all day, but just gets the morning or evening sun, is ideal. Try not to position the pond under trees, otherwise it will fill up with leaves in the autumn.

To begin with, your pond will seem rather bare, but plants will gradually take over, as they do in any natural ecosystem. But plants do take time to get themselves established. You can stock the pond with frog spawn, plants and fish.

It is important to stock your pond with the correct species of plants

Making a pond

1 Excavate a hole in the chosen site. Around the edge, make a shelf for the shallow-water plants, and dig a deep area in the middle for creatures to hide in. Smooth the sides. If any sharp rocks project into the hole, pad these with old newspaper or glass-fibre loft insulation.

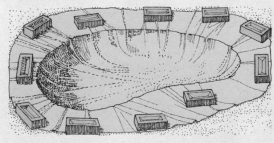

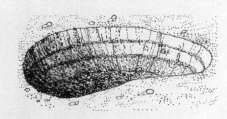

2 Add the lining material, put bricks around the edge to support it, and smooth it into place.

3 Fill your pond with water from a hose, and put turf or paving stones over the edge of the liner. The pond will then be ready to stock with plants.

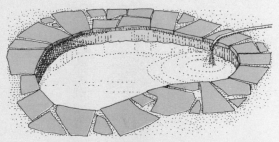

(see page 37). Select plants from the various types available. Oxygenators should be planted first, in the deep area of the pond. These include water starwort, stonewort, water milfoil, and the pondweeds. These species are are valuable because they keep the pond supplied with oxygen. You should also have some of the floating plants, such as frogbit, duckweed and water lettuce. Finally, choose some of the many species that will grow around the side of the pond, in the shallow area. As well as the various sedges, reeds and rushes these include the iris, yellow flag, purple loosestrife and water mint.

Local ponds and streams can provide a good supply of creatures for the pond. You can collect frog, toad and newt spawn, to start a population of amphibians. The larvae of insects such as dragonflies and caddisflies are easily found, as are snails. In addition you will be able to catch freshwater fish.

BREEDING BUTTERFLIES AND MOTHS

It is extremely interesting and quite easy to rear butterflies and moths from their larvae or pupae. It is also an excellent way of obtaining specimens for your collection while helping the species in question. This is because if you take one creature for your collection and release back into the wild the others you have bred, you have protected them against predation. Of those that you rear, it is worth taking two for your collection. You can then mount one specimen to show the undersides of the wings, and the other to show the top.

The table opposite shows several common species of butterflies and moths, and gives details of when you can expect to find them in the various stages of their life cycle.

Do not breed a species and then introduce it into a place where it does not already live without the expert advice of your local natural history society. This sort of carelessness has resulted in the grey squirrel being introduced in England where it has caused considerable damage to trees.

Sources of butterflies and moths

In previous chapters of this book I have described several methods of catching these creatures, such as netting butterflies, catching moths by light traps, beating branches for the larvae, and digging for pupae underground.

The specimens that you catch in the wild may more often than not have parasites. But if you collect a number of them, some will be healthy, while others will give you the opportunity of studying the parasites.

Caring for larvae

Larvae must have the correct plant to feed on, otherwise they will not grow or develop. It is best to put the plant in a pot, to ensure a ready and healthy supply of fresh food. If this is not practicable, you can enclose a naturally growing part of the plant in a muslin bag. This restrains the larvae and protects them from predators. At the next stage many larvae need a particular

Home-made cage

Purpose-built cage

Rearing in cages

Purpose-made cages for butterflies are made of clear plastic to allow a good view of the inhabitants. A muslin-topped cap gives good air circulation, while the base can hold compost for plant food

if required. A home-made cage should be well-constructed, preferably of smooth wood, with few crevices. Keep the cage tall, to accommodate plants, and include muslin panels for ventilation.

position in which to pupate. The usual position is hanging from the lid of the cage or from the muslin bag, or in the soil. You should watch out for this noticing whether the leaves of the food plant project up towards the cage lid, preventing the larvae from getting into the right position for pupating.

Always try to keep your larvae in conditions that are as natural as possible. A cold conservatory or outhouse often gives the right light and temperature.

Looking after the pupae

Pupae are just as delicate as larvae, so keep them in conditions as near as possible to natural. Many pupae

last right through the winter and the adults do not emerge until spring.

Some species allow you to sex them. This is important, because breeding and hatching pupae is frequently the only way to get good-condition virgin adults that you can breed or preserve.

Breeding pairs

You can sex some adult butterflies and moths, but the most reliable way of producing breeding pairs is to start with the pupae. Keep the adults together in a suitable cage with a supply of food. There should also be a food plant on which they can lay their eggs. If you have some newly hatched female moths, you can use them to attract breeding males by their scent—the naturalist Fabre used this technique (see page 65).

If there is no fresh food available for your adult moths or butterflies, you can use a nectar substitute. This can be a combination of honey and water (or sugar and water) on a pad of cotton-wool. Place it among the twigs in the cage so that the insects can feed on it.

The cage must not have any cracks, even tiny ones, since when the minute caterpillars hatch they must not be allowed to escape. When they hatch, put in fresh food leaves. The caterpillars will usually crawl on to these and settle there.

Butterflies are quite selective about where they lay their eggs, whereas many species of moth will simply scatter their eggs anywhere. If you possess a greenhouse you can keep your butterflies and food plants there.

Sexing pupae

With some species of butterfly and moth, it is impossible to tell the sex of the adults. But you can distinguish the sex of the pupae. The male has two small bumps at the tip, and the female has no bumps. This information is useful when breeding.

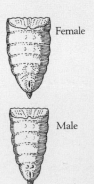

Female

Male

BUTTERFLIES AND MOTHS — THEIR HABITAT AND DIET

Butterflies	When and where eggs are laid	Where larvae pupate	Food (larva and adult)
Brimstone	May–June Buckthorn	Amongst rough vegetation on stem or under leaf	Buckthorn
Comma	April–June Stinging nettle, hop	On food plant	Stinging nettle, hop, currant, elm, gooseberry
Common blue	January, June, September Bird's-foot trefoil, rest harrow	Lower stems on food plant	Bird's-foot trefoil rest harrow, clover
Large white	April–May, July–August Cabbage, nasturtium	On food plant, walls, fences, windows	Cabbage, nasturtium
Peacock	May Stinging nettle	On food plant, or in undergrowth some distance away	Stinging nettle, hop
Red admiral	May–June Stinging nettle	On underside of leaf surrounded by foliage	Stinging nettle, pellitory, hop
Small tortoiseshell	May, August Stinging nettle	Stems, twigs, fences	Stinging nettle
Wall brown	April–July Grass	Among meadow grass	Meadow grass
Moths			
Dun-bar	March–May On food plant	On or just below ground	Oak, birch, elm, sallow. Larvae feed on other caterpillars
Green arches	June–July On food plant	In underground chamber	Dock, plantain, bramble, sallow
Poplar hawk moth	April–June Poplar or sallow leaves	Just below ground	Poplar, sallow, willow, aspen
Red underwing	March–April On bark of poplar	Amongst dead leaves, under loose bark	Poplar, willow
Silver Y	May–June On food plant	Under leaves or debris, on or near food plant	Clover, peas
Vapourer	April–July On food plant	Tree bark crevices, under eaves, on fences	Most deciduous trees and shrubs
Wood tiger	June–August On food plant	On or near ground	Plantain, wild violet, dandelion

KEEPING AQUATIC INSECTS

When walking and collecting near ponds and streams, one of the most common things you will see are aquatic insects such as dragonflies, caddisflies and mayflies. There are many species that are interesting to study, and you can keep their larvae in an aquarium.

The eggs may be difficult to identify, but it will be easy to find the larvae. Caddisfly larvae are particularly interesting, since some species enclose themselves in a protective tube made of leaf fragments, sand grains, twigs, or other material available in the pond. Most species of caddisfly spend the majority of their year-long life cycle as larvae. Mayfly nymphs live even longer before reaching the adult stage.

To keep these insects you should simulate the natural environment as closely as possible. A wide tank, giving a generous area of water surface, is best. Use a sand and gravel base, and keep the water quite shallow. You will also require a good supply of plants.

Some caddisfly larvae are vegetarian. They feed from decaying plant litter on the pond bottom. Introduce some of this material from the natural habitat. Other caddisflies as well as dragonfly and mayfly larvae are carnivorous, eating tadpoles and small fish. You will have to give them a supply of meat, but be careful not to put in too much or you will foul the water.

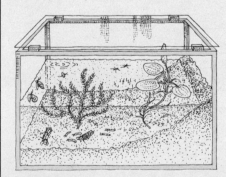

Aquatic environment By sloping the sand and gravel in the base you can imitate the shallow area of a pond or stream. Include plants from a local pond and litter from the pond bottom.

SMALL-ANIMAL TERRARIA

There are several small invertebrates, such as worms, ants and snails, which are interesting and easy to keep. They are all cold-blooded creatures, so they will be most active if the temperature is warm. But be careful they do not dry out. It is best to keep these creatures for a while, observe them and do experiments, and then let them go. If you keep them without looking at them regularly, you might forget their existence and allow them to die. You should also be careful that these creatures do not escape. They can do a lot of damage.

Breeding mealworms

The larvae of the small beetle called *Tenebrio molitor* used to infest grain stores, and came to be known as mealworms. Provided they have a warm dry area to live in, they are easy to keep, and provide a good food source for many birds and reptiles. Larvae, pupae and adult beetles all make excellent food, but allow them to breed for about six months to establish the colony first.

Any secure ventilated container will do for holding the mealworms. A biscuit tin with a perforated lid or an old goldfish bowl with a muslin top will suffice. Whatever the container, keep it in the dark. Feed the mealworms every week with a little dry bran mixed with a selection of flour, oatmeal, chopped dried bread and biscuits. The insects need only a little moisture. An apple core or sliced carrot, replaced every few days, will be sufficient. When you want to take out some worms, use a sieve. Replace the sieved mixture back in the container, as it will contain eggs.

Keeping dermestid beetle larvae

These can be very useful creatures to keep, especially if you intend to clean and preserve the skeletons of small mammals. This is because they eat meat, and will strip the flesh from a skeleton very efficiently (see page 273).

Biological suppliers sell the beetles and you can breed them in a large sealed jar in a warm cupboard or in a container similar to that used for mealworms. Put the carcass in the bottom and they will remove the flesh.

Studying earthworms

It is possible to make a wormery to observe the life of earthworms. You need a transparent container shaped like a narrow box. A wooden frame

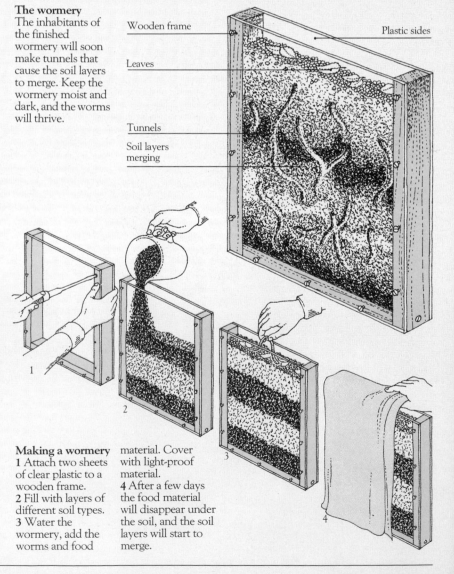

The wormery
The inhabitants of the finished wormery will soon make tunnels that cause the soil layers to merge. Keep the wormery moist and dark, and the worms will thrive.

Wooden frame

Plastic sides

Leaves

Tunnels

Soil layers merging

Keeping mealworms
Put a piece of dry sackcloth or a 10 mm layer of dry sawdust in the container. Cover this with a similar depth of bran or oatmeal and about 20 mm of dried bread and biscuits crumbled and mixed with flour. Top this with another layer of bran, and add about 100 mealworms.

Making a wormery
1 Attach two sheets of clear plastic to a wooden frame.
2 Fill with layers of different soil types.
3 Water the wormery, add the worms and food material. Cover with light-proof material.
4 After a few days the food material will disappear under the soil, and the soil layers will start to merge.

with clear plastic sides is ideal. Fill this with layers of different soil types, such as garden soil, sand and peat, and water the wormery. You can then introduce the worms. Ten or twelve mature specimens will be sufficient for a wormery about 30 centimetres square. You can tell when specimens are fully grown by looking for the saddle, which contains the reproductive organs. Place dead leaves and grass on the surface of the soil as food for the worms, and cover the wormery to simulate the underground darkness.

The formicarium
This should have a tightly fitting glass lid and light-proof cover. A flexible tube connects with the outside world. You can use the end chamber to introduce food.

How to make a formicarium
1 Screw four lengths of wood to the baseboard and position the glass with one edge touching one of them.
2 Add the connecting tube, moulding it in with plasticine.

After a few days all the worms will have tunnelled through the soil, disturbing the soil layers, which will eventually merge. The leaves will have been drawn down into the soil. This is how worms enrich the soil, because they do not eat all the leaves they pull down. In addition, their tunnels help to provide air for the plant roots.

Slugs and snails
You can keep slugs and snails in a tank with a lid. Make sure they are cool and moist. It is best to feed them with

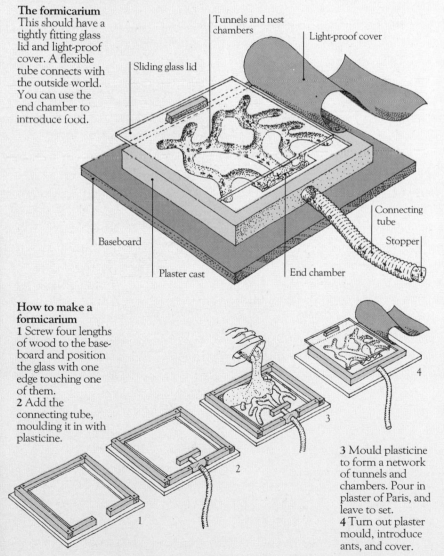

Tunnels and nest chambers

Light-proof cover

Sliding glass lid

Baseboard

Plaster cast

Connecting tube

Stopper

End chamber

3 Mould plasticine to form a network of tunnels and chambers. Pour in plaster of Paris, and leave to set.
4 Turn out plaster mould, introduce ants, and cover.

plant matter, such as vegetable fish food, providing them with water on a moist piece of cotton wool. If you leave some soft fat for them to eat, you will be able to see their minute tooth-marks by examining the surface of the fat with the aid of a hand lens.

Setting up a formicarium
Making an ant farm or formicarium is a simple business, and it enables you to study the habits of a colony of ants at close quarters. Of the various species of ant, the best to keep are the black garden ants, the slightly smaller yellow ants, or red ants. Be sure to obtain a good selection of adults, larvae and pupae. It is also essential to find the queen, who is several times larger than the other ants.

Keep the formicarium dark except when you are observing the ants, and even then use as low a light level as possible. You can feed ants on virtually anything organic—kitchen scraps, bits of meat, fruit and vegetables. Give them a dampened piece of sponge as a water source.

Experiments with the formicarium
Once the formicarium is established, you can set up a variety of experiments. With typist's correction fluid you can mark individual ants and work out their activities during the day. You can calculate an ant's average life span by isolating individual pupae and marking the adults when they hatch out.

Introducing a stem from a rose bush that is thickly covered with aphids will allow you to observe a particularly interesting pattern of behaviour. Within a short time the ants will start to "milk" the aphids by stroking them with their antennae, to obtain the sweet sticky substance they secrete. The aphids are, in fact, like a dairy herd to the ants.

Like mealworms, the ants that you breed in the formicarium are also a valuable food source. Birds and fish will readily take "ants' eggs" (which are sold in pet shops under this name, but which are really ants' pupae) while some reptiles and amphibians will even eat the adults.

REARING FROGS AND TOADS FROM SPAWN

Collecting spawn is an excellent way of studying the amazing change (known as *metamorphosis*) from the tadpole to the adult creature. Both frogs and toads develop in this way, and you can collect either. But do not mix spawn from frogs and toads in the same container. Keeping spawn can be a very useful conservation exercise, since once you have reared the tiny frogs, you can release a number of them that might otherwise have been eaten by predators. This is especially valuable because sadly the so-called common frog is on the decline.

Frogs lay their spawn in a great cloudy mass, whereas toad spawn appears in long strings. (You may also see newt spawn, which is found as single eggs, each surrounded by jelly.) I once went to Cambridge to give a lecture, and afterwards the students (who were all young biologists or hopeful veterinary surgeons) invited me back to their rooms for a drink. During the course of conversation, I discovered to my amazement that none of them knew the difference between frog spawn and toad spawn. I was telling this story to a German scientist I know, saying how it amazed me that these students did not know such a simple thing. I learned the difference by looking in ponds and ditches when I was about eight years old. "Ah," said my German friend, "the trouble is that now they don't let them look in ponds and ditches." So you can see how important field work is to all naturalists.

Collect the spawn as soon as it appears in your local pond. This will probably be in April, or whenever the weather starts to get warmer after the end of winter. Take only a small portion of spawn (about a handful), and some water weed. You should leave the bulk of the spawn in its natural habitat. Keep the spawn you have collected in an aquarium, or even in a dish or bowl. Change the water every two or three days, and keep the container out of direct sunlight. You can now observe the transformations

that occur as your specimens change from eggs to tadpoles, and from tadpoles to frogs or toads. You will notice that in addition to the changes in body shape, the tadpoles will also alter their diet. At first they are vegetarians and feed on water weed. But as their back legs start to grow, they begin to eat meat. Cat or dog food is ideal.

As their tails shrink and the froglets lose their gills and begin to breathe air, they will start to crawl out of the water. You must provide them with small stones, on to which they will climb. You will be surprised how well they jump, so cover the aquarium.

When the young frogs have developed to this stage, it is advisable to release them back into the wild. Only in their natural habitat will they be able to find the constant supply of insects that they need. Be sure to put them back in the pond or lake where

you found the original spawn. This will ensure that the adult creatures have the correct environment.

Experiments with tadpoles

There are two simple experiments that will show the effects of varying conditions on the growth of tadpoles. If you keep one batch of spawn indoors in a warm place, and another in a shed or outhouse, you will discover the effect of temperature on its development. The indoor batch should develop much more quickly.

Another factor that influences the growth and development of these creatures is the amount of iodine in the water. This substance is important for the production of the hormones that stimulate the change from tadpole to adult. By adding just a few drops of iodine to one batch of tadpoles, you will be able to see how vital it is.

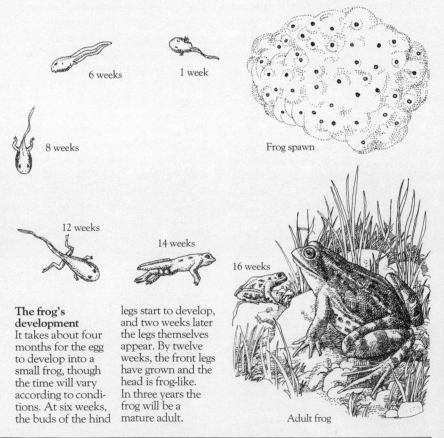

6 weeks 1 week

8 weeks

Frog spawn

12 weeks

14 weeks

16 weeks

Adult frog

The frog's development

It takes about four months for the egg to develop into a small frog, though the time will vary according to conditions. At six weeks, the buds of the hind legs start to develop, and two weeks later the legs themselves appear. By twelve weeks, the front legs have grown and the head is frog-like. In three years the frog will be a mature adult.

KEEPING WILD MAMMALS AND REPTILES

These creatures are extremely interesting to study. But it is best to keep them only as long as is necessary for your observations. Look upon them as though they have come to you for a holiday. Release them after a few weeks, or otherwise let them go in the autumn, so that they can hibernate.

Suitable cages vary according to the species you want to keep. One of the best containers is an aquarium. This is ideal for rodents such as mice and voles (as well as amphibians and reptiles) since they will gnaw through the sides of a wooden cage in no time at all. Always remember to keep the aquarium covered with a piece of gauze or a sheet of glass with a small space for ventilation. This will allow air to circulate, while keeping the creatures in. Position the cage so that the animals receive suitable amounts of light and warmth—most reptiles like a sunny window ledge, whereas mice or voles

prefer a cool shady situation. Of course it is essential to give them a supply of fresh food and water.

It is fascinating to watch and note down the behaviour of your captives. You will find that just as you go to bed, get up, have breakfast, and follow a daily routine, animals also have their own particular ways of spending their lives. Many small mammals, especially wild mice, have delightful habits that the domestic white mouse and other common pets seem to have lost. Try keeping a wild mouse and a domestic mouse side by side in separate cages, and make notes on their habits.

Which animals can you keep?

In Britain you cannot catch any wild bird unless you have a special licence.

This is a very important law, essential for the protection of bird species. Some of the most suitable creatures for the naturalist to keep are the amphibians, particularly the various species of newts, and the reptiles such as lizards and slow-worms. (The latter is in fact a legless lizard.) The smaller rodents such as mice and voles are also rewarding and easy to keep.

Before you decide to catch any of these creatures it is vital to make sure that you can provide it with the correct food and give a constant supply. With the herbivorous animals this is usually quite easy. It is possible to grow food (see page 262), or find it readily. You can supply some of the carnivores

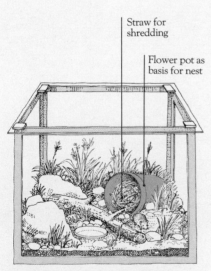

Setting up a vivarium
How you stock a vivarium depends on its inhabitants. Slow-worms, for example, require stones or pieces of wood under which to hide. They prefer a slightly damp atmosphere and will be more active in warm conditions. Mice need material to shred up. Sink their water dish into the compost to stop them knocking it over.

Straw for shredding

Flower pot as basis for nest

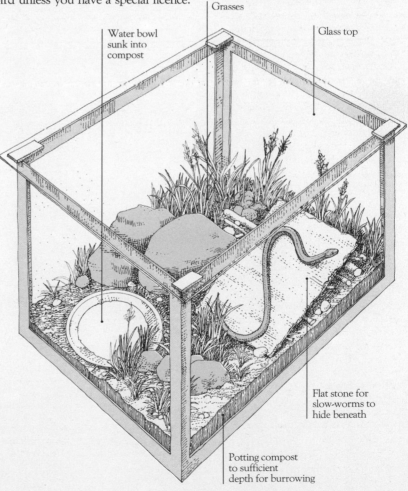

Water bowl sunk into compost

Grasses

Glass top

Flat stone for slow-worms to hide beneath

Potting compost to sufficient depth for burrowing

with things such as blowfly larvae or mealworms (see page 280). But many species of snake, insectivores such as shrews, and some of the small predators like weasels and stoats pose problems. It is difficult to get the right food for them, and they should not be kept by the amateur naturalist. Remember that the welfare of the animal in your charge comes first. If you think that your specimen is suffering in any way as a result of its captivity, you should release it.

Nocturnal behaviour

Many of the creatures you may want to keep are nocturnal, that is to say they sleep during the day (when you want to watch them) and are active at night. In order to observe them, it is possible to reverse their rhythm by turning day into night. It is not difficult to construct a cage that will carry out this process. The light should be controlled by a time switch, and the cage well ventilated. The light comes on at night, so that the creatures think that it is day, and sleep while you sleep. An opaque cover keeps daylight out of the cage during the day, so that the inhabitants think it is night, and become more active. You can watch their behaviour by using a red light, which does not disturb the animals.

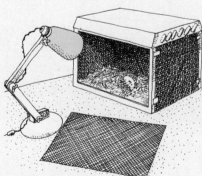

Keeping nocturnal creatures
Use a cage that allows you to block out all the light during the day. You can paint the sides and make a wooden or cardboard cover for the front. To switch the light on and off a time switch is ideal, though you can control it manually. Look at the animals under a red bulb.

KEEPING AND FEEDING LIVE ANIMALS

	Diet	Conditions	Remarks
Newt	Small invertebrates e.g. slugs, worms, insects; small fish	Damp habitat, plus water to swim in	Release adults in natural habitat
Frog	Tadpole—pondweed Froglet—meat e.g. pet food Adult—flies, small beetles, slugs, worms	Water, weed and stone platform. Adults need pond	
Toad	Tadpole—pondweed Toadlet—meat e.g. pet food Adult—snails, young newts, frogs and slow-worms	Damp environment, water, weed and stone platform. Adults need pond	Release adults in natural habitat
Lizard	Small insects e.g. mealworms, blowflies, grasshoppers. Also spiders	Sandy soil, turf, log for basking	Release in autumn for hibernation. Warm environment encourages activity
Slow-worm	Small slugs, snails, worms and insects	Turf, board or stones under which to rest. Light environment, slightly damp	Long living
Grass snake	Tadpoles, frogs, toads, small mammals	Prefers warm environment	Hard to feed; very difficult to keep any snake
Mouse	Wheat. Also grass, fruit, mealworms, nuts	Potting compost, moss, turf, log or stone. Straw for shredding. Don't use wooden container—aquarium best	Generally nocturnal
Vole	Grass. Also wheat, fruit, mealworms	As for mouse	Generally nocturnal
Shrew	Pet food, mealworms, maggots. Need continuous food supply as they eat voraciously	As for mouse	Very difficult to keep: will only eat live or very fresh food. Keep singly or they will fight

WOUNDED AND ORPHANED CREATURES

Caring for wounded or orphaned animals is a full-time job. You must not undertake it lightly. Sometimes a creature that seems in need of care is in fact quite safe. For example, you sometimes see baby birds that look as if they have been deserted by their parents. But these are probably young birds that have just left the nest. Their parents will still be looking after them. So if you find a baby bird, keep it under observation, but do not touch it or move it from the place where you found it, since its parents may still be caring for it. If you decide, after a time, that it has really been deserted, take it and hand-rear it yourself. But remember that this is a difficult and time-consuming task. You can make up a mixture for hand-feeding with hardboiled egg and a little raw meat. It is best to do this with tweezers or a dropper. This is hard work because most parent birds give their offspring a little food at regular intervals through the day. There are similar difficulties with orphaned mammals. There are several milk substitutes that you can use, and you can feed the animals with an eye dropper. But again this is a time-consuming task.

When hand-rearing mammals or birds, over-feeding can be as dangerous as under-feeding. I remember once I was rearing some baby hedgehogs. I left them for the weekend under the care of my sister, with strict instructions as to how much food they should have at each mealtime. Because young hedgehogs, like most baby mammals, are very greedy, my sister was under the impression that I was starving them to death. So she gave them twice as much as I had told her, and all the baby hedgehogs died due to enteritis brought on by over-feeding.

If you find a wounded creature, make sure you are doing the kindest thing by trying to save it. Your local veterinary surgeon or the RSPCA can advise you what the most humane course of action will be.

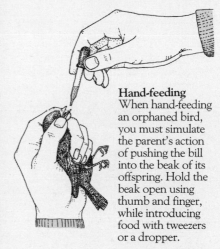

Hand-feeding
When hand-feeding an orphaned bird, you must simulate the parent's action of pushing the bill into the beak of its offspring. Hold the beak open using thumb and finger, while introducing food with tweezers or a dropper.

Window victims
Unconscious birds that have flown into glass should be kept dark and warm until they move. Then release them. Take a bird with broken bones to a vet.

THE ECOLOGY OF A SMALL MAMMAL

When you have collected a lot of information and specimens connected with a particular mammal, it is often worthwhile making a display to show clearly what you have discovered about the species. The centrepiece of the display should be the mammal itself—either its skin, or a drawing or photograph. Around this show its signs—drawings or castings of the tracks and droppings—and a map of its territory. You can also include a photograph or sketch of the mammal's home.

On the edge of the display show the subject's food and its predators—the species that eat it. This will demonstrate the mammal's place in the environment, show how many other species depend on it for a food supply, and how many things it relies on to survive.

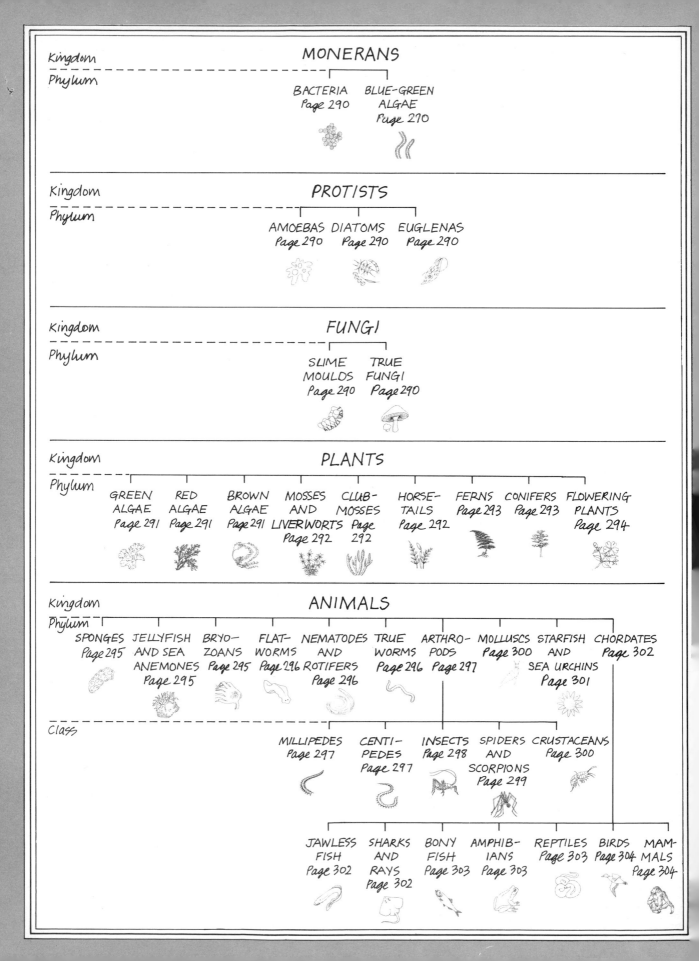

Kingdom

MONERANS

Phylum

BACTERIA
Page 290

BLUE-GREEN
ALGAE
Page 270

Kingdom

PROTISTS

Phylum

AMOEBAS
Page 290

DIATOMS
Page 290

EUGLENAS
Page 290

Kingdom

FUNGI

Phylum

SLIME
MOULDS
Page 290

TRUE
FUNGI
Page 290

Kingdom

PLANTS

Phylum

GREEN
ALGAE
Page 291

RED
ALGAE
Page 291

BROWN
ALGAE
Page 291

MOSSES
AND
LIVERWORTS
Page 292

CLUB-
MOSSES
Page
292

HORSE-
TAILS
Page 292

FERNS
Page 293

CONIFERS
Page 293

FLOWERING
PLANTS
Page 294

Kingdom

ANIMALS

Phylum

SPONGES
Page 295

JELLYFISH
AND SEA
ANEMONES
Page 295

BRYO-
ZOANS
Page 295

FLAT-
WORMS
Page 296

NEMATODES
AND
ROTIFERS
Page 296

TRUE
WORMS
Page 296

ARTHRO-
PODS
Page 297

MOLLUSCS
Page 300

STARFISH
AND
SEA URCHINS
Page 301

CHORDATES
Page 302

Class

MILLIPEDES
Page 297

CENTI-
PEDES
Page 297

INSECTS
Page 298

SPIDERS
AND
SCORPIONS
Page 299

CRUSTACEANS
Page 300

JAWLESS
FISH
Page 302

SHARKS
AND
RAYS
Page 302

BONY
FISH
Page 303

AMPHIB-
IANS
Page 303

REPTILES
Page 303

BIRDS
Page 304

MAM-
MALS
Page 304

WHAT'S WHAT

Despite the bewildering and exhilarating diversity of life we see around us, it is still possible to discern some sort of order and organization in nature. Over the centuries, in their exploration of the living world, naturalists such as Linnaeus have assigned organisms to groups so that the members of each group have something in common with one another—just as, for example, in your class at school everyone is about the same age, or in the yellow pages of the telephone directory the shops are classified together according to what they sell. The study of this grouping and classifying of living organisms is called taxonomy.

On the left you can see one general scheme into which we can fit all living things. This particular version is accepted by most naturalists, although there are several other versions, and opinions differ as to which is the best one. But whichever you prefer, all schemes start with main groups; each main group is divided into sub-groups; each sub-group is broken down still further into smaller sub-groups; and so on. You go on doing this until you reach the final group, the species.

Early naturalists used to group organisms as either plants or animals, but the more we know about the natural world the more we see that this old two-kingdom system does not work. The modern system has five kingdoms. Plants are defined as organisms which, by means of *photosynthesis*, trap the energy in sunlight to make their food. To do this they use the green pigment chlorophyll or something very similar. Animals, on the other hand, are things that engulf or ingest (that is, eat) their food. Fungi are regarded as a third kingdom—they "digest" their food externally and then absorb it in liquid form through the body wall. The other two kingdoms both consist of microscopic single-celled creatures. Protists are organisms each of which has a nucleus; they can't be thought of as plants or animals since some can switch from photosynthesis to eating and back again, depending on the conditions. Members of the fifth kingdom, monerans, don't even have a nucleus. They are the simplest organisms and are probably similar to the very early forms of life that evolved on Earth.

The importance of finding out what's what
The grouping and naming of plants and animals is important to a naturalist for two reasons. First, we must remember what Linnaeus was talking about. Classification provides a universal scientific language so that naturalists and biologists all over the world, although perhaps speaking in their own native tongues, can refer to a group or an organism and know that they are all talking about the same one. The second reason is that classification reveals evolution at work and its results. Before the theory of evolution was generally accepted, Linnaeus and other naturalists assigned plants or animals to groups because of structural

How to use the "What's What"
Everyone uses the classification system automatically to some extent. You know the difference between a bluebell and a hawk, for example, and you know that a bluebell is a flowering plant and that a hawk is a bird. So to start with, look at the key chart (opposite) and turn to the page indicated for the group you think your specimen belongs to. There, listed under the heading for the group, are certain characteristics that only members of that group have. Examine your specimen for these characteristics. If it has them, then you have turned to the correct group and should read on for more information. If it's the wrong group, however, you may get a cross-reference to another page for easily confused and very similar organisms. If you still have no luck then return to the key chart, study your specimen more closely and select another likely group.

features they had in common with one another. For example, all the flowering plants were grouped together, as were all the creatures with six legs. But before Darwin, no one had fully realized that the similarity between organisms in a group was due to the fact they were related to each other—that is to say, they were all descended from a common ancestor. It must have been rather like grouping together people called Durrell just because their names were Durrell, and not realizing that they were all related to each other. Darwin's magnificent theory revealed the true reasons for similarities between certain plants or animals—they were related through the process of evolution.

The modern taxonomist continues to classify organisms on the basis of characteristics that they have in common, but he always has to keep in mind that the most important characteristics are those which show how organisms are related to each other. Going back to the analogy with names, it is quite possible that someone else has "invented" the name Durrell quite independently of my family, so that although they have the same name as us they are not related to us. The same thing can happen in nature, and the taxonomist must always be on the lookout for it. Take, for example, bats (which are mammals) and birds. Both have wings, but taxonomists would not consider "having wings" to be a good reason for putting birds and bats in the same group. They know from fossil records that although both bats and birds have a very ancient common ancestor (since mammals and birds descended from the reptiles), their wings developed only *after* the birds and the mammals started their evolution as distinct and separate groups.

A good classification scheme helps any naturalist to construct an evolutionary tree, and if you have a book on fossils you can draw one for yourself. The height of the whole tree represents time. Below is a tree

The evolution of the vertebrates
The tree shape is often used by naturalists and biologists to show how different groups of organisms are thought to have evolved from a common ancestor. But do not read too much into these trees. They make one particular path of evolution seem an established fact, which it isn't. Our knowledge of evolution and the construction of evolutionary trees rely to some extent on fossils, and we admit that the fossil record is pretty sparse and that there are large gaps. The discovery of one new fossilized tooth can cause whole trees to be re-drawn. Also, most trees do not show all the groups that become extinct along the way—they usually have room to show only those branches that lead to extant organisms (those alive today). But despite these drawbacks, trees are invaluable in helping us to picture the probable course of evolution, and they give us a framework to help in our classification of living animals and plants.

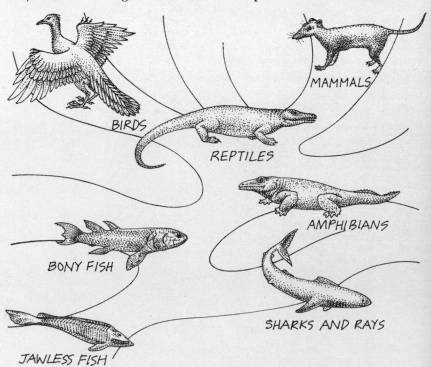

that covers the 500 million years since animals with backbones (the vertebrates) first appeared in our fossil records. Each limb and branch coming off the main trunk represents a distinct group evolving, and shows how long ago and from what ancestors it came. You can see for example that the first vertebrate was a strange jawless fish very similar to the lampreys of today. You can also see that the ancestors of today's amphibians (frogs, toads and salamanders) evolved from a fish-like creature about 400 million years ago, and that a primitive amphibian gave rise to the whole range of reptiles about 50 million years later. Reptiles continue today, of course, in the shape of turtles and snakes and lizards. The dinosaurs have long since become extinct. But the reptiles gave rise to the birds and mammals about 180 million years ago. A tree such as this just shows the evolution of the major classes of vertebrates but you can take any branch—let's say the mammals—and study the fine twigs that split off from it. Now one can see the evolution of bats and primates, rodents, carnivores and ungulates. The extreme tips of the twigs on the top of the evolutionary tree represent the species that are living in the world at present.

By knowing how to use the system of classification a naturalist can identify the species of a particular organism he is studying; by understanding how the system was set up he can appreciate the organism's place in the evolutionary history of life. For example, some people still think that Darwin said we were descended from chimpanzees. In actual fact, he said nothing of the kind. What he said was that we and the other present-day great apes have a common ape-like ancestor, and from that common ancestor we branched out in different directions—some to become chimpanzees, some to evolve into gorillas, and some to become human beings.

2 Leaves with stipules, these sometimes joined to the leaf-
  stalks; trees, shrubs or herbs:
  3 Carpels more than 2, or if few, then fruits with numerous
    hooks on the outside; leaves simple or more often pinnate,
    trifoliolate or 5-foliolate; calyx often with an epicalyx
                                              Rosaceae (4–43)
  3a Carpels 2, often partially united below; herbs:
    4 Leaves pinnate . . . . . . Rosaceae (4–43)
    4a Leaves not pinnate . . . . Saxifragaceae (425–437)
2a Leaves without stipules; mostly herbs:
  5 Carpels not immersed in or enclosed by the floral axis:
    6 Herbs:
      7 Carpels usually numerous, or if few the petals often
        modified into tubes or spurs and the leaves much
        divided . . . . Ranunculaceae (214–241)
      7a Carpels few; petals never modified; leaves fleshy, un-
        divided . . . . . . Crassulaceae (421–424)
    6a Shrubs . . . . Rosaceae (4–43)
Carpels immersed in the floral axis or enclosed by the
  calyx-tube:
                  and showy, solitary; aquatic plants
                                      Nymphaeaceae (242, 243)
                  spike-like panicles
                                      Rosaceae (4–43)

Taxonomic keys
For the accurate identification so important to a serious naturalist you need what is called a "taxonomic key". This is a book or leaflet which asks a series of questions about your specimen. By answering "Yes" or "No" to each question in turn you go on to the next one, and so on until you end up at the correct identification for your organism. To use the real keys employed by experienced zoologists and botanists you must be familiar with the anatomy of your organism and the naming of its various parts. This can be quite complicated, especially in plants, and sometimes it is difficult even for the professional botanist to classify a specimen using a key. If you want to have a go at keys for your plants or animals, ask at your local natural history society or library, or contact an organization such as the British Museum (Natural History).

THE CLASSIFICATION HIERARCHY AT WORK

Linnaeus and other naturalists developed a hierarchy system to classify organisms. The idea is similar to someone posting you a letter: first the letter goes to the right country, then to the county or state, then the town, then the street name, and finally to your house. In the natural world, the country has its equivalent in the "kingdom", which is the largest group (there are only 5 kingdoms for all living things). The next group down is the phylum, then class, order, family, genus, and finally the species. This diagram shows the system at work in classifying a weasel.

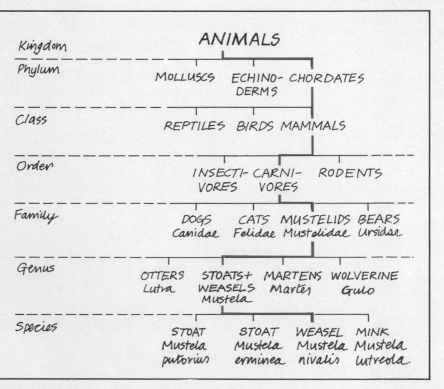

Kingdom — ANIMALS

Phylum — MOLLUSCS ECHINO-DERMS CHORDATES

Class — REPTILES BIRDS MAMMALS

Order — INSECTIVORES CARNIVORES RODENTS

Family — DOGS Canidae CATS Felidae MUSTELIDS Mustelidae BEARS Ursidae

Genus — OTTERS Lutra STOATS + WEASELS Mustela MARTENS Martes WOLVERINE Gulo

Species — STOAT Mustela putorius STOAT Mustela erminea WEASEL Mustela nivalis MINK Mustela lutreola

MONERANS
Kingdom: *Monera*

PROTISTS
Kingdom: *Protista*

FUNGI
Kingdom: *Fungi*

○ Microscopic
○ Single cells (if many individuals joined, then each one is exactly the same)
○ No nucleus ("control centre") in the cell (if there is a nucleus, see Protists, right)

Monerans are very simple organisms, being neither plants nor animals. There are two main groups, bacteria and blue-green algae, and they are found by the billions in water, soil, dust, air, other organisms—virtually everywhere. About 4,000 species.

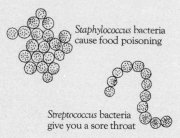

Staphylococcus bacteria cause food poisoning

Streptococcus bacteria give you a sore throat

Bacteria are extremely small, down to one thousandth of a millimetre across. Some are spherical, some are rod-shaped, others are corkscrew-shaped. They are responsible for decomposition and decay, some types of fermentation, and many diseases in plants and animals (disease-causing bacteria are called *pathogens*).

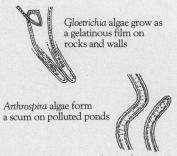

Gloetrichia algae grow as a gelatinous film on rocks and walls

Arthrospira algae form a scum on polluted ponds

Blue-green algae can colonize bare rock or concrete (where they form inky streaks), tree trunks, the sub-zero Arctic and the 75°C water of hot springs. Like plants, they get their energy from sunlight.

○ Microscopic
○ Single cells (if many individuals joined together, then each one is exactly the same)
○ Nucleus ("control centre") in each cell (if there isn't a nucleus, see Monerans, left)

These minute water-dwelling organisms were formerly divided into two main groups: the protozoa were single-celled animals, while the protophyta were single-celled plants. This division is not much use, since some protists can change from plant to animal and back again. There are about 11 groups of protists, of which three common types are shown here. They live everywhere there is water, and form the bulk of the plankton in the sea. At least 50,000 species.

Amoeba engulfs a food particle

Amoebas (Phylum *Sarcodina*) are fairly large protists about one-tenth of a millimetre long. They move by flowing their jelly-like bodies over the ground, and pursue and engulf minute food particles.

Diatoms come in many shapes and sizes

Diatoms (Phylum *Chrysophyta*) make up a large proportion of oceanic plankton. They contain chlorophyll like plants, and their cell walls are made of silica and sculpted into wondrous shapes.

Euglena swims by lashing its flagellum

Euglenas (Phylum *Euglenophyta*) usually get their energy from sunlight, like plants, but if kept in the dark can absorb food through their body walls, like fungi.

○ Feed by dissolving and absorbing nutrients from dead animal and plant matter
○ Main body is a mass of unsegmented threads
○ Reproduce by spores

The fungi include mushrooms, toadstools and moulds. The main body of a fungus is a mass of threads called hyphae which interlace and form a network called the mycelium. The threads secrete chemicals called enzymes that dissolve organic matter, and the resulting nutrient-rich liquid is absorbed through the fungal body wall as food. The mushroom or toadstool is the spore-producing part. About 100,000 species.

Slime moulds are common on the damp woodland floor in autumn

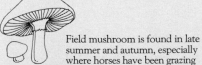

Slime moulds (Phylum *Myxomycophyta*) are unusual jelly-like fungi that look similar to slugs and can even "crawl" slowly. Common on dead wood, most slime moulds are small (about 15 mm across) and brightly coloured.

Field mushroom is found in late summer and autumn, especially where horses have been grazing

True fungi (Phylum *Eucomycophyta*) are of many different kinds. Mildews, rusts and choke fungi are visible as white or brown powdery marks on the host, usually a plant. Pin-moulds are white cottonwool-like growths on dung, old bread and vegetable refuse. Sac fungi are so named because spores are borne in a sac-shaped structure, and they include yeasts (used in fermentation), moulds (used in cheese-making), and edible "mushrooms" like morels and truffles. Club fungi bear their spores in club-shaped structures and include most mushrooms and toadstools, also puffballs and stinkhorns.

GREEN ALGAE
Kingdom: *Plantae*
Phylum: *Chlorophyta*

o Simple plants, single-celled or long threads or frond-like
o Under the microscope, most cells look similar to each other
o Chlorophyll not masked by other pigments, so green in colour (if red or brown, see right)

About 90 per cent of the 6,000 or so species of green algae live in fresh water, mostly as microscopic swimming forms or fixed filamentous types such as slimy green pondweeds. Most familiar, however, are those green algae found along the shoreline. They are very simple plants with hardly any specialized structures.

Sea lettuce has bright green fronds but no stem or roots

Sea lettuce (*Ulva lactuca*) is a green alga about 30 centimetres long. It is common on rocky shores and salt marshes, and is tolerant of fresh water for a time.

RED ALGAE
Kingdom: *Plantae*
Phylum: *Rhodophyta*

o Small or medium-sized plants, usually with fine frond-like branches, living in the sea
o Always fixed by a "holdfast" or similar structure
o Red pigment phycoerythrin masks normal green chlorophyll, so red or purple in colour (if brown, see Brown algae, right)

Red algae are abundant in warmer seas, especially in the deep water of tropical seas. They form chalky encrustations in rock pools and on the bed of the Mediterranean. Rare in colder northern and southern seas. About 4,000 species.

Crimson-tuft alga can withstand being uncovered by the tide for short periods

Crimson-tuft seaweed (*Plumaria elegans*) is up to ten centimetres long. It grows as deep-red delicate fronds hanging from rocks on the middle and lower shore, and often forms attachment for purse sponges.

BROWN ALGAE
Kingdom: *Plantae*
Phylum: *Phaeophyta*

o Simple plants, mainly large and many-celled, living in the sea
o Always fixed by a "holdfast" or similar structure
o Brown colour due to pigment fucoxanthin, which masks the normal green of chlorophyll (if red, see Red algae, left)

Nearly all 2,000 species of brown algae are marine seaweeds, ranging from minute branching forms to massive giant kelps many metres long. The body of the plant shows separate parts—the stipe (stem), fronds (leaves) and holdfast (roots). They are common in colder seas and shores, and show marked zonation on the intertidal regions of rocky shores.

Sugar kelp is one of the larger algae to be found on rocky shores

Sugar kelp (*Laminaria saccharina*) grows up to three metres long. Its several fronds look like a crinkled ribbon, carried on a slender stalk and with a two-tiered holdfast. This brown alga grows on sheltered rocky shores from the low water mark down into shallow water.

MOSSES AND LIVERWORTS

Kingdom: *Plantae*
Phylum: *Bryophyta*

o Small green flowerless land plants (if large then see Clubmosses, right)
o Cushion-like growth of small pointed leaves (mosses) or flat lobed and fleshy (liverworts)
o Breed by spores

Mosses and liverworts are fairly simple low-growing plants, with the minimum of specialized structures. They do not have flowers but breed instead by spores, and there are no roots to speak of. Both types of plant are usually found in damp or wet places, though a few types of moss can grow on drier soil.

Bog moss
Sphagnum plumulosum forms large brownish-green hummocks on wet heaths, bogs and moors

Mosses (Class *Musci*) are common in damp places. Species tend to be restricted to different soils, either acid or alkaline. They grow as low springy cushions of small pointed leaves arranged spirally around a stem, and though the individual plants are small they can spread to cover large areas. In the Arctic tundra they form great "blankets" of vegetation, and large numbers are found in humid tropics. Mosses breed by spores carried in capsules on stalks. About 10,000 species.

Pellia liverwort has a thallus one centimetre across, and grows on the sides of ditches and moist banks by streams

Liverworts (Class *Hepaticae*) are flat dark-green plants that grow almost flat on the ground. The body of the plant, which looks like a veinless leaf, is called a thallus and is lobed (like a liver—hence the name). Liverworts breed by spores carried on capsules which are short-lived. They are always found in very damp situations such as wet bogs and the sides of streams. About 14,000 species.

CLUBMOSSES

Kingdom: *Plantae*
Phylum: *Lycodophyta*

o Green flowerless land plants
o Look like a large moss (if small then see Mosses and Liverworts, left)
o Breed by spores

Clubmosses are an ancient group which 250 million years ago included giant trees. The 400 living species are found in damp places; they are numerous in the tropics, but in temperate regions are found chiefly on wet heaths and mountainsides.

Clubmosses resemble ordinary mosses but are much larger

Stagshorn clubmoss (*Lycopodium clavatum*) grows to three metres long. Its leaves are finely toothed at the edges, and its spores form a fine yellow powder known as lycopodium powder.

HORSETAILS

Kingdom: *Plantae*
Phylum: *Arthrophyta*

o Large green plants without flowers or leaves
o Jointed tubular stem with rough feel
o Joints of stem have crown of toothed sheaths
o Spores produced in egg-shaped cones (if a cone-producer but a woody stem then see Conifers, right)

The 30 or so species of horsetails are survivors of a huge group of plants that included the giant fern-like trees growing in coal forests 250 million years ago. Several species are common on marshy ground, others on waste land or hedgerows. They are quite spectacular when covering large areas, and resemble no other type of plant.

Horsetails have distinctive whorls of long pointed leaf-like spikes at intervals along the stem

Common horsetail (*Equisetum arvense*) is about half a metre tall and is common in woods and fields on light soil. It can be a serious weed of cultivated land.

FERNS

Kingdom: *Plantae*
Phylum: *Pteridophyta*

o Large green land plants, no flowers
o Frond-like leaves unwind and expand as they grow
o Button-like spore sacs on undersides of leaves

Ferns are an ancient group of plants which are still flourishing today. The majority (over 90 per cent) are large tropical tree ferns up to 15 metres tall; there are a few aquatic types, and of course the familiar woodland ones found mainly on rich moist acid soils. About 12,000 species.

Most ferns have distinctively elegant lace-like leaves

Buckler fern (*Dryopteris dilatata*) is about one and a half metres high. Its fronds form crown-like tufts and its stalks are clothed in brown pointed scales. Like many ferns it spreads by underground stems called rhizomes, and it is common in woodlands.

CONIFERS

Kingdom: *Plantae*
Phylum: *Coniferophyta*

o Shrubs or trees, with woody trunks and branches
o Leaves usually needle-like or scaly
o Seeds borne in cones (if a non-woody cone producer then see Horsetails, left)

Coniferous trees are most widespread in the Northern Hemisphere. They are resistant to cold, and all except the larches are evergreen. Conifers were the dominant vegetation on Earth some 150 to 200 million years ago. About 300 species.

Although conifers bear flowers of a sort, there are technical differences between conifer flowers and the flowers of a "flowering plant" which mean that conifers are not included in the group called Flowering plants (see page 294). Also, not all conifers are evergreen—larches are conifers that shed their leaves in autumn, and so they are deciduous conifers. And, strangely enough, not all conifers bear cones; the juniper is one that bears soft fleshy berries instead.

Scots pine loses its lower branches as the tree gains height

Scots pine (*Pinus sylvestris*) grows up to 30 metres high. The bark of the lower trunk is rough reddish-brown, but on the upper parts is light-red to orange. Old specimens have flat crowns. Once widespread in Scotland, this tree has been severely depleted by man but it is still widespread in Northern Europe.

Italian cypress has its branches sharply unswept to give a spike shape

Italian cypress (*Cupressus sempervirens*) is a narrow columnar tree that reaches 30 metres high and has scale-like leaves with adjacent pairs at right angles. The cones are round and knobbly, like dark scaly berries.

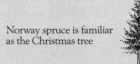

Norway spruce is familiar as the Christmas tree

Norway spruce (*Picea abies*) grows up to 20 metres and has a conical shape with short branches. It bears long cylindrical heavy cones.

FLOWERING PLANTS

Kingdom: *Plantae*
Phylum: *Anthophyta*

o Green plants bearing flowers, which may be large, colourful and scented or small and inconspicuous
o Well-defined leaves, stems and roots
o The microscope reveals unique pipe-like water-conducting tissue called xylem

Flowering plants are the dominant and most varied present-day vegetation. Apart from the plants normally thought of as "flowers" this group includes grasses, deciduous and tropical trees, insectivorous plants and cacti. There are over 250,000 species, with new ones being discovered regularly.

The class Anthophyta has two main subdivisions, depending on whether the seeds of the plant each have one or two cotyledons (seed-leaves—see the comparison table below). Beyond this, flowering plants are divided into families. There are dozens of common families—for example, the *Ranunculaceae* family includes the meadow buttercup, creeping buttercup, lesser celandine, meadow-rue, marsh marigold and stinking hellebore. It is often quite difficult to assign an unfamiliar flower or tree to a particular family or species—one needs to examine the plant very carefully and know several technical names for the various parts of the flower. Some examples are shown right to illustrate the principles of flower classification.

Rushes are grass-like plants of damp places

Soft rush (*Juncus effusus*) is a monocotyledon, as shown by its parallel leaf veins and flower parts in threes. The three sepals of each tiny flower resemble petals, and there are no leaves; rushes belong to the family *Juncaceae*.

Pyramidal orchid grows in dry grassy places

Pyramidal orchid (*Anacamptis pyramidalis*) is a monocotyledon belonging to the distinctive *Orchidaceae* family, which have three-petalled flowers with the lower middle petal forming a lip or keel.

Red campion is a plant of hedge and wood

Red campion (*Silene dioica*), a dicotyledon, has simple leaves in opposite pairs, five separate petals and not more than ten stamens. It is a member of the *Caryophyllaceae* family.

Daisy is common on short grass everywhere

Daisy (*Bellis perennis*) is a member of the *Compositae*, the largest plant family. This dicotyledon has many tiny self-contained flowers called florets packed on to one flower head.

Broom has bright yellow insect-pollinated flowers

Broom (*Cytisus scoparius*) is a dicotyledon of the *Leguminosae* family. It has bilaterally symmetrical flowers, only one carpel and its stamens united into a sheath.

Beech has a tall smooth trunk and dense crown

Beech (*Fagus sylvatica*) flowers are tiny catkins with no petals, and are either male or female. This dicotyledonous tree belongs to the family *Fagaceae*.

COTYLEDONS—ONE OR TWO?	
Monocotyledons (*Monocotyledonae*)	**Dicotyledons** (*Dicotyledonae*)
One cotyledon (seed-leaf) to nourish germinating embryo	Two cotyledons (seed-leaves) to nourish germinating embryo
Leaf veins run parallel to each other	Leaf veins form branching network
Water- and sap-conducting tubes in stem scattered about irregularly	Water- and sap-conducting tubes in stem arranged in neat regular bundles
Flower parts usually in threes or multiples of three	Flower parts usually in fours or fives or multiples of fours or fives
Grasses, sedges, rushes, some pondweeds, lilies, irises, orchids, palm trees. About 55,000 species.	Vast majority of present-day plants—flowers, trees, shrubs and herbs. About 200,000 species.

SPONGES

Kingdom: *Animalia*
Phylum: *Porifera*

o Body soft but not smooth due to supporting spicules or fibres
o Body anchored to a surface and may be encrusting, globular or vase-like
o System of passages through which water flows into and out of the body

Sponges are immobile plant-like organisms but the fact that they "eat" tiny food particles in the water means that they are animals. The sponge body is perforated with small pores through which water is drawn in, and food in the form of minute suspended particles is extracted; the waste water is then expelled through a single large opening. Sponges are mainly marine, distributed from the lower shore to very deep water. About 10,000 species.

Pieces of breadcrumb sponge often break from their rocky base and are cast up on the beach

Breadcrumb sponge (*Halichondria panicea*) is up to 20 centimetres across and its body is made up of many small conical projections which look like miniature volcanoes. It encrusts shells and rocks from midshore to deep water.

JELLYFISH AND SEA ANEMONES

Kingdom: *Animalia*
Phylum: *Coelenterata*

o Simple soft-bodied animals, radially symmetrical ("circular")
o Single body opening surrounded by stinging tentacles
o Body wall made of two cell layers sandwiching a jelly-like middle layer

The coelenterates include hydras, sea anemones, corals and jellyfish. They live their lives in two distinct phases— a fixed cylindrical polyp (the hydra stage) and a free-swimming bell-shaped medusa (the jellyfish stage). All 10,000 species are aquatic, and most are marine.

Common jellyfish has four pink horseshoe-shaped reproductive organs

Jellyfish (Class *Scyphozoa*) are mainly free-floating semi-transparent animals with long stinging tentacles dangling below a bell-shaped body.

Dahlia anemone reverts to a blob of jelly when tide goes out

Sea anemones and corals (Class *Anthozoa*) are immobile animals which might be mistaken for plants. Sea anemones capture prey with their grasping tentacles, which possess numerous powerful stinging cells. Coral animals are like small sea anemones but with hard skeletons, and in the tropics their dead remains build into huge reefs.

Freshwater *Hydra* is over one centimetre tall

Hydras (Class *Hydrozoa*) are elongated cylinders, usually fixed at one end and with an array of waving tentacles at the other end.

BRYOZOANS

Kingdom: *Animalia*
Phylum: *Bryozoa*

o Small animals with tentacles, anchored together in colonies
o Each individual is enclosed in a capsule or embedded in a jelly-like substance
o Each has a separate mouth and anus

Bryozoans are small colonial creatures that in salt water are called "sea mats" or "corallines" and in fresh water are called "moss animals". Most species are marine and occur as film-like growths on weeds, shells and rocks. About 4,000 species.

A sea mat colony shows up as a white lacy encrustation on a frond of seaweed

Sea mat (*Membranipora*) forms flat whitish encrustations on stalks and fronds of brown seaweed. The individual animals are about one millimetre long and are rectangular in shape with a blunt spine on each side. Found from middle shore down to shallow water.

FLATWORMS

Kingdom: *Animalia*
Phylum: *Platyhelminthes*

o Small soft flattened body
o Not segmented (if segmented then see True worms, right)
o Aquatic or damp-living

Flatworms can be distinguished from most other worm-like creatures by their flattened or ribbon-like profile. Most are small and inconspicuous, though the parasitic flukes and tapeworms can cause wasting disease in animals. Flatworms are hermaphrodites, and often have a long complex life cycle. About 25,000 species.

Free-living flatworm (planarian) hunts small prey on the stream bed

Free-living flatworms (Class *Turbellaria*) are mostly marine bottom-dwellers, with some freshwater species in ponds and streams. They usually have several pairs of eyes at the head end and a "mouth" in the centre of the underside, through which a muscular tube, the pharynx, is protruded to grasp small animal prey.

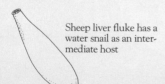

Sheep liver fluke has a water snail as an intermediate host

Flukes (Class *Trematoda*) are leaf-like translucent creatures that are parasitic on many vertebrates, most commonly fish. They have suckers near the mouth and tail to grip the host. It is quite common for a fluke to have two (or more) hosts in its life cycle.

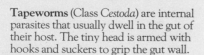

The pork tapeworm, which lives in the human gut, can grow to over 3 metres in length

Tapeworms (Class *Cestoda*) are internal parasites that usually dwell in the gut of their host. The tiny head is armed with hooks and suckers to grip the gut wall.

NEMATODES AND ROTIFERS

Kingdom: *Animalia*
Phylum: *Aschelminthes*

o Small, translucent, and either worm-like (nematodes) or tubby cylinders (rotifers)
o Not segmented (if segmented then see True worms, right)
o Move by sinuous writhing (nematodes) or leech-like looping movements (rotifers)

Nematodes (also known as roundworms) are one of the most numerous organisms on earth; in rich soil there may be 100 million per square metre. Most are very small and can only be seen under the microscope. Both free-living and parasitic forms are common; the latter include threadworm (pinworm) of children and the pig hookworm. About 20,000 species. Rotifers are also small and incredibly numerous. They are called "wheel-animacules" because of the ring of tiny hairs around the mouth. About 1,500 species.

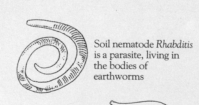

Soil nematode *Rhabditis* is a parasite, living in the bodies of earthworms

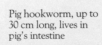

Pig hookworm, up to 30 cm long, lives in pig's intestine

Rotifers are cylindrical semi-transparent creatures seldom more than a millimetre long

TRUE WORMS

Kingdom: *Animalia*
Phylum: *Annelida*

o Tubular worm-like or leech-shaped body
o Body divided into many similar segments
o Bristles called chaetae protrude from sides of body, except leeches, which have suckers at the front and back (if body flattened with no segments or chaetae then see Flatworms, left; if small and translucent with no segments or chaetae then see Nematodes and rotifers, left)

True worms are a very varied group of creatures which have colonized many different habitats, but they all have in common the fact that they are segmented. Usually the segments at the head and tail ends are different in structure, but outwardly very similar. About 14,000 species.

Earthworm *Lumbricus* feeds on organic material in soil

Earthworms and bloodworms (Class *Oligochaeta*) conform closely to the basic worm shape, and their chaetae (bristles) are very small. They are hermaphrodite, and move by alternately lengthening and shortening the body.

King ragworm *Nereis* is up to 20 cm long and its powerful mouthparts can give you a nasty nip

Marine worms (Class *Polychaeta*) such as ragworms and paddleworms have large chaetae, often borne on paddle-like side extensions of the body, and usually swim by S-shaped wriggles and waving of the chaetae.

Medicinal leech, about 10 cm long, feeds on mammalian blood

Leeches (Class *Hirudinea*) are flat and without chaetae. They can move either by undulating the body and "flapping" along or by "looping" using the suckers under the head and tail.

ARTHROPODS

Kingdom: *Animalia*
Phylum: *Arthropoda*

Arthropods are a group of animals whose name means "joint-legged", and this is what sets them apart from all other invertebrates. To qualify as an arthropod you must possess jointed legs and a hard outer skeleton called an "exoskeleton"— and, of course, you must not possess a backbone.

The phylum *Arthropoda* is the largest phylum in the natural world. It is estimated there are over one and a half million species of arthropod, ranging from microscopic springtails crawling in the soil to giant crabs whose legs are a metre long. The phylum is split up into five classes: insects (themselves the biggest single group of animals), crustaceans, millipedes, centipedes, and spiders, scorpions and other arachnids.

MILLIPEDES

Kingdom: *Animalia*
Phylum: *Arthropoda*
Class: *Diplopoda*

o Long cylindrical segmented body
o Two pairs of jointed legs per segment (if one pair of legs per segment then see Centipedes, right)
o Head bears short antennae often bent at an angle (if long antennae then see Centipedes, right)

Millipedes tend to live in damp places, burrowing in the surface layers of soil and in leaf litter for their food of decaying vegetation. They are slow-moving and possess numerous "stink glands" along the body which produce a repellent substance. Some millipedes are squat and can roll up into a ball, in a way similar to woodlice. About 7,000 species.

Spotted snake millipede has "stink glands" which show up as orange-red spots on each body segment

CENTIPEDES

Kingdom: *Animalia*
Phylum: *Arthropoda*
Class: *Chilopoda*

o Long flattened segmented body
o One pair of jointed legs per segment (if two pairs of legs per segment then see Millipedes, left)
o Head bears long antennae (if short antennae then see Millipedes, left)

Centipedes are mostly fast-moving hunters of other small invertebrates, active mainly at night in damp places such as leaf litter and other rotting vegetation. At its head end each centipede has a pair of poison claws. About 1,500 species.

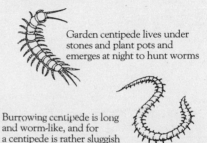

Garden centipede lives under stones and plant pots and emerges at night to hunt worms

Burrowing centipede is long and worm-like, and for a centipede is rather sluggish

INSECTS

Kingdom: *Animalia*
Phylum: *Arthropoda*
Class: *Insecta*

○ Three pairs of jointed legs at some stage in life cycle (if four pairs then see Spiders and scorpions, right; if five pairs then see Crustaceans, page 300)
○ Hard outer skeleton
○ Body divided into three parts
○ One or two pairs of wings

Insects are by far the largest class of living things. They have invaded and colonized almost every habitat you care to think of—the land, air, fresh water, under the ground, and there are even some in the sea. They range in size from almost microscopic spring-tails to 15 centimetre Goliath beetles. About one million species.

Springtail

Springtails (Order *Collembola*) are tiny pale wingless insects that live in soil and leaf litter. They jump by flicking their abdomens, hence their name.

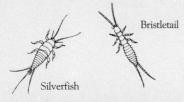

Bristletail

Silverfish

Silverfish, firebrats and bristletails (Order *Thysanura*) are small and wingless, with two or three prongs for a tail. It is often difficult to make out the three body divisions.

Mayfly

Mayflies (Order *Ephemeroptera*) live only a few days as adults but the aquatic larvae may live for years. The adults look like small dainty dragonflies; they rest with their wings together over the back (like butterflies), forewings are larger than hind-wings, and there are long tail prongs.

Damselfly

Dragonfly

Dragonflies and damselflies (Order *Odonata*) have two roughly equal pairs of large wings with networks of veins, a long slender body and powerful flight. They are voracious hunters as are their aquatic larvae. Damselflies, like mayflies, hold their wings together over their backs at rest; dragonflies spread theirs sideways.

Stonefly

Stoneflies (Order *Plecoptera*) have aquatic larvae; the adults are similar to mayflies but hold their wings out sideways at rest, and have shorter tail prongs.

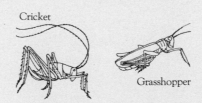

Cricket

Grasshopper

Grasshoppers, locusts and crickets (Order *Orthoptera*) are medium-sized to large insects, herbivorous and with powerful folded back legs for jumping. Grasshoppers have short stubby antennae, crickets have long wavy ones.

Stick insect

Stick insects (Order *Phasmida*) have long thin twig-like bodies and legs, and their wings may resemble leaves.

Earwig

Earwigs (Order *Dermaptera*) resemble thin beetles, with the front pair of wings hardened into a case-like covering to protect the rear pair. They have pincer-like tails.

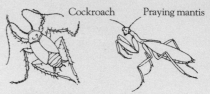

Cockroach

Praying mantis

Cockroaches and mantises (Order *Dictyoptera*) are fairly primitive insects. Cockroaches are beetle-like with very long antennae and long spiky legs. Mantises have their clutching forelegs doubled up, as though praying—hence their name.

Termite

Termites (Order *Isoptera*) are small, pale and often eyeless and wingless ant-like creatures. They live in colonies with a queen as the only breeding member.

Biting louse

Biting lice (Order *Mallophaga*) are parasites on birds and few mammals. They are small and wingless, and have claw-like legs for clinging.

Sucking louse

Sucking lice (Order *Anoplura*) are similar to biting lice, but have tube-like mouth-parts for sucking up body fluids.

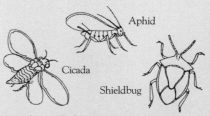

Aphid

Cicada

Shieldbug

True bugs (Order *Hemiptera*) include aphids, shieldbugs, groundbugs, capsids, cicadas, hoppers and pondskaters. They are small compact insects whose beak-like mouthparts are designed for piercing and sucking plant sap or small animal prey.

Lacewing

Lacewings and ant lions (Order *Neuroptera*) possess small slender bodies, two pairs of large lacy wings and rest on undersides of leaves by day. They are carnivorous.

Scorpion fly

Scorpion flies (Order *Mecoptera*) are similar to lacewings, but their wings are narrower and longer and the tail is bent over in a scorpion-like sting. The beak-shaped mouthparts are adapted for biting and chewing.

Caddisfly

Caddisflies (Order *Trichoptera*) have well-known aquatic nymph larvae which build themselves protective cases out of gravel or plant debris. The adult caddisfly resembles a cross between a moth and a fly, with hairy wings and no mouthparts (so they cannot feed).

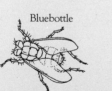

Bluebottle

Cranefly

True flies (Order *Diptera*) encompass houseflies, craneflies, mosquitoes, horseflies and blowflies. The larvae are legless maggots, and the adults have one pair of well-developed wings.

Flea

Fleas (Order *Siphonaptera*) are small wingless parasites, with large back legs for leaping and tube-like mouthparts for bloodsucking.

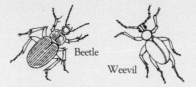

Beetle

Weevil

Beetles and weevils (Order *Coleoptera*) have the front pair of wings modified into hard case-like covers for the back pair. Weevils have an elongated snout-like beak on the head.

Butterfly

Moth

Butterflies and moths (Order *Lepidoptera*) possess two pairs of large, often colourful wings. The larvae are called caterpillars, the pupae chrysalises. Moth species outnumber butterfly species by about 30 to 1.

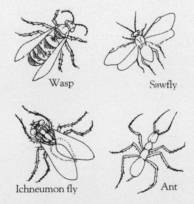

Wasp

Sawfly

Ichneumon fly

Ant

Bees, wasps, sawflies, ichneumon flies and ants (Order *Hymenoptera*) have two pairs of transparent membraneous wings (except for worker ants, which are wingless), and the bees, wasps or ants are often social and live in colonies.

SPIDERS AND SCORPIONS

Kingdom: *Animalia*
Phylum: *Arthropoda*
Class: *Arachnida*

o Four pairs of walking legs (if three pairs then see Insects, left)
o Body usually in two parts
o Hard outer skeleton
o Jaws bear fangs

A large and varied group of arthropods, most of which are land-living hunters. The arachnids include spiders, scorpions, mites, ticks, also various false scorpions, whip scorpions, false whip scorpions and so on, and the horseshoe (king) crabs. Approaching 100,000 species.

The tremble spider *Pholcus*, when disturbed, vibrates its web rapidly and becomes virtually invisible

Spiders (Order *Araneae*) are hunters of invertebrates. Some hunt by stealth, some run down prey, some spin webs. A typical spider has well-separated head and body, and eight pairs of "eyes" plus powerful fangs.

The common harvestman *Phalangium* is less nocturnal than other species

Harvestmen (Order *Opiliones*), unusually for arachnids, have abdomen and body joined, long lanky legs and a single pair of eyes on a "periscope" at the front.

Most scorpions are ground-dwellers and prefer to hide in dark corners during the day

Scorpions (Order *Scorpiones*) are large arachnids with lobster-like pincers. The flexible abdomen has 12 segments.

Harvest mite inserts its blade-like mouthparts through thin skin and causes intense irritation

Mites and ticks (Order *Acarina*) have small round globular bodies with projecting head parts. Many are parasitic.

CRUSTACEANS

Kingdom: *Animalia*
Phylum: *Arthropoda*
Class: *Crustacea*

- Body covered with hard shell, as jointed plates or a large carapace
- Several (often five) pairs of jointed legs (if four pairs then see Spiders and scorpions, page 299; if three pairs then see Insects, page 298)
- Two pairs of feelers or similar appendages in front of mouth
- Body divided into head, thorax and abdomen (though it is often difficult to make out these divisions)

Crustaceans are a numerous and very variable group of creatures. Nearly all live in water; most are marine (crabs, prawns and barnacles), some are freshwater (water fleas, crayfish and shrimps) and a few are land-dwelling (woodlice and robber crabs). They are especially numerous in the sea where, as a constituent of plankton, and as krill, they form an important food source for other animals. The classification of crustaceans is very complicated, especially since there are many different small or almost microscopic kinds which are divided up according to technical details. Included here are some of the more familiar crustacean groups. About 25,000 species.

Water flea *Leptodora* swims by rowing along with its oar-like antennae

Water fleas and similar creatures (Orders *Branchiopoda*, *Ostracoda*) have their bodies enveloped in a two-part shell and swim by means of their long hairy antennae. The copepods (Order *Copepoda*) are also small swimming crustaceans, who look like crosses between woodlice and lobsters. Copepods are a major part of oceanic plankton and are one of the most numerous kinds of creature on earth.

Goose barnacle *Lepas* enclosed in a two-valved shell and anchored by a fleshy wrinkled stalk

Barnacles (Order *Cirripedia*) are protected by calcareous plates and have six pairs of limbs that act like long fingers to sweep food into their mouths. They are usually fixed in one place; some, like the acorn barnacle, have plates welded to the rock; others are attached by a fleshy stalk.

Fish louse *Argulus* can swim from one host to another

Fish lice (Order *Branchiura*) are small transparent crustaceans, flattened and with two large suckers plus hooks and spines so they can cling to their fish hosts.

Sea slater *Ligia* scavenges for dead fragments on the upper shore

Woodlice, water lice and sea slaters (Order *Isopoda*) are flattened and have segmented shells and seven similar pairs of legs. The woodlouse is one of the few crustaceans to conquer land.

Freshwater shrimp *Gammarus* swims on its side, often with the smaller female on the male's back

Freshwater shrimps and sandhoppers (Order *Amphipoda*) are thin (being flattened side-to-side) and fast-moving, feeding on decaying organic matter.

Squat lobster *Galathea* is about 8 cm long and will not hesitate to use its claws in defence

Crabs, lobsters, prawns, shrimps and crayfish (Order *Decapoda*) are the largest crustaceans and typically have five main pairs of limbs, although the front pair may be pincers rather than legs.

MOLLUSCS

Kingdom: *Animalia*
Phylum: *Mollusca*

- A sheet of soft tissue, the mantle, covers all or part of body
- The mantle secretes a shell which lies within the body or on the outside
- Body divided into head, muscular foot and "visceral mass" containing guts and reproductive organs

Molluscs are the second-largest group of animals (after insects). They are mostly water-dwelling and many live in the sea; the familiar land-dwelling slugs and snails make up a relatively small proportion of the total number of mollusc species. They are distinguished from other similar invertebrates by the lack of jointed legs. Molluscs are varied in appearance and range from tiny snails half a millimetre long to the giant squid which may have a body six metres long, with tentacles twice as long again. About 100,000 species.

Plaited door snail has eyes on the tips of its two longer tentacles

Slugs and snails (Class *Gastropoda*) are an enormous group of molluscs which have colonized almost every habitat. Most have a single shell, which can be coiled (like a snail), cone-shaped (like a limpet) or very small or even absent (as in some slugs). Besides the familiar garden slugs and snails, gastropods include pond snails, sea snails such as whelks and periwinkles, and sea slugs (nudibranchs) such as the sea lemon. The name "gastropod" comes from the observation that slugs and snails appear to be walking on their stomachs.

Cockle filters food from water passed in and out of its body through siphons

Bivalves (Class *Bivalvia*) include the familiar two-shelled molluscs such as mussels, oysters, cockles, clams and razorshells. The two shells are nearly always of equal size. These molluscs are mostly filter-feeders and burrow into sand or mud (or even stone or wood), or are fixed by short anchoring threads.

STARFISH AND SEA URCHINS

Kingdom: *Animalia*
Phylum: *Echinodermata*

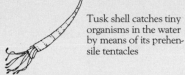

Tusk shell catches tiny organisms in the water by means of its prehensile tentacles

Tusk shells or tooth shells (Class *Scaphopoda*) each have a single shell in the shape of tapering cylinder, like an elephant's tusk, which is open at both ends. They live in the sea, burrowing in sand, and are up to 13 centimetres long.

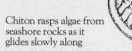

Chiton rasps algae from seashore rocks as it glides slowly along

Chitons (Class *Amphineura*) are also called "coat-of-mail" shells; they are small flattened creatures, rather like woodlice, that adhere tenaciously to rocks on the intertidal shore. The shell is made of eight linked plates and the creature has a broad muscular foot.

Common octopus is an active predator of fish, shellfish and crustaceans

Octopus, cuttlefish and squid (Class *Cephalopoda*) are large complicated molluscs which live in the sea. They have large well-developed brains, good eyesight and long tentacles surrounding the mouth. Cuttlefish and squid have small internal shells (familiar as "cuttlebone"), octopuses do not have a shell.

○ Radially symmetrical ("circular") with arms or body divisions in fives or multiples of five
○ Hard skeleton made up of bone-like plates
○ Move by means of tiny tube feet

Echinoderms, which include the familiar starfish and sea urchins, are a distinct group of creatures with a long fossil record. They live on the shore or sea bed and can creep along by waving their rows of hydraulically operated tube feet to and fro. Most echinoderms are able to regrow broken-off arms. About 6,000 species.

Some starfish, like the sun star, have more than 5 arms

Starfish and cushion stars (Class *Asteroidea*) mostly have the typical five-armed body. Their tube feet have suckers which are used in relays to prise open mussels, oysters and other bivalves. In cushion stars the arms are not very prominent and some are pentagon-shaped.

The brittlestar can catch small worms with its long agile arms

Brittlestars (Class *Ophiuroidea*) have a small central body and long flexible arms which, as their name suggests, are easily broken off. They are found in deeper offshore waters, often in large numbers.

The edible sea urchin's long tube feet extend beyond the spines

Sea urchins (Class *Echinoidea*) have spherical bodies with the mouth on the underside. The ball-shaped skeleton ("test") has long spines and holes through which stick long tube feet. The spines can be tilted by muscles and act with the tube feet to move the creature along, and in some species the spines are poison-tipped.

The sea cucumber ejects a mass of sticky white threads from one end if molested

Sea cucumbers (Class *Holothuroidea*) are aptly named. They have soft cylindrical bodies, small inner skeletons and three rows of tube feet on the underside which are used for locomotion. The 20 or so long feathery tube feet at one end sweep organic particles into the mouth.

The feather star hauls itself along the sea bed using its arms

Sea lilies and feather stars (Class *Crinoidea*) have long delicately-feathered arms and, in the case of the sea lily, the creature is attached by a stalk to the sea bed. Their mouths, unlike other echinoderms, are on the upper side.

CHORDATES

Kingdom: *Animalia*
Phylum: *Chordata*

There is no simple gradation from the vast assemblage of "lower" animals, popularly known as invertebrates (animals without backbones) to the familiar "higher" ones. There is, of course, no creature with only half a backbone. But there are curious animals who possess what is called a "notochord"—a tough yet flexible rod of tissue that runs down inside the back and gives the same kind of support as the string of vertebrae that make up the spine. The significant point is that the vertebrate animals, during a very early stage in their development, before the backbone has appeared, go through a phase of having a notochord. Because of this, biologists consider that the backbone may well have evolved from the notochord, and therefore taxonomists group creatures who possess a notochord along with those who possess a backbone, rather than leaving them with the invertebrates. This grouping has been given the name *Chordata* and has superceded the old group *Vertebrata*. So the situation now is that there are several phyla of invertebrates—things like worms, molluscs and arthropods—and one phylum *Chordata* which encompasses both creatures with backbones (fish, amphibians, reptiles, birds and mammals) and creatures with notochords, like the larva of the sea squirt and small fish-like lancelet.

JAWLESS FISH

Kingdom: *Animalia*
Phylum: *Chordata*
Class: *Agnatha*

o Body slender and fish-like
o No jaws, but mouth is a sucker armed with many small teeth (if jawed then see Sharks and rays or Bony fish, right)
o No pelvic fins

The small group of primitive jawless fish live mainly as parasites on other fish or as bottom scavengers. The two best-known kinds are the lampreys and the hagfishes. The fins and sense organs of these fish are poorly developed, and they swim by clumsy thrashing movements. About 60 species.

North Atlantic hagfish lives in shallow water, feeding on other fish which it envelops with slime

The eel-like lamprey has small eyes and a large oval sucker on its head for clinging to its host

SHARKS AND RAYS

Kingdom: *Animalia*
Phylum: *Chordata*
Class: *Chondrichthyes*

o Fish whose internal skeleton is made of cartilage, not bone (if bony then see Bony fish, right)
o Skin rough and covered with toothed scales (if flat scales then see Bony fish, right)
o No gill covers, so gill slits clearly visible
o Unequally-lobed tail fin

"Cartilaginous" fish, as the sharks, dogfish, skates and rays are known, are mostly predatory fish with sandpapery skin. They lack a swim bladder so are either continually on the move or spend much time hunting or resting on the sea bed. They are marine, and their empty egg capsules washed up on the shore are known as "mermaid's purses". About 600 species.

Lesser-spotted dogfish feeds on molluscs and other bottom-dwelling invertebrates

The common ray's pectoral fins form wing-like flaps on its flattened body

BONY FISH

Kingdom: *Animalia*
Phylum: *Chordata*
Class: *Osteichthyes*

- Fish whose internal skeleton is made of bone (if cartilage then see Sharks and rays, left)
- Body covered by bony plates or scales (if rough toothed scales see Sharks and rays, left)
- Gills covered by single flap of skin, the "operculum"
- Tail fin usually has equal lobes

The vast majority of modern-day fish are in this group. They live in both fresh water and in the sea, and some (like mud-skippers) are able to live for some time away from water. Bony fish possess swim bladders and have an extremely wide range of body shapes and sizes, from the snake-like eel to the tall thin sunfish. Over 20,000 species.

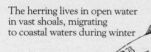

The herring lives in open water in vast shoals, migrating to coastal waters during winter

The slender eel has long fins on its back and underside

The bream is a strongly built fish with a deep body and small fins

AMPHIBIANS

Kingdom: *Animalia*
Phylum: *Chordata*
Class: *Amphibia*

- Soft scaleless skin
- Eggs jelly-like, hatch into aquatic larvae called tadpoles
- Usually four-limbed with webbed feet

Amphibians are not as dependent on water as are fish, but they are more tied to it than reptiles, birds and mammals. An amphibian needs water for the development of its eggs through the larval stage, the tadpole, which breathes by gills. Also the adult needs moisture to keep its skin damp since it "breathes" partly through the skin, though it has lungs and pumps air in and out of them by raising and lowering the floor of the mouth. The skin contains glands that secrete a distasteful, sometimes poisonous substance. About 3,500 species.

Tree frog has sucker-like tips to its fingers and toes

Frogs and toads (Order *Anura*) have short front legs, squat bodies, no tails (hence their name Anura) and long powerful back legs which enable them to leap many times their own length. Toads have drier wartier skins than frogs and tend to walk rather than hop or jump.

Common newt is usually found in water only in the breeding season

Salamanders (Order *Urodela*), which include newts and mud puppies, are lizard-like with long bodies and tails. Some live in moist places on land, others are fully aquatic and retain their larval gills.

Caecilian burrows for small soil or litter invertebrates

Caecilians (Order *Apoda*) are worm-like burrowing amphibians confined to the tropics. They are blind and legless and usually pale in colour.

REPTILES

Kingdom: *Animalia*
Phylum: *Chordata*
Class: *Reptilia*

- Skin dry with horny scales or bony plates
- Large heavily-yolked eggs with leathery shell
- No larval stage

The present-day reptiles are the remains of a much larger group of animals that dominated life on earth some 200 million years ago. The large eggs have their own food and water supplies which means that, unlike amphibians, reptiles are not dependent on water to complete their life cycle. Although modern forms look primitive, as a group the reptiles are successful in living on land, in fresh water and in the sea. About 5,000 species.

Snapping turtle of North America lurks in weeds waiting for prey

Tortoises and turtles (Order *Chelonia*) have a body enclosed in a shell made of bony plates covered with horny scales. The upper shell is the carapace, the lower one the plastron. The ribs are fused to the shell. There are no teeth, but a horny powerful beak.

Grass snake hunts for frogs and small mammals

Snakes and lizards (Order *Squamata*) have elongated bodies, and skins made of small overlapping scales with broader scales along the belly. The mouth is large and gaping, and tongue notched or forked. Usually, lizards are four-legged while snakes are legless, but there are lizards with tiny functionless legs (such as some skinks) and also legless lizards (such as the slow-worm).

American alligator is distinguished from a crocodile by its broader shorter head

Crocodiles and alligators (Order *Crocodilia*) are large lizard-like reptiles covered with thick hard square scales. They have large mouths filled with rows of similar-looking teeth, and most have their eyes and nostrils high on the head.

BIRDS

Kingdom: *Animalia*
Phylum: *Chordata*
Class: *Aves*

o Warm-blooded
o Skin bears feathers (except on legs, which have horny scales)
o Front pair of limbs formed into wings
o Lay large yolky eggs with chalky shells

Birds are a large, successful and very distinctive group of creatures. They mastered the power of flight early in their evolutionary history (though some, like the ostrich, have since become flightless) and they have complex mating and nesting behaviour which often involves displaying plumage and producing songs or calls. Parental care is well developed. The class *Aves* is divided into about 30 orders (whose names end in *-formes*) and each order is divided into several families (which end in *-idae*). Some of the more familiar orders are mentioned here. About 9,000 species.

Hawk
Eagle

Eagles, hawks and vultures (Order *Falconiformes*) are birds of prey. Buzzards, eagles, hawks, harriers and kites belong to the *Accipitridae* while the kestrel, merlin and hobby are *Falconidae*.

Pheasant
Grouse

Game birds (Order *Galliformes*) include grouse, partridge, quail and pheasant. They are stout-bodied land birds who run on their strong legs rather than fly.

Owl

Owls (Order *Strigiformes*) are nocturnal predators with silent flight, flat faces and large forward-facing eyes for night vision.

Storm petrel
Shearwater

Petrels and shearwaters (Order *Procellariiformes*) are chiefly ocean birds who nest on coasts.

Swan
Mallard

Ducks, geese and swans (Order *Anseriformes*) are generally large semi-aquatic birds, collectively called wildfowl. They fly powerfully and often in formation and have webbed feet.

Lark
Crow

Perching birds (Order *Passeriformes*) contains about half of all bird species. Their four-toed feet are modified for a secure grip, usually with three claws pointing forward and one back. Within this group are the swallows (*Hirundinidae*); crows, rooks and jays (*Corvidae*); tits (*Paridae*); finches (*Fringillidae*); mockingbirds (*Mimidae*); shrikes (*Laniidae*); larks (*Alaudidae*); and thrushes (*Turdidae*), which are a large and mixed family taking in the chats, redstarts, robin, wheatears, nightingale, blackbird and redwing.

Sandpiper
Redshank

Gulls and waders (Order *Charadriiformes*) are a large group including auks such as puffins and guillemots. Most waders are fairly similar-looking birds with long spindly legs, compact bodies and long thin probing bills.

MAMMALS

Kingdom: *Animalia*
Phylum: *Chordata*
Class: *Mammalia*

o Warm-blooded
o Skin bears fur or hair
o Usually four-limbed
o Female suckles young on milk from mammary glands

Mammals are a successful and numerous group, and with their large brains, well-developed sense organs and complex behaviour they are often regarded as the most "advanced" members of the animal world. There are three subclasses of the mammal class. The monotremes, such as the duck-billed platypus and spiny anteater, are egg-layers. The second sub-class are marsupials like the kangaroo, koala and opossum, where the young are born at a very early stage of development and continue their growth in the mother's pouch. In the third sub-class, the placentals (which make up the majority of mammals), the offspring are nourished by the mother's body through the placenta. There are about 4,000 species of mammal, of which one half are rodents and one quarter are bats.

Shrew

Mole

Shrews, moles and hedgehogs (Order *Insectivora*) are generally small mammals with small similar-looking teeth for crunching up their food of small invertebrates. Shrews are, with bats, the smallest mammals.

Bat

Bats (Order *Chiroptera*) are the only mammals truly to conquer flight, though some others can glide fairly well. Most bats are small insect-eaters and have similar-looking teeth, but the larger ones are fruit-eaters and some, the "vampire" bats, are blood-suckers.

Armadillo

Armadillos, anteaters and sloths (Order *Edentata*). Although their name means "toothless", armadillos and sloths have small simple back teeth. They are exclusively New World creatures, and the armadillo's fur is supplemented by strong horny plates which are jointed and in some species it can roll into a ball.

Dormouse

Porcupine

Rodents (Order *Rodentia*) include rats, mice, voles, squirrels, beavers, gophers, porcupines and guinea-pigs. The name means "gnawing animal" and the large projecting front teeth are used for gnawing at food, nest material and all manner of other substances.

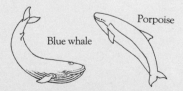

Hare

Pika

Rabbits, hares and pikas (Order *Lagomorpha*) resemble rodents with their large gnawing front teeth. The rabbits and hares have long ears, large prominent eyes for 360-degree vision and well-developed back legs for jumping.

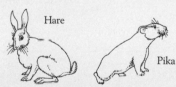

Porpoise

Blue whale

Whales, dolphins and porpoises (Order *Cetacea*) are large marine mammals sometimes mistakenly called fish. Like all mammals they are warm-blooded and they even have a few remaining hairs, usually on the nose. Most are hunters, but the world's largest mammal—the blue whale—and other "baleen" whales have hair modified into sieve-like whalebone plates in the mouth that filter tiny crustaceans (krill) from the water.

Cougar

Jackal

Carnivores (Order *Carnivora*) include hunting flesh-eaters such as the cats, the dogs (foxes, wolves and jackals), bears, sea-lions and seals, racoons, pandas and the mustelids (weasels and stoats). Most members have four long pointed canine teeth (eye-teeth or dog-teeth) and are large powerful predators that pounce on or run down their prey.

Bushbaby

Chimpanzee

Primates (Order *Primates*) are the lemurs, bushbabies, monkeys, apes and humans. A primate's hand is supple and has an opposable thumb, which allows it to pick up a small object between thumb and finger. The primates are physically fairly unspecialized, a horse or whale being much more distinctive in its anatomy. But for many biologists the primate's large and complex brain and resulting intelligence compensate for any lack of physical specialization.

Elephant

Elephants (Order *Proboscidea*) have long sensitive trunks and upper incisor teeth that grow into the characteristic tusks. The African bull elephant is the largest land mammal, but even so is less than one-twentieth the weight of a blue whale.

Zebra

Rhino

Odd-toed ungulates (Order *Perissodactyla*) include the horses, asses and zebras, who have one functional toe on each foot, and rhinos and tapirs with their three-toed feet (though tapirs have four toes on each of their front feet).

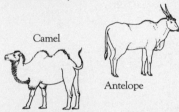

Camel

Antelope

Even-toed ungulates (Order *Artiodactyla*), also known as cloven-hooved ungulates, are one of the most numerous groups of mammals. Included here are antelopes, deer, giraffe, camels, hippos, and the wild sheep, pigs and cattle as well as their domesticated relatives. Like the odd-toed ungulates they are large herbivorous mammals with well-developed back (molar) teeth for chewing plant material.

THE NATURALIST AND
THE FUTURE

In the years ahead, naturalists have a vital role to play in the future of the world. Next to professional scientists they are one of the most important groups of people, for with their help in learning about the world, and in protecting it and persuading others to protect it, they will become the guardians of our planet and of our welfare as a species.

Let us take a look at the future as it is predicted today and see how naturalists can help repair some of the damage done so far. It is a fairly gloomy picture. To begin with, the world is grossly over-populated and growing more so. It is estimated that in the next 20 years the numbers of human beings in the world will rocket from 4,000 million to 6,000 million—half as many again. Any other animal species in this situation would face what is called a "crash", which means that the species outstrips its food supply and the population literally eats itself out of existence. Only a few individuals survive; the rest die, so reducing the numbers to a level that nature's food supply can cope with. Already, with starvation and malnutrition in many parts of the world and with other vital resources in dwindling supply, the signs of a "crash" for our species are evident. Although man is a clever species and has adapted himself to almost every environment in the world, since his tenancy he has treated each environment in such a profligate manner that, without immediate action, he faces a grim future.

During the next 20 years, as our numbers increase by half, it is believed that one-third of the world's arable land may be destroyed in one way or another. Constructing huge cities to house the flood of humanity, building roads and airports—all these activities swallow up good land. Our appalling agricultural methods are also to blame, degrading what rich agricultural land is left, and the over-grazing by our domestic creatures causes deserts to spread over once-fertile areas like a deadening sandy tide. Our treatment of the forests and seas is no better. Half the tropical forests will be gone in 20 years, and with them their many treasures. The seas of the world are treated as giant cesspools instead of living breathing organisms—they have become the world's dustbin for everything from nerve gas to sewage.

One of the greatest problems is that many people, especially those in government and those working in the huge businesses that exploit natural resources all over the world, are under the impression that these things don't matter. They turn a blind eye to the problems, or they try to give the impression that if there are any problems then our brilliant technology can cope. This is untrue; in spite of the cleverness of man, our world is being destroyed by our much-vaunted technology more quickly than technology can repair the damage—if it can ever be repaired.

All the facets of the great jig-saw puzzle of nature interlock, however remotely, with each other. What is being done to the rain forests, for example, affects us all, not just people living in and around them. The rain forests in Central America are the vital winter home of hundreds of thousands of warblers. In the spring and summer, these warblers spend their time in North America where they act as a much more efficient form of insect control than any man-made insecticides. However, the tropical forests, the winter resorts of these birds, are being felled to create grasslands for raising beef cattle. Where does this meat go? It goes to North America and Europe to make hamburgers. What is worse, the grasslands newly-created from the felled forest are only temporary, since the thin top soil is deprived of its protective green blanket and is soon eroded away. So, eventually, not only will the man in New York or London pay more for his hamburger but he will also pay more for—or have to do without—the foodstuffs the warblers would have protected.

The role of the future naturalist

The naturalists of the future will play a more important part in the development and protection of the planet than did the naturalists of years ago, simply because the task is so urgent. Our extraordinary technology can be turned to the naturalist's aid: video and still cameras to record with, tape recorders for sound, aqualungs and submarines to study the sea and balloons or helicopters to study the air. Even satellites are being used, to help monitor vegetation changes over vast areas. But remember that in the great task of recording and studying no tool is too humble; a microscope is better than a magnifying glass, but the magnifying glass lets you take the first step towards knowledge. Likewise, it is essential to keep the sea clean, but it is important not to pollute the village duckpond.

Whether you look through a microscope or a magnifying glass, whether you study a tropical forest or a tiny city garden, you are gaining knowledge about how the world works. With this knowledge comes understanding; then, as a naturalist who understands the workings of ecosystems (our life-support systems, in fact), you can explain to others what is going wrong and, most important, why it is going wrong. You can help teach others how to live within nature's rules. In the past, by bending or disregarding these rules, we have put our future and the future of many other species in jeopardy.

The naturalists of today are lucky. With the age of cheap air travel they can explore parts of the world which their forebears never could, and this alone shows how small the world has become and how important the planet is to us. What would have taken months of travel even 60 years ago can now be accomplished almost overnight. This is wonderful, but also hazardous, for it means that more and more people (many of them uncaring) are getting to places which hitherto have remained inviolate. These people must be taught (if they are not naturalists) to observe and respect rather than destroy or violate. It is up to the naturalists of the world to show them how to enjoy and protect that heritage. It matters not where they come from—a naturalist, when he looks at an atlas, knows that national boundaries are merely political and not biological. Unfortunately, the political borders have come to matter and they cause a tremendous amount of strife in the world. They mark out territories for the different groups of our species, and we spend much time and effort squabbling over them; yet the naturalist knows that, in reality, the world and its future belong to all living things, not just to one species.

GLOSSARY

Adaptation Any feature of a living organism that allows it to deal with its environment efficiently, and so improves its chances of survival. The long bill of the curlew, which can probe for deep-buried worms, is a good example.

Aerobic Term used to describe an organism that can live and be active only in the presence of oxygen.

Anaerobic Term used to describe an organism that is living or active in the absence of oxygen.

Annual Plant that completes its life cycle in one growing season, so that each individual lives for only one year.

Antennae Paired sense organs found on arthropods and other invertebrates. Antennae normally take the form of long segmented "feelers" on the animal's head.

Anther Pollen-producing part of the male organ of a flower.

Anthocyanin Group of red, violet and blue plant pigments. They appear in flowers, fruits, stems and leaves, and are responsible for autumn tints.

Arboreal Tree-dwelling.

Arthropod Member of the most numerous animal group, which includes insects, spiders, crabs and centipedes. Their bodies have pairs of jointed limbs and are usually covered with a hard skeleton. The word arthropod literally means "joint-legged".

Artificial selection Choosing and breeding certain individuals to give offspring with the particular characteristics that the breeder wants.

Bio- (From the Greek word *bios*, life) Prefix used in a number of terms, including biology, the study of living things, and biomass, the total weight of all the living things in a given area.

Botany The study of plants.

Bract Modified leaf which bears a flower.

Byssus Mass of threads formed by the hardening of the sticky secretion produced by certain molluscs such as the common mussel. This enables the animal to attach itself firmly to stones, a great advantage in tidal waters.

Caecilians Legless worm-like burrowing amphibians of the tropics.

Carnivore Used as a term of classification, this word denotes a meat-eating mammal of the order *Carnivora*. The word is more generally used of any creature whose diet is mainly flesh.

Carotenids Yellow, orange and red plant pigments (carotene is found in carrots). They assist in the process of photosynthesis, by absorbing light and passing on energy to chlorophyll.

Catkin Reproductive part of a plant, usually a tree. Catkins are made up of many small flowers. They are always either male or female, and sometimes the different sexes grow on separate trees, as on the goat willow, so that one tree is male while another is female.

Chitin Chemical substance which gives the hardness to the outer covering of insects and other arthropods.

Chlorophyll Green pigments present in most plants. The absorption of light by chlorophyll provides energy that is used in the process of photosynthesis, enabling plants to manufacture energy-containing food substances (sugars) from carbon dioxide and water.

Chrysalis Popular term for the pupa of a butterfly or moth.

Cilia Small numerous hair-like projections from a cell, which flick from side to side to produce movement.

Class Grouping of living things, made up of a number of orders. Several classes together form a phylum.

Climax The form of community ultimately resulting from the process of succession, when an ecosystem is allowed to develop naturally.

Cloaca Single opening at the back of the body, into which the kidney and reproductive ducts open. It is found in fish, amphibians, reptiles, birds, and a few mammals, such as the spiny anteater.

Cocoon Case made partly or completely of silk, constructed by an insect larva. The cocoon will protect the insect's "resting" stage, which is known as the pupa.

Commensal Term describing species that associate closely without apparent mutual advantage. Commensal creatures may share the same burrow or shell. The word commensal literally means "at the same table".

Community Group of organisms living and interacting within a given area. The community and the non-living environment together make up the ecosystem.

Competition Contest between living organisms to get essential commodities such as space, food or light which are in limited supply.

Compositae The largest family of flowering plants, which includes daisies and thistles.

Convergent evolution Process through which creatures or plants evolve similar features independently of each other. The process is the result of the two species living in similar habitats, and therefore playing similar roles in nature. Thus a dolphin, which is a mammal, has evolved a shape similar to a fish, because both creatures move through water.

Coppice Small wood or thicket grown for the purpose of periodically cutting the underwood for domestic use. To coppice a tree is to cut or "prune" the tree close to ground level, so that several trunks grow from the same set of roots.

Cotyledon Specialized leaf forming part of the embryo in a seed, containing reserves of food that are used in germination.

Crepuscular Term describing behaviour that takes place during twilight (dusk or dawn).

Cruciferae Large family of flowering plants, many of which, like the cabbage, radish and mustard, are important food sources for mankind.

Deciduous Term describing plants that shed their leaves at a certain season. A familiar example is the oak tree.

Diatom Single-celled water-living protist. Diatoms are photosynthetic, and are found in enormous numbers near the water surface, where there is most light.

Dioecious Plant species that has male reproductive organs (flowers) on one individual, female on another.

Diurnal Term describing behaviour or activity that takes place during daylight.

Earth The burrow or burrows of a fox family.

Ecdysis Moulting, or shedding an outer layer of skin (as in insects).

Ecology The study of living organisms in relation to their surroundings. Ecology is a wide-ranging science that includes subjects from animal behaviour and the study of predation, to geology and soil science.

Ecosystem Community of organisms along with their environment. Ecosystems are usually very large—such as the arctic tundra—but may be as small as a wood or pond.

Ectoparasite Organism that lives parasitically on the outside of another, for example the tick that lives on a sheep.

Endoparasite Organism that lives parasitically inside another, for example the fluke (a kind of flatworm) that lives in the liver of a sheep.

Entomology The study of insects.

Enzyme Substance produced by a living cell, which controls the speed of certain chemical reactions.

Epiphragm Membrane (composed of calcium phosphate and mucus) secreted by snails before periods of cold or heat. It seals the shell opening, to prevent water loss.

Epiphyte Plant that grows on the trunk or branches of another plant simply in order to gain support, using the support plant as a parking place.

Errant Term used to describe marine worms that are free-living and active, in contrast to the tube-dwelling types. The ragworm is an example of an errant worm.

Ethology The study of animal behaviour.

Evolution The accumulation of changes in a species over generations, by processes such as natural selection.

Family Grouping of living things, made up of a number of genera. Several families together form an order.

Fertilization The combining of male and female reproductive cells to produce another cell (called a zygote), which develops into a new individual.

Filter feeding The active straining of minute food particles suspended in the water by aquatic animals such as bivalves, crustaceans and sea squirts.

Flagellum Long whip-like thread protruding from certain single-celled creatures. The creature moves by lashing the flagellum.

Foetus The embryo of a mammal once it takes on recognizable features. The human embryo, for example, is termed a foetus two months after fertilization.

Food chain Chain of events in which, for example, a plant is eaten by a herbivore, the herbivore is eaten by a carnivore, then that carnivore is eaten by a second carnivore, and so on. In nature there are seldom more than five or six links in a chain. Several food chains combine to make up a food web.

Food web Network combining several food chains, where one plant or animal may be eaten by a range of consumers.

Fossorial Adapted for digging, such as the spade-like front feet of the mole, or referring to a burrowing creature.

Gall Abnormal plant growth caused by the activity of insects, mites, nematodes or fungi. The spangle galls found on oak leaves, for example, are caused by the presence of the gall wasp larvae developing inside the galls.

Gallinaceous Term used to describe birds of the order Galliformes. These are often known as game birds, and include grouse, partridge and quail.

Garrigue Low scrub vegetation found in southern France.

Gene The "unit" of inheritance, which controls whether a person has blue or brown eyes, for example.

Genetics The study of heredity (how an organism inherits characteristics from its parents) and the variations and resemblances of different generations of organisms.

Genus (plural genera) The second smallest grouping of living things, made up of a number of similar species. Several genera together form a family.

Habitat Distinctive and characteristic surroundings, such as a deciduous woodland or a pond. A habitat is determined chiefly by the vegetation.

Halophyte Plant that can live in salty conditions, for example, the marsh samphire, which lives on salt marshes.

Hanger Wood growing on the side of a steep hill, usually in southern England.

Hemiptera Order of insects with sucking mouthparts, such as the aphid and the leafhopper. Hemipterans are usually called bugs.

Herbivore Animal whose diet consists exclusively or mainly of plants, for example the deer and the rabbit.

Hermaphrodite Creature that possesses both male and female reproductive organs, such as many worms.

Herpetology The study of amphibians and reptiles.

Hymenoptera Order of insects that often live in colonies. They include bees, wasps and ants.

Hypha (plural hyphae) Thread-like strand or filament making up a fungal body.

Ichthyology The study of fish.

Imago The fourth, or adult, stage in the life of certain insects, such as the butterfly or moth.

Imprinting The tendency of young animals to follow the first moving thing they see. In the wild, this is usually the mother, but in captivity, where the mother may not be the first thing they see, they might follow a human being or any moving object.

Insectivore Used as a term of classification, this word denotes an insect-eating mammal of the order Insectivora (such as the shrew). The word is more generally used of any insectivorous creature (such as the bat)—a creature that eats mainly insects.

Instinct Inborn or innate behaviour. For example, many newly hatched birds act in a frightened manner when they see the shape of a hawk for the first time, even though the parents have not taught them that the hawk is dangerous.

Invertebrate Animal without a backbone. Invertebrates make up over 90 per cent of all animal species.

Labellum (plural labella) Lobe or pad at the end of an insect's proboscis. In flies the labellum is used to soak or sponge up liquid food.

Labium In insects, the lower lip. The labium is made up of several parts. In plants, a petal or petals shaped like a lip, as in some orchids.

Larva Active but immature or juvenile stage in the life cycle of an insect or similar creature (or an amphibian). The larva does not normally look like the adult. Common examples are caterpillars (larvae of a butterfly) and tadpoles (larvae of a frog).

Lek Area in which breeding males compete for territories. The males who occupy territories in the centre of the lek manage to mate with the highest proportion of available females.

Lepidoptera Order of insects, made up of the butterflies and moths.

Limnology The study of freshwater plants and animals.

Littoral Term used to describe a shore area or a creature or plant that lives on the shore.

Metamorphosis Change in body form between larva and adult, mainly in insects, other arthropods, and amphibians. There are two types: in complete metamorphosis there are distinct changes, such as from egg to caterpillar to chrysalis to adult butterfly; in incomplete metamorphosis the changes are less marked, such as from dragonfly larva to adult dragonfly.

Mimicry Similarity in appearance or behaviour between one organism and another, which confers protection against predators. In Batesian mimicry a harmless creature mimics a dangerous or distasteful one and so gains protection. In Mullerian mimicry several dangerous species have similar markings (called warning coloration), and a predator who tackles one unsuccessfully learns to avoid all similar creatures in the future.

Monoecious Plant species that has male and female reproductive organs (flowers) on the same individual.

Mustelidae Family of mammals including the stoat, weasel, mink and badger.

Mycelium The main body of a fungus, made up of a network or maze of threads called hyphae.

Mycology The study of fungi.

Mycorrhiza Association of a fungus with the roots of a higher plant, usually a tree.

Natural selection The survival of an organism to reproduce successfully owing to some characteristic which has helped it in the struggle for existence.

Nectar Sugary substance secreted by flowers to attract insects, thereby aiding pollination.

Neoteny Persistence of larval features throughout an individual's life. Neoteny is common in amphibians—for example, the adult alpine salamander retains the tadpole's larval gills.

Niche The "role" of a species in nature—what it eats, what eats it, where it lives, and so on.

Nocturnal Term describing behaviour or activity that takes place during the hours of darkness.

Nutrient cycle The travels of a chemical substance, such as carbon or phosphorus, back and forth between the living and non-living worlds.

Nymph Young insect resembling a small wingless adult. This stage occurs in insects with an aquatic larval stage, such as dragonflies, stoneflies and mayflies, as well as grasshoppers and bugs.

Omnivore Animal that eats plants and animals.

Order Grouping of living things made up of a number of families. Several orders together form a class.

Organic In chemistry, containing the substance carbon in some form or other. In nature, connected with living (or ex-living) material.

Ornithology The study of birds.

Oviparous Egg-laying, as in most animals except mammals.

Ovoviviparous Forming eggs which are retained in the mother's body until they hatch, then giving birth to the young.

Ovule Female part of a flower that, when fertilized by pollen, develops into a seed.

Parasite Organism living on or in another organism called the host. The parasite obtains food or some other resource at the expense of the host.

Parthenogenesis Form of reproduction in which eggs develop without fertilization. It is common in insects such as aphids, stick insects and gall wasps, as well as some fish.

Pathogen A disease-causing organism such as certain bacteria.

Perennial Plant that survives as an individual year after year in some form—for example as a bulb.

Pheromone Chemical substance released by an animal, which influences the behaviour or development of other individuals of the same species. The wind-borne sex attractant chemical of the female emperor moth is a good example.

Phloem Pipe-like tissue through which nutrient-containing sap is distributed to the various parts of a plant.

Phoresy Condition in which one animal is carried around by another. Ground beetles carry mites in this way.

Photoperiodism Response of a plant or animal to the relative lengths of day and night. Some birds, for example, come into breeding condition only when the days get to a certain length in the spring.

Photosynthesis The manufacture of food substances (mainly energy-containing sugars) by green plants using the energy in sunlight, carbon dioxide gas from the atmosphere and water.

Phylum Grouping of living things made up of a number of classes. Several phyla together form a kingdom.

Phyto- Prefix used to denote connection with plants. For example, phytoplankton means the tiny plants in plankton.

Plankton Small organisms that float and drift in the surface layers of seas and lakes. Plankton are mostly small (less than a millimetre) and include larval types of larger animals.

Pollard To cut a tree some way above the ground, to promote the sprouting of thin branches to give shade or that can be used for fencing or firewood. This method of cultivation is often used with willows.

Pollen Small or microscopic bodies produced by flowering plants, containing the male reproductive cells.

Pollination Process by which pollen is transferred from the male part of a flower to the female part, to form the fertilized ovule which will grow into a seed.

Precocial Term used to describe birds, such as the plover, whose young are alert and active immediately after hatching from the egg.

Proboscis Trunk-like projecting mouthparts or snout, such as the tube-like sucking mouthparts of a butterfly.

Pupa The "resting" stage in the life cycle of certain insects, in between the larva and the adult. The pupa of a butterfly or moth is popularly known as a chrysalis.

Quadrat Square of vegetation (usually 1 metre square) chosen at random to study the composition of vegetation in a survey of a selected area.

Queen The only fertile female of social insects such as the wasp, bee or ant.

Sap Fluid containing sugars and other high-energy substances made by photosynthesis, which is transported from the leaves to all parts of a plant through the pipe-like phloem tubes.

Saprophyte Organism which derives nourishment from decayed organic matter. Most fungi are saprophytes, as are certain flowering plants, such as the bird's nest orchid.

Satyridae Family of butterflies commonly called "browns". Most have pale-centred eye spots and plain bands in the veins of their fore-wings.

Seed Product of the fertilization of the female ovule by the male pollen. The seed can often withstand adverse conditions and germinate into a new plant when conditions become favourable.

Selection See Artificial selection and Natural selection.

Sett The burrows of a badger family.

Species The smallest grouping of living things, whose members can interbreed to produce fertile offspring. The members of a species all look very similar. Several species together form a genus.

Sporangium Spore case, such as that found on the underside of fern leaves.

Spore Single cell enclosed in a protective coating which can withstand adverse conditions such as extreme cold or dryness, and which can grow into a new individual when conditions become favourable. Unlike a seed, a spore is not the result of fertilization.

Stamen The male organ of a flower, consisting of a stalk bearing at its tip an anther which produces pollen.

Stigma The part of the female reproductive structure of a flower which receives the pollen.

Stomata Tiny holes in the surface of a leaf or stem which allow air to circulate to the cells inside the plant.

Succession The sequence in which different groups of plants and animals colonize an area, each community replacing the former until the climax community is reached.

Symbiosis Different species living together to their mutual benefit. The alga produces food that is used by the fungus, and the fungus provides protection for the alga.

Taiga The large belt of coniferous forest across northern Europe and Asia.

Taxonomy The study of the classification of living things.

Territory Area occupied by one or a pair of animals, and actively defended against others of the same species and sometimes against intruders of different species.

Thallus The main body of a simple plant, which is not divided into stem or leaves, for example the liverwort's fleshy lobes.

Turdidae Family of perching birds, including thrushes and blackbirds.

Viviparous Giving birth to formed young, as in most mammals and some reptiles.

Worker Infertile member of a colony of social insects such as bees, wasps or ants.

Xerophyte Plant that can survive in conditions of water shortage. Xerophytes range from cacti to plants of heathland, such as heather.

Xylem Pipe-like tissue through which water and some dissolved nutrients are carried from the roots to the various parts of a plant.

Zoo- Prefix denoting a term connected with animals, as in zoology, the study of animals.

INDEX

Page numbers in **bold** type indicate a photograph or illustration.

ACKNOWLEDGMENTS

In writing a book like this, of course you do not labour alone. Lee and I would like to put on record our grateful thanks to the following people, without whose help, encouragement and expertise the completion of this work would not have been possible:

Steve Parker, Neville Graham, David Black, Stuart Jackman, Phil Wilkinson, Mark Richards, Joss Pearson, Rosamund Gendle and everyone else at Dorling Kindersley, for their patience and hard work.

Philip Dowell, and his assistant Andy Butler, for the miraculously beautiful photographs of the items we collected.

Eric Thomas, the meticulous and inspired artist who is responsible for many of the remarkable illustrations.

John Stidworthy, who with patience and good humour answered our many and varied questions.

John Hartley, my assistant, who as usual put up with us, fetched and carried, and even cooked meals for us.

Renata Vassaillou, who typed the early chapters in Provence.

Joan Benn, our ever-cheerful and long-suffering secretary, for her magnificent shorthand and typing, and for dealing so competently with the text.

All our friends who, without rancour, put up with us being thoroughly anti-social during the two years that we worked on the text.

The authors of the many books that we ransacked in pursuit of answers to our numerous queries.

And last but not least, all the naturalists, both amateur and professional, that we have met and corresponded with around the world, for their understanding help at all times. In particular, the following people helped us and the Dorling Kindersley team with our collecting and offered their expert advice on identification:

Juliet Bailey, The Wildfowl Trust, Slimbridge, Gloucestershire; Bob Britton, Station de la Tour du Valat, Camargue, France; Eric Groves, British Museum (Natural History) Department of Botany, London SW7; Bob Moseley, Calshot Activities Centre, Southampton; Eric Newrith, Margaret McMillan House Field Centre, Wrotham, Kent; John Perry, Victoria College, Jersey; Ian Swinney, Bookham Commons Trust, Surrey; Tony Thomas and Graham Hobbs, Slapton Ley Field Centre, Devon; Nigel Webb, Furzebrook Research Station, Wareham, Dorset; Derek Wells, Nature Conservancy Council, Huntingdon.

Dorling Kindersley Limited would like to thank the following people for their assistance:

Richard Dawes, Martin Dohrn, Sue Gooders, Billy Hall, Nigel Haselden, John Huxley, Angela Jackson, Malcolm Rush, Jane Parker, Jim Scott, Bob Smiles, Andrew Stanger, Mary Trewby, Alan Ward, Peter Ward.

Illustrators
Marrion Appleton
Brian Craker
Rosamund Gendle
Sheila Hadley
Anthony Maynard
Robert Micklewright
Eric Thomas
Ken Wood
David Worth

Photographers
Abbreviations: b bottom, c centre, l left, r right, t top.

Jacket photographs Philip Dowell 1–6 Philip Dowell 8 Lawrence Durrell 10t Francois Gohier/Ardea 10c Rod Williams/Bruce Coleman 10b Hans Reinhard/Bruce Coleman 12t Mary Evans Picture Library 12c&b Mansell Collection 14 Mansell Collection 15 Philip Dowell 17 David and Katie Urry/Ardea 20 Philip Dowell 24 J A Bailey/Ardea 28 Philip Dowell 33 Philip Dowell 40 Stephen Dalton/Natural History Photographic Agency 45t Bob Gibbons/Ardea 45bl P A Bowman/Natural Science Photos 45rc Martin Dohrn/Science Photo Library 45br R C Revels/Natural Science Photos 48 Ian Beames/ Ardea 50–1 Philip Dowell 54tl Eric Herbert/Natural Science Photos 54bl D Bonsall/Natural Science Photos 54r Avon and Tilford/Ardea 58–9 Philip Dowell 62 Jane Burton/Bruce Coleman 66–7 Philip Dowell 70–1 Philip Dowell 74–5 Philip Dowell 79 John Mason/Ardea 81 Ian Beames/Ardea 84t Ian Beames/Ardea 84b Rod Williams/Bruce Coleman 85 Jeff Foott/Bruce Coleman 88 Bruce Coleman 96 Wayne Lankinen/Bruce Coleman 100t Charlie Ott/Bruce Coleman 100b R. Balharry/Natural History Photographic Agency 101 Jane Burton/Bruce Coleman 105 Ian Beames/Ardea 108–9 Philip Dowell 113t&bl N A Callow/Natural History Photographic Agency 113br A and E Bomford/Ardea 117 L Campbell/Natural History Photographic Agency 124–5 Philip Dowell 128 Geoffrey Kinns/Natural Science Photos 130 Hans Reinhard/Bruce Coleman 134–5 Philip Dowell 139 Ake Lindan/Ardea 142 Martin W Grosnick/Ardea 146 Alain Compost/Bruce Coleman 151 C B Frith/Bruce Coleman 155 Pekka Helo/Bruce Coleman 158–9 Philip Dowell 161 Bruce Coleman 164 Avon and Tilford/Ardea 168–9 Philip Dowell 173t&b Martin Dohrn/Science Photo Library 176–7 Philip Dowell 180 Martin Dohrn/Science Photo Library 184–5 Philip Dowell 188 David George 192–3 Philip Dowell 197 R J C Blewitt/Ardea 200 Martin Dohrn 204–5 Philip Dowell 208 Philip Dowell 210 Jane Burton/Bruce Coleman 214–5 Philip Dowell 219t C B Frith/Bruce Coleman 219bl&br Ron and Valerie Taylor/Ardea 223 David George 226–7 Philip Dowell 231 Martin Dohrn 234 Ron and Valerie Taylor/Ardea 239t W W F and A I Giddings/Bruce Coleman 239b Jan and Des Bartlett/Bruce Coleman 240 Philip Dowell

Photographic Services
Negs

Typesetting
Advanced Filmsetters (Glasgow) Limited

Reproduction
A Mondadori, Verona

FURTHER READING

A true naturalist is always hungry for information about his subject. Of course, first-hand knowledge obtained by observation, recording and other practical work is irreplaceable, but there is much to be gained by simply reading the works of the great naturalists.

You can learn a lot about the art of meticulous observation from writers in earlier centuries, so try to procure the books of Charles Darwin, Henri Fabre, Gilbert White and James Audubon. A useful reference work, which can still be picked up second-hand, is Lydekker's *Royal Natural History*. Although fairly ancient (the okapi had not been discovered) it is filled with information and beautiful line drawings by some of the best animal illustrators of the time.

Read also the works of the early explorers—quite apart from containing a lot of useful information, they are great adventure stories. Especially good are the early writings of Bates, Wallace, von Humboldt, Hudson and Waterton when exploring in South America; Chapman, Selous and Baker in Africa, and Webber and others in India. Try to read also the works of naturalists who studied and explored in the early years of this century—people like William Beebe in the Americas and also Roy Chapman Andrews in Asia.

There are so many modern works that one cannot possibly list them all, but here is a short list to act as a springboard (many of these books have bibliographies which will introduce you to further works on the subject): H C de Wit on plants; Edwin Way Teale on insects (mainly American, but the methods of study are the same); Robert Mertens and H W Parker on the reptiles of the world; David Lack on bird behaviour, and especially his classic *The Life of the Robin*; George Schaller's extraordinary books on the lives of lions, tigers and gorillas; and David Fleay's numerous works on the fascinating and remarkable fauna of Australia.

Some of the classics in animal behaviour which you should try to read include Konrad Lorenz's *King Solomon's Ring*, *Animal Architecture* by Karl von Frisch, and Donald Griffin's *Listening in the Dark*. And good works on conservation include Aldo Leopold's *A Sand County Almanac*, Rachel Carson's *Silent Spring* and *Extinction* by Paul and Anne Erlich.

As general works of reference, you cannot do better than beg, borrow or steal Bernhard Grzimek's *Animal Life Encyclo-paedia* and H C de Wit's *Plants of the World*. They are expensive, but will repay you a thousand-fold. William Collins have published in the past, and continue to publish, excellent handbooks and field guides on natural history, not only of Britain and Europe but of the whole world, and it is well worth collecting these for your library.

There are many bookshops, both new and second-hand, where you can discover treasures, but if you want the most helpful firm, who have an enormous stock of books on every aspect of natural history, Wheldon & Wesley Ltd are the people (their address is on the right).

USEFUL ADDRESSES

Organizations

Amateur Entomologists' Society
4 Steep Close, Orpington, Kent
Arranges field trips and holds a large annual exhibition.

Botanical Society of the British Isles
c/o Department of Botany, British Museum (Natural History), Cromwell Road, London SW7
Organizes field trips, exhibitions and conferences; produces the journal "Watsonia". Societies dealing specifically with lichens, mosses and ferns are also based at the museum.

British Butterfly Conservation Society
Tudor House, Quorn, Leicestershire
Involved in butterfly-watching, conservation work to improve the habitat and monitoring of populations.

British Museum (Natural History)
Cromwell Road, London SW7
Provides information on all aspects of natural history, and a centre for identification of specimens.

British Trust for Ornithology
Beech Grove, Tring, Hertfordshire
Arranges field meetings, annual bird-watchers' conference, and publishes "Bird Study" and "BTO News".

Royal Society for the Protection of Birds
The Lodge, Sandy, Bedfordshire
Maintains a large network of bird reserves, also research and the magazine "Birds".

World Wildlife Fund
11–13 Ochford Road, Godalming, Surrey
Raises money to help nature conservation on an international basis.

Royal Society for Nature Conservation
The Green, Nettleham, Lincolnshire
Coordinates over 40 county and regional Nature Conservation Trusts which own and manage their own nature reserves. Produces the magazine "Natural World".

Equipment and supplies

Watkins & Doncaster
Four Throws, Hawkhurst, Kent
Wide range of naturalists' equipment and specialist books.

Gallenkamp & Co Ltd,
PO Box 290, Technico House,
Christopher Street, London EC2
Laboratory equipment, including mounting boards and dissecting kits.

L Christie
129 Franciscan Road, Tooting,
London SW17
Entomological supplies and specimens, cabinets and cases, books (postal only).

Specialist bookshops

F W Classey Ltd
PO Box 93, Faringdon, Oxfordshire
Specializes in entomology, but deals with all aspects.

Wheldon & Wesley Ltd
Lytton Lodge, Codicote, Hertfordshire
All aspects of natural history, with enormous stocks.

A FINAL WORD

We hope that you enjoy this book and that, like naturalists everywhere, you appreciate the fact that the world and its wildlife are being unthinkingly destroyed by what is called progress. If so, would you like to help in the work we are doing at the Jersey Wildlife Preservation Trust? The Trust endeavours to save and build up colonies of almost extinct animal species, and to train people from different parts of the world in the arts of captive breeding and controlled reintroduction. We plead on behalf of the wild plants and creatures of the world because they cannot plead for themselves, and after all it is your world you are helping to preserve. If you wish to know more, write to me at our headquarters and we would be delighted to send you full details:

Jersey Wildlife Preservation Trust
Les Augres Manor, Trinity, Jersey,
Channel Islands.

CODES FOR THE NATURALIST

In our crowded and busy world it is vital that everyone respects and cares for the countryside and its wildlife. On this page are some codes of behaviour for the naturalist, though of course such guidelines apply to anyone who comes into contact with nature. For these I have used bits and pieces from a number of codes suggested by various nature organizations (see page 319). If you are ever in doubt about a certain activity, consult one of these bodies or refer to the *Wildlife and Countryside Act 1981* (HMSO). A list of the 39 creatures protected in Britain is contained in Schedule 5 of the Act. This list includes bats, a number of butterflies and moths, some newts and toads, and several lizards and snakes. Schedule 8 of the Act contains a list of 61 protected plants. And remember that, as a naturalist, it is part of your job to encourage your family and friends to follow the codes.

CODE FOR THE COLLECTOR

1 Take no more specimens than are necessary for your purpose.
2 Do not take the same species in numbers from the same place year after year.
3 Predators or parasites of whatever you collect should not be destroyed.
4 Do not disturb all potential home sites in your search for a species. For example, don't turn over all the rotting logs in a wood or investigate all holes in every tree.
5 Re-site traps that are catching creatures unnecessarily.
6 When taking plant samples, cut the material cleanly with a knife or secateurs—do not break it off.
7 For local or rare species, take only one or two specimens and avoid collecting in well-worked or over-worked areas.
8 Tell the authorities of unusual finds that you make, and if you are collecting in an area of special interest to conservationists, then supply a list of species found.
9 Never collect for commercial gain.
10 In taking specimens for your permanent collection, especially of rare species, wherever possible you should take the creatures into captivity, breed them and release those surplus to your requirements.
11 Do not attempt to reintroduce species or reinforce endangered populations without the advice and consent of the proper authorities.
12 When taking animals for breeding or rearing at home, make sure you take no more than you can support by the available food supply. Be sure that you are leaving enough food for your specimens' wild relatives. And remember that the specimens you take may themselves be the food source of another species.
13 Never disturb a member of an endangered species. Captive breeding of endangered species is an acceptable practice, and indeed is the only hope for some species since the numbers in the wild are too low to recover naturally. But it should be done only by experts, and only after much discussion with both conservationists and the governmental authorities of the country concerned.

CODE FOR THE COUNTRYSIDE

1 Leave things as you found them. This means stones or rocks you have overturned or weeds that you have dragged in from a lake.
2 Don't litter, especially in a water supply.
3 Guard against starting a fire. Every year a vast area of forest or scrubland is laid waste by carelessly lighted fires started by campers or picnickers.
4 In farming areas keep to the paths—don't walk straight across fields.
5 Drive or cycle carefully, and keep pets under control, for a slow tractor or a herd of cows might be around the bend.
6 Obtain permission to enter private lands and, where necessary, public nature reserves.
7 Don't leave tell-tale signs that might give away the location of a nest or burrow to predators (both human and otherwise).
8 Make yourself aware of the laws in your country pertaining to wild areas and their inhabitants. In Britain, for example, it is illegal to uproot any plant unless it is on your own land and it is illegal to disturb any nesting bird for whatever purpose. Also, permits are required for photography for some of the rarer birds.
9 Do not tell just anybody where you find something interesting. Remember that a secret shared in many cases no longer remains a secret—only tell people whom you know will respect the information.

CODE FOR PHOTOGRAPHERS AND TRAPPERS

1 Permits may be necessary to photograph or to trap and release some species. Check with your local natural history society.
2 If you use baits or lures, do not overdo things so that your subjects become used to their alternative food source.
3 In your trapping activities, never use substances or methods that can be detrimental to your catch or to other organisms in the area, like drugs or sticky materials.
4 If you want to keep your catch for a while, make sure that you provide the necessary conditions to keep it healthy and happy until its release.
5 Release your captive in appropriate surroundings, either in its original location or away from the trap site if you no longer want to trap that creature for your studies.
6 The welfare of the subject or captive is always more important than your photograph or trapping project.